HEALTHCARE TECHNOLOGIES SERIES 41

Innovations in Healthcare Informatics

Other volumes in this series:

Volume 1	**Nanobiosensors for Personalized and Onsite Biomedical Diagnosis** P. Chandra (Editor)
Volume 2	**Machine Learning for Healthcare Technologies** D.A. Clifton (Editor)
Volume 3	**Portable Biosensors and Point-of-Care Systems** S.E. Kintzios (Editor)
Volume 4	**Biomedical Nanomaterials: From design to implementation** T.J Webster and H. Yazici (Editors)
Volume 6	**Active and Assisted Living: Technologies and applications** F. Florez-Revuelta and A.A Chaaraoui (Editors)
Volume 7	**Semiconductor Lasers and Diode-based Light Sources for Biophotonics** P. E Andersen and P.M Petersen (Editors)
Volume 9	**Human Monitoring, Smart Health and Assisted Living: Techniques and technologies** S. Longhi, A. Monteriù and A. Freddi (Editors)
Volume 13	**Handbook of Speckle Filtering and Tracking in Cardiovascular Ultrasound Imaging and Video** C.P. Loizou, C.S. Pattichis and J. D'hooge (Editors)
Volume 14	**Soft Robots for Healthcare Applications: Design, modelling, and control** S. Xie, M. Zhang and W. Meng
Volume 16	**EEG Signal Processing: Feature extraction, selection and classification methods** W. Leong
Volume 17	**Patient-Centered Digital Healthcare Technology: Novel applications for next generation healthcare systems** L Goldschmidt and R.M. Relova (Editors)
Volume 19	**Neurotechnology: Methods, advances and applications** V. de Albuquerque, A. Athanasiou and S. Ribeiro (Editors)
Volume 20	**Security and Privacy of Electronic Healthcare Records: Concepts, paradigms and solutions** S. Tanwar, S. Tyagi and N. Kumar (Editors)
Volume 23	**Advances in Telemedicine for Health Monitoring: Technologies, design and applications** T.A. Rashid, C. Chakraborty and K. Fraser
Volume 24	**Mobile Technologies for Delivering Healthcare in Remote, Rural or Developing Regions** P. Ray, N. Nakashima, A. Ahmed, S. Ro and Y. Soshino (Editors)
Volume 26	**Wireless Medical Sensor Networks for IoT-based eHealth** F. Al-Turjman (Editor)
Volume 29	**Blockchain and Machine Learning for e-Healthcare Systems** B. Balusamy, N. Chilamkurti, L.A. Beena and P. Thangamuthu (Editors)
Volume 31	**Technologies and Techniques in Gait Analysis: Past, present and future** Nachiappan Chockalingam (Editor)
Volume 33	**Electromagnetic Waves and Antennas for Biomedical Applications** L. Wang
Volume 34	**The Internet of Medical Things: Enabling technologies and emerging applications** S.K. Pani, P. Patra, G. Ferrari, R. Kraleva and H. Wang (Editors)
Volume 36	**Applications of Machine Learning in Digital Healthcare** M.H. Silveira and S.-S. Ang (Editors)
Volume 37	**Technology-Enabled Motion Sensing and Activity Tracking for Rehabilitation** W. Zhoa
Volume 38	**Healthcare Monitoring and Data Analysis using IoT: Technologies and applications** V. Jain, J.M. Chatterjee, P. Kumar and U. Kose (Editors)
Volume 39	**Digital Tools and Methods to Support Healthy Ageing** P.K. Ray, S.-T. Liaw and J.A. Serano (Editors)
Volume 40	**Applications of Artificial Intelligence in E-Healthcare Systems** Munish Sabharwal, B. Balamurugan Baluswamy, S. Rakesh Kumar, N. Gayathri, and Shakhzod Suvanov (Editors)
Volume 42	**Smart Health Technologies for the COVID-19 Pandemic: Internet of medical things perspectives** C. Chakraborty and J.J.P.C. Rodrigues (Editors)
Volume 44	**Medical Information Processing and Security: Techniques and applications** A. Kumar Singh and H. Zhou (Editors)
Volume 46	**Digital Twin Technologies for Healthcare 4.0** R.K. Dhanaraj, S. Murugesan, B. Balusamy, and V.E. Balas
Volume 50	**Explainable Artificial Intelligence in Medical Decision Support Systems** Agbotiname Lucky Imoize, Jude Hemanth, Dinh-Thuan Do and Samarendra Nath Sur (Editors)

Innovations in Healthcare Informatics

From interoperability to data analysis

Edited by
Mohamed Abouhawwash, Sudeep Tanwar,
Anand Nayyar and Mohd Naved

The Institution of Engineering and Technology

Published by The Institution of Engineering and Technology, London, United Kingdom

The Institution of Engineering and Technology is registered as a Charity in England & Wales (no. 211014) and Scotland (no. SC038698).

First published 2023

The Institution of Engineering and Technology
Futures Place
Kings Way, Stevenage
Hertfordshire SG1 2UA, United Kingdom

www.theiet.org

British Library Cataloguing in Publication Data
A catalogue record for this product is available from the British Library

ISBN 978-1-83953-458-4 (hardback)
ISBN 978-1-83953-459-1 (PDF)

Typeset in India by MPS Limited

Cover Image: Laurence Dutton/E+ via Getty Images

Contents

Preface

The existing solutions for healthcare data analytics environments are restricted to scalability and high response time, which may not apply to real-time response generation for various healthcare informatics applications. Hence, quick decisions must be taken to perform data analytics on the generated data for the applications above. In this context, deep learning can play an essential role in the successful execution of various models and techniques to make quick decisions and adapt to an increase in data generation from different smart devices. Some of the challenges with these systems are as follows.

- *Interoperability*: It is the ability of different information systems, devices, and applications ("systems") to access, exchange, integrate, and cooperatively use data in a coordinated manner within and across organizational, regional, and national boundaries. This supports the provision of timely and seamless information portability and optimizes the health of individuals and populations globally. Health data exchange architectures, application interfaces, and standards enable data to be accessed and shared appropriately and securely across the complete spectrum of care, within all applicable settings and with relevant stakeholders, including the individual.
- *Consumerization*: Today, many of us still cannot go online to make an appointment with our primary care physician (PCP) for a physical or a sick visit or see the average wait time for the ER. When today's consumer-centric options, such as a local pharmacy or urgent care center, do allow for this level of visibility and ease, people who grew up digitally will question the value of maintaining a relationship with a PCP who does not provide this level of visibility. With this in mind, this edited book addresses how providers and others use existing and developing technologies to support open, proactive, two-way communication between all segments of the healthcare world. This includes hospitals, clinicians, patients, vendors, and anyone else who plays a role in the community.
- *Health data analytics*: The digitization of healthcare systems in clinical settings, combined with the explosion of personal data collection devices, provides the opportunity to use data for revolutionizing approaches to care at all levels with an emphasis on precision medicine and person-centered care. This big data opportunity requires expertise in health informatics, data science, and computational modeling. To address this, we have added a course on Introduction to Health Data Analytics. In addition, other courses have been

added or are being proposed to enhance the way we address this issue, including Patient Engagement Informatics and Analytics and Claims Data Analysis.

As mentioned above, we have seen tremendous growth in healthcare data generation from different IoT-enabled intelligent devices in the recent era. It generates the need to explore new models to process this data to infer new results. Hence, this edited book aims to present advances in healthcare informatics technologies by highlighting improvements over traditional technologies to introduce emerging technologies and platforms for big data analytics for different applications in healthcare. This book will explore all the methodologies of machine learning (basics of deep learning), algebra (mathematics behind deep learning), and models and techniques used in deep learning for big data analytics in healthcare informatics applications. We will also emphasize advanced-built applications using deep learning models. Moreover, it will describe advances in leveraging specialized platforms such as Hadoop, Spark, and others for data analytics in healthcare informatics networks. We will focus on the methodologies, theories, tools, applications, trends, challenges, and case studies of healthcare informatics. Recent related efforts in the areas will also be investigated, especially in academia and industry. We are committed to making the different chapters of the book in a synchronized manner to read by all readers. We hope that it will be a valuable reference for students, instructors, researchers, industry practitioners, and related government agencies staff.

This book is the first to address the issues and their solutions in healthcare informatics. The deep learning models, such as basic NN, CNN, and RNN, will be explained in detail for different applications. Also, how GAN can be used to design encoders and decoders is explained in detail in the proposal. There is a tremendous amount of information flow between different healthcare networks using various encoding and decoding techniques. But, how to handle such a massive amount of data from the end users' perspective is a critical challenge in front of the research communities. Hence, the need for deep learning models is inevitable in this environment. So, in summary, to the best of our knowledge, this book is the first attempt to analyze the healthcare informatics using deep learning models and techniques.

The book contains 11 chapters and brief summary of each chapter is as follows.

The chapter "Introduction to healthcare informatics: fundamentals and historical background" provides information on data, knowledge, wisdom to handling health data, electronic health records, health information technology, key players, the importance of data analytics, healthcare standards, and ethics to be followed in the health sectors. The chapter also discusses various M-Health technologies used in the healthcare sector.

The chapter "Healthcare informatics: an overview of privacy and security" aims to increase cybersecurity awareness among patients, healthcare staff, and organizations and presents countermeasures for future research. It also focuses on several methods to protect healthcare information, including information security risk management, data encryption, and data protection laws/regulations. In addition, the COVID-19 pandemic highlighted the crucial need to enhance IT systems security. It also offers students, scholars, and scientists a comprehensive background concerning healthcare informatics on two relevant topics: privacy and security.

The chapter "Health informatics and its contribution to health sectors" discusses informatics, related definitions, Health Informatics (HI), and its relation with other disciplines. The chapter also provides an educational overview of the evolution of HI, different HI technologies, benefits, and challenges of HI to its various stakeholders. Finally, it ends with some thoughts on HI's future growth.

The chapter "Role of Internet of Things and Artificial Intelligence for healthcare informatics: an overview" proposes a framework that uses IoT and AI as critical technologies to deliver safe medical services to patients to give a thorough analysis of current healthcare systems. According to the framework, combining IoT and AI offers several features, including data protection, data management, effective diagnosis, and real-time health monitoring, which solve the shortcomings of the existing hospital information system. The hospital environment is improved by including such features in the present system. The chapter is concluded with a thorough analysis and suggestions.

The chapter "Blockchain for security and privacy in healthcare informatics" explores the different aspects of developing blockchain-based security solutions for healthcare informatics. It also identifies the security requirements and challenges in the healthcare system. It also describes the security aspects of centralized/decentralized healthcare systems and elaborates on adaptions and the impact of Blockchain Technology in Healthcare Systems. Security attributes and architecture of blockchain and its applicability in developing secure frameworks for healthcare informatics have also been explained here. How the issues of identity and trust management can be addressed in healthcare using blockchain has also been presented in this chapter. The security of big data in healthcare using blockchain, along with the technologies like IoT and IoE, has also been addressed in this chapter. Finally, the chapter highlighted blockchain-based healthcare systems' challenges, opportunities, and future insights.

The chapter "Unification of machine learning and blockchain technology in healthcare industry" discusses a detailed comparison of intelligent applications in the Healthcare sector, which mainly includes applications like smart grids, smart cities, etc. Data security increases by combining both machine learning and blockchain.

The chapter "Robotics and process automation technologies for healthcare informatics" discusses the various challenges of using robots in the healthcare industry. The challenges are like what training the robots should get to work perfectly in the given environment and how the robots assist in the scarcity of nursing, doctors, and other healthcare providers by supporting the providers with repetitive duties, allowing the practitioner to undertake higher-level functions, how the robots minimize the need for "outsourcing production" management operations in healthcare. These intense observations and literature conclude that the health robotics categorization outlines how to offer high-quality patient care, effective medical operations, and a comfortable workplace for doctors and healthcare staff. The chapter concludes with the case study where the robot protects the privacy and confidentiality of the patient's data.

The chapter "Evolving technologies: IoT and Artificial Intelligence for healthcare informatics" focuses on the proliferation of AI and IoT in various healthcare segments, clinical and non-clinical, along with the context of machine learning and deep learning. The chapter includes domains of healthcare benefit, AI and machine learning algorithms and techniques involved at various stages, use of IoT in healthcare, challenges faced, their limitations, and future scope.

The chapter "Confidentiality, integrity, authentication (CIA) model for secured storage and processing of personal health records using Merkle hash tree and recurrent neural networks (RNN)" used RNN deep learning technique and the cloud environment is set up with OpenNebula cloud. The algorithms for confidentiality, integrity, and authentication algorithms are implemented in java. Cryptanalysis against different cryptographic attacks proves that our proposed method contributes a secure and proficient mechanism for storing health information. Furthermore, after comparing the encryption time of our system with the standard encryption algorithms, it is inferred that this CIA model incurs less overhead for encryption. And also, better performance has been observed on the factors like an error rate of 2% and an accuracy rate of 98%, minimum overhead on computation, communication, and storage costs.

The chapter "Blockchain technology integrity in health informatics using hyper ledger platform" proposed a conceptual framework and implementation of blockchain technologies with IoT using hyper ledger on potential applications in healthcare. A novel framework was proposed based on the most popular Ethereum and hyper-ledger of fabric platforms/frameworks in this domain. Briefly discussed the present status of IoT-based healthcare models in the domain and analysis of IoT and blockchain applications in healthcare. IoT and BCT improve medical image quality and storage or transmission for healthcare application challenges and issues. Medical image diagnosis benefits from the current future direction in this domain. In conclusion, the use of BCT in healthcare is growing exponentially. There are some areas of the healthcare domain where BCT has the potential to have a significant impact.

The chapter "A recent study based on prediction approach over Coronavirus Disease-2019 cases in India using regression modeling of Artificial Intelligence" shows an analysis might help political leaders and health authorities to allocate enough resources, including personnel, beds, and intensive care facilities, to manage the situation in the next few days and weeks.

Dr Mohamed Abouhawwash, USA
Dr Sudeep Tanwar, Ahmedabad, India
Dr Anand Nayyar, Da Nang, Viet Nam
Dr Mohd Naved, Delhi, India

About the editors

Mohamed Abouhawwash received the BSc and MSc degrees in statistics and computer science from Mansoura University, Mansoura, Egypt, in 2005 and 2011, respectively. He finished his PhD in statistics and computer science, 2015, in a channel program between Michigan State University, USA, and Mansoura University, Egypt. He is at Computational Mathematics, Science, and Engineering (CMSE), Biomedical Engineering (BME) and Radiology, Institute for Quantitative Health Science & Engineering (IQ), Michigan State University, East Lansing, MI 48824, USA. He is an assistant professor with the Department of Mathematics, Faculty of Science, Mansoura University, Egypt. In 2018, Dr Abouhawwash is a visiting scholar with the Department of Mathematics and Statistics, Faculty of Science, Thompson Rivers University, Kamloops, BC, Canada. His current research interests include evolutionary algorithms, machine learning, image reconstruction, and mathematical optimization. Dr Abouhawwash was a recipient of the best master's and PhD thesis awards from Mansoura University in 2012 and 2018, respectively.

Sudeep Tanwar is working as a full professor at the Nirma University, India. He is also a visiting professor with Jan Wyzykowski University, Poland, WSG University in Bydgoszcz, Poland, and the University of Pitesti, Romania. He received BTech in 2002 from Kurukshetra University, India, MTech (Honors) in 2009 from Guru Gobind Singh Indraprastha University, Delhi, India and PhD in 2016 with specialization in Wireless Sensor Network. He has authored 4 books and edited 24 books, more than 400 technical articles, including top cited journals and conferences, such as *IEEE TNSE, IEEE TVT, IEEE TII, IEEE TGCN, IEEE TCSC, IEEE IoTJ, IEEE NETWORKS, IEEE WCM, ICC, IWCMC, GLOBECOM, CITS*, and *INFOCOM*. He initiated the research field of blockchain technology adoption in various verticals in the year 2017. His H-index as per Google Scholar and Scopus is 64 and 53, respectively. His research interests include blockchain technology, wireless sensor networks, fog computing, smart grid, and the IoT. He is a member of the Technical Committee on Tactile Internet of IEEE Communication Society. Recently, he has been awarded cash prize of Rs, 50,000 for publishing papers with 5+ Impact factor and publication of books with Springer, IET and CRC under the scheme of "Faculty Awards and Incentives" of Nirma University for the year 2019–2020. He has been awarded the Best Research Paper Awards from IEEE IWCMC-2021, IEEE ICCCA-2021, IEEE GLOBECOM 2018, IEEE ICC 2019, and Springer ICRIC-2019. He has won Dr KW Wong Annual Best Paper Prize (with 750 USD) for 2021 sponsored by

Elsevier (publishers of JISA). He has served many international conferences as a member of the Organizing Committee, such as the Publication Chair for FTNCT-2020, ICCIC 2020, and WiMob2019, and a General Chair for IC4S 2019, 2020, 2021, 2022, ICCSDF 2020, FTNCT 2021. He is also serving the editorial boards of COMCOM-Elsevier, IJCS-Wiley, Cyber Security and Applications – Elsevier, Frontiers of blockchain, SPY, Wiley, IJMIS Journal of Inderscience, JCCE, and JSSS. He is also leading the ST Research Laboratory, where group members are working on the latest cutting-edge technologies.

Anand Nayyar received PhD (Computer Science) from Desh Bhagat University in 2017 in the area of Wireless Sensor Networks, Swarm Intelligence and Network Simulation. He is currently working in School of Computer Science-Duy Tan University, Da Nang, Vietnam as Professor, Scientist, Vice-Chairman (Research) and Director – IoT and Intelligent Systems Lab. A Certified Professional with 125+ Professional certificates from CISCO, Microsoft, Amazon, EC-Council, Oracle, Google, Beingcert, EXIN, GAQM, Cyberoam, and many more. Published more than 150+ research papers in various high-quality ISI-SCI/SCIE/SSCI Impact Factor Journals cum Scopus/ESCI indexed Journals, 70+ papers in international conferences indexed with Springer, IEEE and ACM Digital Library, 40+ book chapters in various SCOPUS, WEB OF SCIENCE Indexed Books with Springer, CRC Press, Wiley, IET, Elsevier with citations: 9500+, H-Index: 52 and I-Index: 180. Member of more than 60+ associations as senior and life member including IEEE, ACM. He has authored/co-authored cum edited 40+ books of computer science. Associated with more than 500+ international conferences as program committee/chair/advisory board/review board member. He has 18 Australian Patents, 4 German Patents, 4 Japanese Patents, 11 Indian Design cum Utility Patents, 1 USA Patent, 3 Indian Copyrights, and 2 Canadian Copyrights to his credit in the area of Wireless Communications, Artificial Intelligence, Cloud Computing, IoT, and Image Processing. Awarded 38 Awards for Teaching and Research – Young Scientist, Best Scientist, Best Senior Scientist, Asia Top 50 Academicians and Researchers, Young Researcher Award, Outstanding Researcher Award, Excellence in Teaching, Best Senior Scientist Award, and many more. He is listed in Top 2% Scientists as per Stanford University (2020, 2021, and 2022). He is acting as an associate editor for Wireless Networks (Springer), Computer Communications (Elsevier), International Journal of Sensor Networks (IJSNET) (Inderscience), Frontiers in Computer Science, PeerJ Computer Science, Human Centric Computing and Information Sciences (HCIS), IET-Quantum Communications, IET Wireless Sensor Systems, IET Networks, IJDST, IJISP, IJCINI, and IJGC. He is acting as an editor-in-chief of IGI-Global, USA Journal titled *International Journal of Smart Vehicles and Smart Transportation (IJSVST).* He has reviewed more than 2,500+ articles for diverse Web of Science and Scopus Indexed Journals. He is currently researching in the area of Wireless Sensor Networks, Internet of Things, Swarm Intelligence, Cloud Computing, Artificial Intelligence, Drones, Blockchain, Cyber Security, Network Simulation, Big Data and Wireless Communications.

Mohd Naved is a distinguished Associate Professor with an impressive career spanning over a decade in the fields of Business Analytics, Data Science, and Artificial Intelligence. As an educator, Dr. Naved has consistently demonstrated a commitment to the highest standards of teaching and mentoring, ensuring that his students receive an education that is both cutting-edge and grounded in real-world experience. His dedication to helping students achieve their full potential extends beyond the classroom, as he has been an active participant in the university's Mentor-Mentee Program, providing guidance and support to over 150 undergraduate and postgraduate students. In addition to his teaching prowess, Dr. Naved has excelled in the areas of education management, research, and curriculum development. He has served on various committees and led initiatives related to curriculum development, faculty recruitment and retention, and accreditation, contributing to the institutions he has worked with becoming centers of academic excellence in their respective fields. He has also successfully led the launch of several BBA/MBA programs, resulting in increased admissions and student satisfaction.

As a researcher, Dr. Naved has made significant contributions to the fields of Business Analytics, Data Science, and Artificial Intelligence, with over 80+ publications in reputed scholarly journals and books. His research focuses on the applications of these disciplines in various industries, and he has supervised numerous research projects and dissertations, guiding students to successful outcomes.

Chapter 1

Introduction to healthcare informatics: fundamentals and historical background

V. Pandimurugan[1], Mohammed Abouhawwash[2,3], Rahul Mandviya[4] and Chetaly Mawal[4]

Abstract

Health informatics is new emerging field all over the world, and it provides good quality health services to all the people in the affordable price and also in an accessible manner. There are so many definitions available for health informatics (HI); in simple way, we can say, how the information technology plays a role in the healthcare sector in terms of collection of data, handling data, storage, retrieval, management, etc. Handling data in the healthcare sector can be a very challenging job due to the increasing sophistication and standardization of data in the medical field. Data comes in various forms such as computable electronic data, structured electronically-entered data that cannot be computed by other systems, and unstructured data. HI or medical informatics or clinical informatics manage healthcare data and information through the application of computers and other technologies. It deals with the data that is between the people and health entities, for example, medical informatics data needs to be shared with informaticians to deal with the data in the various aspects like analysis, transfer, knowledge, and innovation based on that data. HI field also interfaces with other fields like computer science specially with artificial intelligence (AI), biomedical engineering, health sciences and public health. This chapter provides the information of data, knowledge, wisdom to handling health data, electronic health record, health information technology key players, importance of data analytics, healthcare standards and ethics to be followed in the health sectors.

Keywords: Health informatics; Information hierarchy; HIT; EHR; Health standards; M-health

[1]School of Computing, Networking and Communications, SRMIST, India
[2]Department of Mathematics, Faculty of Science, Mansoura University, Egypt
[3]Department of Computational Mathematics, Science, and Engineering (CMSE), College of Engineering, Michigan State University, USA
[4]School of Computing Science and Engineering, VIT Bhopal University, India

1.1 Introduction

Technology and healthcare merge in the field of health informatics (HI). Utilizing technology to gather and use data to enhance patient care, healthcare sector operations, and support for research and development in the healthcare sector is both an art and a science.

The world's first hospital was initiated by the leader Benjamin Franklin in 1752, now known as Pennsylvania Hospital. He worked as the secretary of the hospital and maintained the health records of patients including their name, address, disorder, and date of admission and discharge. To raise the standard of medical records, the American Health Information Management Association (AHIMA) was founded in 1928. The electronic health records (EHRs) that are widely used today in the world of informatics were first introduced in the form of electronic medical records in 1949 by Dr Gustov Wagner. He initiated an organization named the Society for Medical Documentation, Informatics, and Statistics. In the late 1960s, the use of electronic medical records was observed, and software and databases were developed in hospitals to securely store the data of the patient.

In 1968, AI Mikhailov created the structure and defined the properties of the information system. Various systems and software began to be released for patient scheduling and later to record and document management. Meanwhile, an organization called the American Health Informatics Association (AMIA) was created to educate health professionals in HI and systems. The US government has invested about 1 million dollars in their military medical department for soldiers. A few years later, the first Windows-based software, EpicCare, was released for EHRs and Mr Obama later introduced the National EHR System [1–3]. However, over time, the use of computer systems increased in the market and the demand for mobile health led to the birth of mobile health (M-health).

In day-to-day life, data plays a vital role, and it transforms from information to knowledge, and wisdom also provides an opportunity in digital technology. To maintain the healthcare activities like therapy, deciding staffing levels, and data exchange, we need an information management system that may be called a clinical or HI system. For creating informatics management, we must design the model and its purpose to develop with three important parameters such as model, measure, and manage. Any informatics management must have the following steps to provide better quality outputs and user needs for real-world problems.

- ✓ Define the goal of the information system.
- ✓ Make a model for the system.
- ✓ Measure the model by collecting and gathering the data.
- ✓ Access the measured model with sample input and outputs.
- ✓ Manage the model and alter it when change is required.

This chapter provides the following information to the reader:

- *Definition of health/bio/medical informatics and its historical highlights*
- *Identify the key players in the health information technology (HIT)*

- *Describe the EHR role in HI*
- *Describe the applications of M-health in HI*
- *State the potential impact of health standards and policies*

1.1.1 Information hierarchy

Information hierarchy means it provides the level of data and its transformation into the different levels based on the one state to another state. Data, information, knowledge, and wisdom are called the information hierarchy.

Data may be values, symbols, and characters based on which operations are performed by the computer, which may be stored or transferred by the communicating media.

Information means how the humans and computer must conclude the data into meaningful for the real-world applications. Knowledge is whether it is right or wrong or true or false which is to be concluded based on the information. Wisdom is a combination of information and knowledge to use the best and innovative solutions that depend on the situation. Figure 1.1 highlights information hierarchy.

1.1.2 Evolution of HI

Last 50 years and before, there is not much more technological development, it was very crucial thing to handle a large amount of data, but now the evolution of computers and new technologies has made easy to manage those data in an efficient manner. Before the evolution of HI, medical data, and lab reports, patient history was handled by the exclusive person in each hospital and clinics. Maintaining patient and medical records in papers is easy but it has many disadvantages like manually they can do some false reports, handwritten by physicians may be

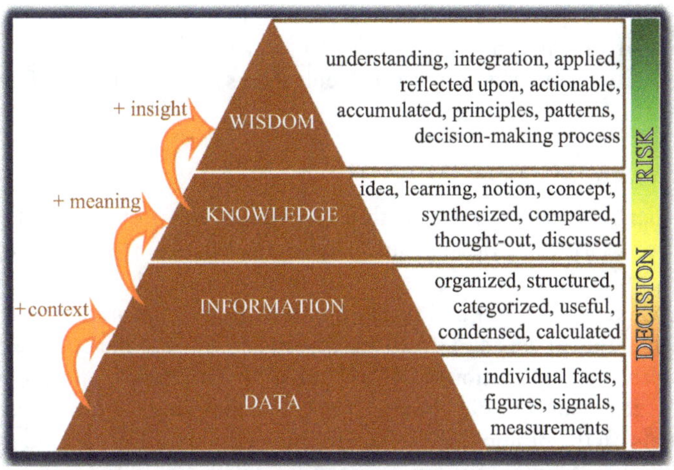

Figure 1.1 Information hierarchy

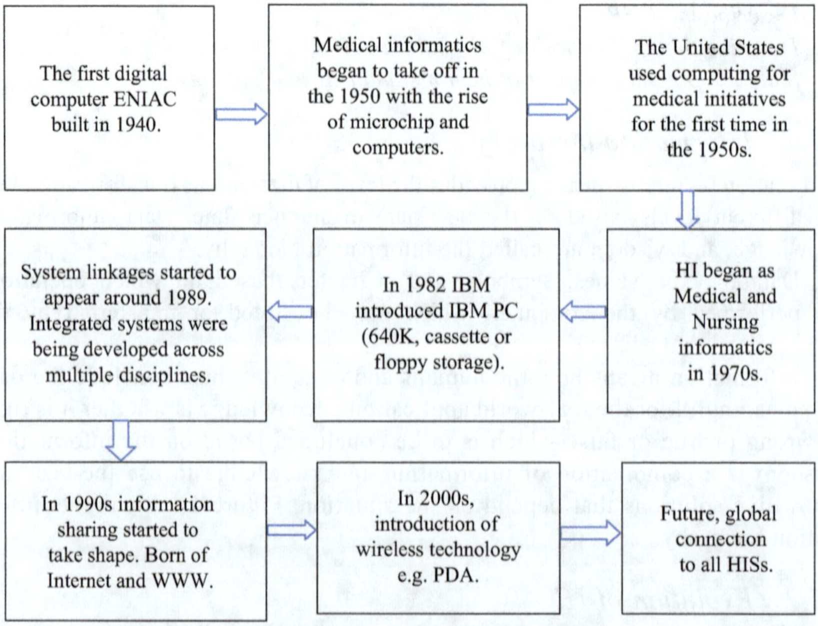

Figure 1.2 Evolution of HI

difficult to understand by the patient caretakers, pharmacies, etc. [4]. Clearly, we all know that without the evolution of computers, there is no health information system. In 1940, the first ENIAC computer was developed, and it required a large amount of space, later IBM launched personal computers in the year 1982. Due to the launch of personal computers, HI systems started to grow with the help of software and system linkages [5–8]. Another milestone for HI, World Wide Web, provides the public access to Internet, due to that sharing and exchanging of information is possible and anywhere, anyone could access the information in the world. Figure 1.2 enlightens the evolution of healthcare informatics.

1.1.3 Definition of HI

HI means handle the health data with a set of rules and protocols that should be followed for storing, retrieving, and sharing the data or present the data for research, clinical aspects, and some innovative activities in the healthcare applications. It is also defined as an applied filed of information science on managing healthcare data and gives intelligent solutions to the complex problem in the healthcare applications with the help of computers and technologies. Bioinformatics is an interdisciplinary field, which applies the innovative methods and tools on the biological data to provide better solutions for the real-world problems [9,10]. Medical informatics can be defined as a field of science and engineering that provides methods and technology for the acquisition of data,

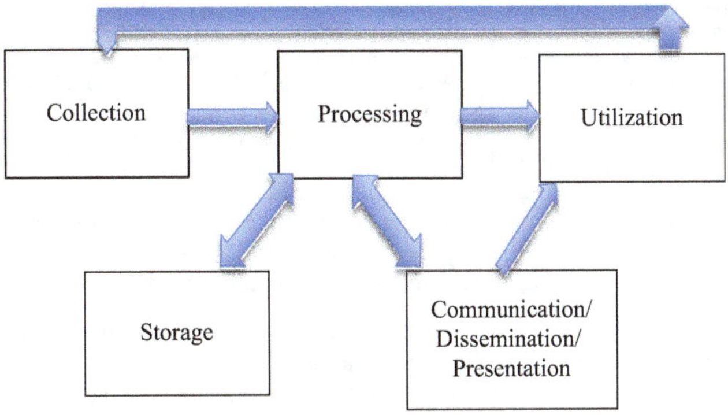

Figure 1.3 General process for information system

personal data, medication information, lab reports to be stored, retrieved, and shared in a secure way for better quality of healthcare to all the peoples. For better understanding, Figure 1.3 shows the general process for any information system.

1.2 Key players of HIT

For implementing information technology tools in healthcare, first we need to identify who are all the important key players for efficient health information systems. The following are the list of key players in health information technology:

- Patients
- Physicians and nurses
- Support staff
- Public health
- Federal and State Governments
- Medical educators
- Insurance companies (payers)
- Hospitals and clinics
- Medical researchers
- Technology vendors

Telemedicine and patient monitoring is also one of the key important factors for HI; it will help the people who are residing in the urban and non-communicable places, they will also get the care at right time through this technology, due to which many developing countries are giving primary health tips and facilities to their citizens [11,12]. Internet of Medical things plays an important role because of IoT devices, many healthcare devices take care of the

Organization	**Functions**
Institute of Medicine	It was created in 1970 by the National Academy of Sciences with the responsibility of assessing healthcare-related policy and offering advice to the Federal Government and the public
Association of American Medical Colleges (AAMC)	Optimize the practice information management. Support resources and tools, innovation, and dissemination of research findings
Healthcare Information Technology Standards Panel	The Department of Health and Human Services (DHHS) established it in 2005, and it offers a blueprint for how standards and specifications should be developed
National eHealth Collaborative	Prioritization of HIT standards to promote interoperability. Program for online education of stakeholders on various HIT issues
Certification Commission for Healthcare Information Technology	Physicians may invest in health information technology (HIT) with less risk if they can assure that it is interoperable, increase the availability of HIT incentives, and expedite the adoption of interoperable HIT
American Recovery and Reinvestment Act	Improve medical quality, patient safety, healthcare efficiency, and health inequities through engaging patients and families, improving care coordination, ensuring proper privacy and security of personal health information, and improving population and public health
Health IT Policy Committee	Set up the legal foundation necessary for the creation and expansion of the national health information exchange. Enrollment, data intermediaries, privacy and security, governance, and quality measures

many patients' lives in a secure way. It is the most important factor in the HI field; IoT makes everything possible in the secure and easy monitoring way of the patient's life, communicating information, sharing, and securing. In 2018 article, 'Health Information Technology Continues to Show Positive Effect on Medical Outcomes: Systematic Review' found that over 86% of the reviewed literature shows positive effects of HIT on the efficiency and effectiveness of medical outcomes [13–15]. Directed exchanges utilize secure messaging to send and receive information between known parties, usually healthcare providers or healthcare facilities. Query-based exchanges initiated by a question or request of personal health information use a data repository or record locater to find patient's EHRs. Consumer-mediated exchanges refer to the process where a patient actively participates in their healthcare delivery by communicating relevant health information.

Big data is a crucial component in the health sectors because in this filed, so much structured and unstructured data are involved for the various operations such as drug delivery, big data analytics through disease prediction, artificial intelligence techniques that can be applied on the data and shows how to store data centrally and securely. It is also used for developing health-related service providing companies and patient–doctor relationship and it's policies.

1.2.1 Organizations involved with HIT

It is important for all the fields to have an exclusive organization to guide the members, employees, and users to monitor the standards and procedures that are to be followed effectively and whatever the new emerging technology has arrived, to train the people through workshop, conference, short-term course, etc.

1.2.2 Barriers to HIT adoption

Building a health information system infrastructure is difficult one in terms of many parameters like time, money, knowledge, Internet, etc. [16,17]. Adopting HIT varies from developed countries to developing countries for the above said parameters.

Inadequate infrastructure: Most developing countries do not spend more money on building healthcare systems for buying efficient systems, software, hardware, etc. Strong infrastructure is required for strong healthcare systems to deliver quality healthcare services to the public.

Poor Internet availability: Internet plays a crucial role in any informatics system for sharing and exchanging information effectively. Without proper Internet facility, it is difficult to achieve the best healthcare service.

Lack of professionals: For operating or working the information system, we need skilled and trained professionals for use of any new technology. When we plan for training the professional, we must spend more money on the resource person and learning the new modern aids. So training requires cost as well as time.

Social and cultural barriers: Digital gap and e-readiness in the healthcare sector are significant social and cultural impediments. Due to stakeholders' lack of desire, curiosity, and fear of embracing and utilizing new technologies, it is complicated to transform health information system from paper based to digital because they are more comfortable with their conventional approach and routing practice.

Organizational barriers: Organizations are hesitating to invest money for new modern technology because there is a lack of evidence like no survey, proof model for better results to show the improvement, and they are convenient with the old method.

1.3 EHR

In health informatics, EHR plays a vital role; without it, HI is not fulfilling. EHR development and promotion initiatives date back more than 35 years. EHRs, however, have only just started to take root in the field of health informatics [18–20]. EHR is the single point of deposition and access for nearly all formal elements of patient's data.

EHR definitions: It defines the health-related information in the electronic format, which are to be stored, retrieved, and shared in the secured manner.

An EHR of a person complies with nationally accepted interoperability standards and may be generated, managed, and viewed by authorized physicians and employees from many healthcare organizations.

1.3.1 History of EHR

The first electronic health record (EHR) systems appeared in the 1960s with the Mayo Clinic being one of the first major health systems to implement an EHR [21,22]. According to the survey conducted by the University of Scranton, in the year 1972, many hospitals started to implement EHRs for the management of patient. To solve standardized difficulties as EHR development advanced, Health Level 7 was established in 1987. According to Greater Than One Labs, a digital communication company was established in New York City, and it now has members in 55 countries. By the year 2000, the Institute of Medicine established a goal for all doctors to regularly utilize the computers.

1.3.2 Key components of EHR

The key components of the EHR can be classified as the administrative components and lab components. Administrative components are important in EHR because it consists of patient registration, admission in and out details, medication, etc. It also includes patient demographics, employer information, chief complaint, etc. [23,24].

Laboratory systems are included in the EHR for providing schedules, billing, results, and other information which is more required for maintaining the patient's history. Rarely laboratory data is fully incorporated into an EHR [25]. When a laboratory information system is created by the same vendor as an EHR, it is not simply integrated with it, as is the case with the Cerner laboratory information system's interface with more than 400 different laboratory tools, many technologies and analyzers are used in the diagnostic laboratory procedure. Figure 1.4 shows key components of EHR.

1.3.3 Benefits of EHRs

EHRs and the capacity to exchange health information electronically can improve your organization's operations and help you in giving better and safer treatment to the patients. Figure 1.5 shows the history of EHR and HIS. EHRs assist clinicians

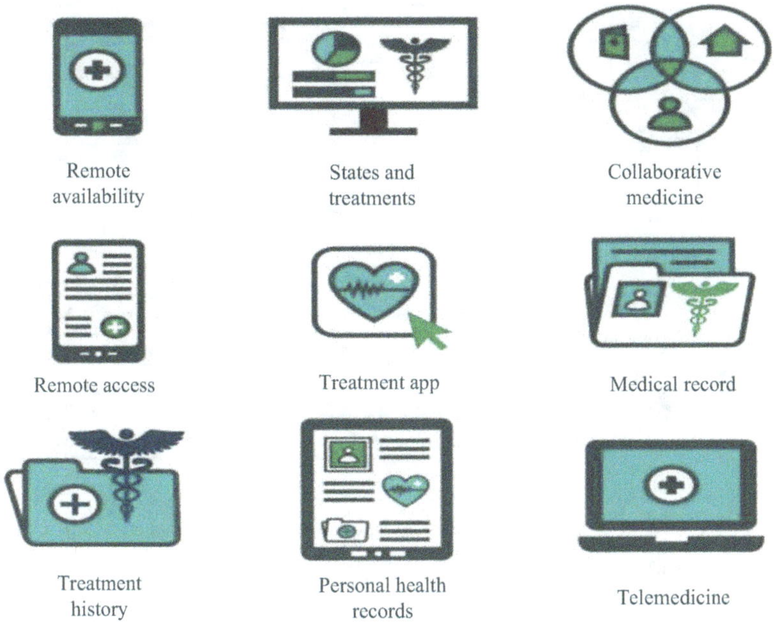

Figure 1.4 Key components of EHR

in improving in the management of patient care and delivery through the following (Figure 1.6 shows the benefits of EHR):

- Patients' information will provide accurate and up to date data for better care.
- For efficient caring, quick access can be given to clinician and physician.
- Transforming information securely between all types of stake holders.
- Reduce the medication and human errors.
- Better health-care convenience with the help of better communication.
- Prescribing can be safer compared to old method (paper).
- Helping promote legible, complete documentation and accurate, streamlined coding and billing
- Patient data is more secure and provide better privacy.

1.3.4 Steps to adopt and implement an EHR

The following are the steps to adopt and implement an EHR system:

Step 1: The patient started to provide the details of the previous medication, allergy, and surgeries. Patients are able to see the online appointments, visits, and prescriptions in their portal itself. Figure 1.7 shows the Stages of implementing EHR.

Step 2: According to the EHR information, system automatically provides the details of next visit and optimal time for better patient management.

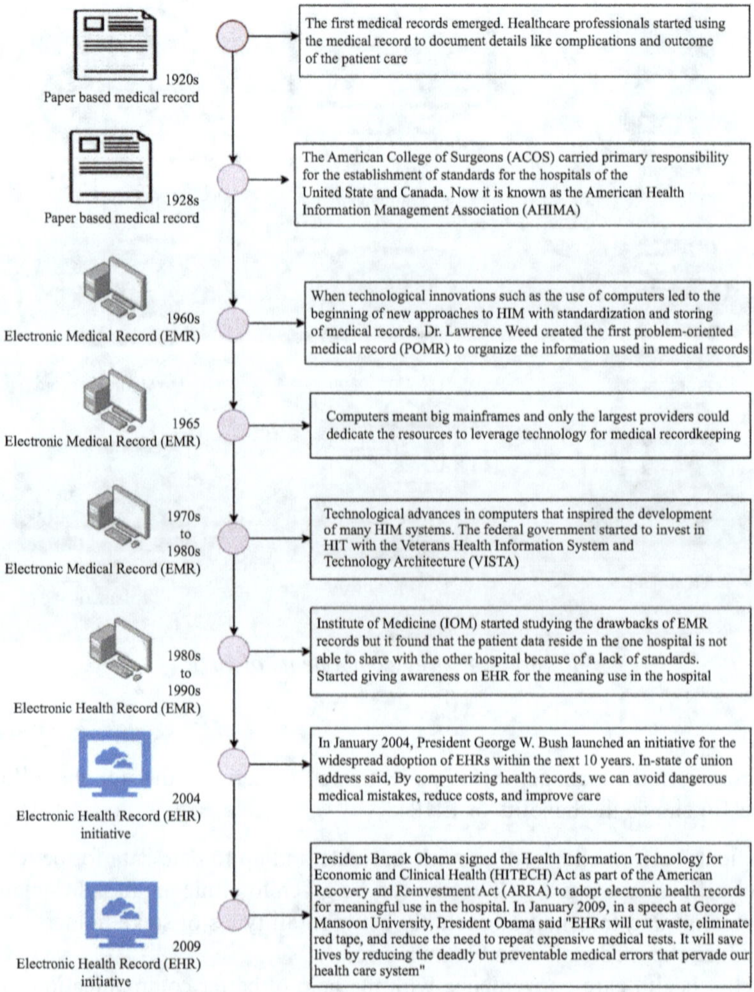

Figure 1.5 History of EHR and HIS

Step 3: Based on the conflicting details, doctor can check the list of appointments and patient details before visiting the patient. Diagnosis and step-by-step medication plan can be easily provided to the patients by the doctor.

Step 4: Transmitting the prescription is possible and its automation, so pharmacy can provide the medicine without any delay and stock maintenance is also easy.

Step 5: The patient receives the bill from the financial division once it is automatically generated by the EHR platform.

Step 6: The system generates the insurance claim, making sure that its structure complies with the requirements of the patient's insurance company.

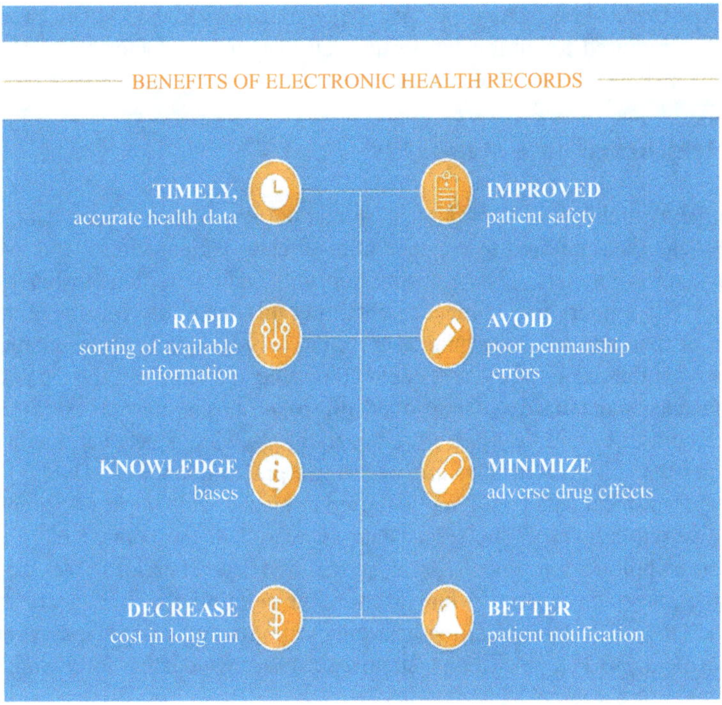

Figure 1.6 Benefits of EHR

Stage	Cumulative Capabilities
Stage 7	Complete EMR; CCD transactions to share data; Data warehousing; Data continuity with ED, ambulatory, OP
Stage 6	Physician documentation (structured templates), full CDSS (variance & compliance), full R-PACS
Stage 5	Closed loop medication administration
Stage 4	CPOE, Clinical Decision Support (clinical protocols)
Stage 3	Nursing/clinical documentation (flow sheets), CDSS (error checking), PACS available outside Radiology
Stage 2	CDR, Controlled Medical Vocabulary, CDS, may have Document Imaging; HIE capable
Stage 1	Ancillaries - Lab, Rad, Pharmacy - All Installed
Stage O	All Three Ancillaries Not Installed

Figure 1.7 Stages of implementing EHR

Step 7: The medical laboratory may also be given access to the EHR in cases when a patient is required to submit lab results. On demand, a doctor can see the test findings.

1.4 Healthcare data standards

Despite the availability of communication technology to enable such data, a large portion of the data required to enhance clinical care, patient safety, and quality is dependent on computers. Data transmission still cannot be done easily or inexpensively. The primary impediment to obtaining this capability was the unintended adoption of data standards for clinical information organization, presentation, and coding so that the receiving system could comprehend and accept the data [26–28]. Common data standards have improved information sharing across commercial clinical laboratories and healthcare organizations, as well as between facilities and pharmacies that fill prescriptions and healthcare practitioners [29]. Due to the lack of standards, clinical data could not be reused to meet a wide range of patient safety and quality requirements. The term "data standard" in the context of healthcare refers to techniques, protocols, and terms and specifications for gathering, exchanging, storing, and retrieving information pertaining to healthcare applications, such as medical records, medications, radiographic images, payments, and reimbursement [30,31]. Health data standardization includes the following:

- **Data element definitions:** Definition of data content to be collected and exchanged.
- **Data exchange format:** A standard format for electronically encoding data elements, such as sequences and error handling. The exchange standard can also include a document architecture for structuring data elements during the exchange and an information model that defines the relationships between data elements in a message.
- **Terminology:** A syntax that describes the medical terms and concepts used to describe, classify, and code data elements and data representation languages.
- **Knowledge representation**: A standard way to electronically represent medical literature, clinical guidelines, etc., to support decision making.

1.4.1 HL7

For the transformation and exchange of EHRs HL7, a non-profit organization that was established in 1987 and was recognized by the American National Standards Institute in 1994 is necessary. It has members in more than 50 nations. A framework for exchanging, integrating, sharing, and retrieving electronic health information is provided by HL7 and to its members. These standards describe the language, structure, and data types necessary for smooth system integration, defining how information is packed and communicated from one party to the next [32,33]. The HL7 standard, which is said to be the most commonly utilized in the world, supports the management, delivery, and evaluation of clinical practice and medical services. By offering instructions on how to put that standard into practice,

HL7 also encourages interoperability in international healthcare. The seventh level of the ISO concept for integrating open systems is referred to as Level 7. This makes it clearer that HL7 messaging is located in the stack's application layer and that the lower layers are components of the overall toolkit [34,35]. Patient Administrative Systems (PAS), electronic practice management, lab systems, dietary, pharmacy, billing, and EHRs all communicate via the HL7 data standard.

1.4.2 Clinical document architecture

To compare the contents of documents produced by information systems with highly varied features, it is necessary to free up crucial clinical content that is now contained in free text clinical notes. Standardized semantics are now long-term available due to the range of clinical signals including structure, underlying information model, level of semantic coding, usage of standard medical words, platform, and give unique qualities [36,37]. Although it is a start in the right direction, the existing Clinical Document Architecture (CDA) standard does not permit these vital goals to be fully enabled. A document markup standard known as CDA defines the semantics and structure of clinical texts. The clinical document includes the following unique elements and serves as a record of observations and performance:

- *Persistence*: Clinical documents do not change for a period determined by regional and regulatory requirements.
- Clinical records are maintained by the person or organization responsible for their care.
- *Certification function*: A clinical document is a collection of legally recognized information. Certification of the entire clinical document applies to the entire document, not part of the document that does not contain the full context of the document.
- Clinical documents are in a human readable format.

The Extensible Mark-up Language (XML) family of HL7 Version 3 standards includes the CDA, which draws semantic content from the HL7 Common Reference Information Model (RIM) and CDA (XML). The EHR, personal health record (PHR), discharge summaries, and progress notes are all kept in the CDA. Version 3 development technique includes the creation of an XML-based RIM implementation. The exact style of XML representation in HL7 is a careful balance of technical, practical, and functional considerations.

1.4.3 Logical observations: identifiers, names, and codes (LOINC)

A key clinical word for clinical test orders and results, the name and code (LOINC) of the logical observation identifier, is a component of a set of established standards for the electronic interchange of clinical health information. The HL7 Standards Development Organization designated LOINC as the recommended code set for test names in exchange among healthcare facilities, laboratories, and test equipment in 1999 [38]. Over 72,000 phrases that are utilized in test results are in the LOINC database. This is required since different laboratories have different, incompatible codes. LOINC applies universal codenames and identifiers to medical terms related to EHRs [39–41]. The purpose is to support the electronic exchange and collection of clinical results such as clinical tests, clinical observations, results management, and research. LOINC has two main parts: Lab LOINC and Clinical LOINC. The Clinical LOINC contains a Document Ontology subdomain that captures clinical reports and document types. Benefits of implementing LOINC include improved communications across the integrated medical services network, improved EHRs across the community, automatic submission of reportable disease cases to public health agencies such as disease management and epidemic detection [42]. There is an improvement in the transmission of information and provides information about the services provided and reduces system errors. This will greatly improve the overall quality of medical care.

1.4.4 Digital imaging and communications in medicine

Digital imaging and communications in medicine (DICOM) is the standard for communicating and managing medical imaging information and related data. DICOM was founded in 1983 by the National Electrical Manufacturers Association (NEMA) and the American College of Radiology (ACR). It is most commonly used for the storage and transfer of medical images and integrates medical imaging equipment such as scanners, servers, workstations, printers, network hardware, image archives from multiple manufacturers, and communication systems (PACS). It is widely adopted in hospitals and is also being adopted in smaller applications such as dental clinics and clinics. DICOM groups information into records. For example, a file containing a chest X-ray may contain the patient ID to prevent the image from being accidentally separated from this information [43,44]. This is similar to embedding tags in an image format such as JPEG to identify and describe the image. DICOM consists of services, most of which involve sending data over the network. The offline media file format will be added to the standard later. The DICOM storage service is used to transfer images or other persistent objects to image archives and communication systems.

1.4.5 Systematized nomenclature of medicine: clinical terminology

SNOMED CT is a collection of medical terminology that can be processed by computers and is systematically arranged. It includes codes, phrases, synonyms, and meanings used in clinical records and reports. It is regarded as the most complete and multilingual phrase for clinical medicine in today's use. The major goal is to effectively gather clinical data and encode the meaning utilized in health information to enhance patient care. It gives the most basic definition of an EHR. Clinical observations, symptoms, diagnoses, treatments, bodily structures, organisms, and other etiologies, as well as chemicals, medications, equipment, and specimens are all included in through reports. SNOMED CT is a collection of medical terminology that can be processed by computers and is systematically arranged. It includes codes, phrases, synonyms, and meanings used in clinical records and reports. It is regarded as the most comprehensive and multilingual phrase for clinical medicine in the entire world [45,46]. The major goal is to effectively gather clinical data and encode the meaning utilized in health information to enhance patient care. It gives the most basic definition of an EHR. Clinical observations, symptoms, diagnoses, treatments, bodily structures, organisms, and other etiologies, as well as chemicals, medications, equipment, and specimens are all included in through reports.

1.5 M-Health

Telehealth is a broad term that encompasses methodologies like remote patient monitoring, reminder services, and other remote care technologies [47]. M-health is a subset of the broader term telehealth. It is a term coined in 2008 by Rockefeller Foundation in the seminal conference "eHealth connect," and this was a clear representation of the evolution of E-health. In layman's terms, M-health is the utilization of wireless technology for delivering health care services and necessary data and information [48–51]. This can be achieved through mobile phones, patient monitoring devices, personal digital assistants, smart monitoring IoT devices, tablets, and any other devices that are capable of wireless connectivity. M-health means using mobile devices to monitor or detect biological changes in the human body, and device management units such as hospitals and clinics. Service providers collect clinical data and use it for care provision.

1.5.1 Different types of M-health technology in the health sector

There has been a tremendous increase in the availability of wireless connectivity and devices capable of utilizing it. This has led to a boost in the development of M-health technology where mobile and other wireless devices will be used in the healthcare industry. Mobile technology has been used for storing medical information, patient monitoring, clinical decision support making, and other medical purposes. Accordingly, this is leading to the growth of M-health worldwide and from this, we come to our next point of various types of M-health technologies used in the health sector.

The following are the various types of technologies used in M-Health:

1. Tablets
2. Smartphones and Apps
3. Wearables
4. Implantables

Tablets

Since the launch of the very first tablet by Apple in the global market, the healthcare industry has been discovering innovative solutions for taking advantage of this technology to aid doctors, nurses, and other medical staff. In earlier days, nurses or medical staff used to jot down every observation of a patient or any research but since there has been the growth of technology in the healthcare industry as well, now they can use tablets for such clinical work. Tablets are wireless electronic devices that are widespread worldwide due to their portability. It can collect, observe, analyze, and access information anytime and anywhere. Let us consider the case of hospitals where nurses, physicians, doctors, and other supporting staff members have started using tablets instead of pen-paper. Earlier they used to check the patients, note down their symptoms and observations. But this process was prone to basic human errors causing trouble to the patients [49,52]. Many times it was observed that the medical reports were misplaced or the patient and pharmacist were unable to read them for later reference. So, to avoid this, tablets are preferable as they help in improving point of care documentation. To access a patient's medical history, according to recent laboratory reports, the usage of EHR can be considered for the real-time coordination between patient and doctor. Post-surgery (if any), the tablets can be used for educating the patient about future medications and online therapies for recovery. Tablets also promote telemedicine, Internet-based communication from different parts of the country or even the world easily for treating them from anywhere.

Smartphones and Apps

Smartphones play a vital role in today's world and are now being used by people all around the world. This smartphone revolution has enabled the healthcare industry to come up with notable innovative products which not only include simple mobile applications but also hardware applications. Healthcare companies like FreeStyle Libre and Accu-Chek have come up with innovative solutions for Continuous Glucose Monitoring (CGM) systems. CGM automatically keeps track of a patient's glucose level throughout the day and night through the data received from a small sensor worn on the back of the upper arm. Blood-glucose meters display glucose levels only but CGM is capable of providing detailed insights like where glucose level was, where it is at present, and where it will be headed to. All this can be simply provided on a mobile app. A mobile app is capable of providing real-time glucose information like current glucose level, trend arrow, history of glucose levels, tracking food, and helping decide what food should be eaten to control glucose level. Advanced techniques of NLP like text-to-speech and speech-to-text can simplify the usage of applications for senior citizens. Many fitness apps help track the number of calories taken in and calories burned based on reading received from the gyroscope of a smartphone [53,54]. Recently, considering the COVID-19 pandemic situation, many people have faced problems related to mental health. This is where apps like MoodKit, Talkspace, and Better Stop Suicide played a major role in improving the mental health of patients. These apps provide access to a large variety of stress buster activities, features to track progress, directly work with mental health professionals virtually, 24 × 7 access to different therapists, recorded audio clips that contain stories to be heard in stressful situations to calm the mind, exercises for mood-boosting, etc. Many apps like QUP in collaboration with doctors came up with features like online appointment booking to avoid long waiting queues and crowds outside clinics, online consultation for diseases which does not require the physical presence of a doctor and also helped manage health records easily as it provides all prescriptions and reports on the application itself. A few pieces of research are being done on detecting deadly diseases like oral cancer, skin cancers, based on the images taken by smartphones using advanced deep artificial neural networks for early diagnosis. Many medical awareness campaigns can be carried out through social media platforms to target audiences based on region and the medical threats in that region.

Wearables

Wearables are generating a boom in the healthcare industry as they are gaining popularity in the fitness space. The fitness freaks are now demanding to keep a track of their health by monitoring through wearables. Wearable technology includes electronic devices which can be worn on various body parts according to the device. There is an increasing demand for products like Fitbits, Smartwatches, Virtual Reality (VR), etc. recently due to great influence on consumers due to their features such as easy to use and portability. These days people carry their wearables along with smartphones in their pockets anywhere they go. Since there is a great demand for wearables, now even sports athletes are using these to keep a count of their oxygen level, blood glucose level, pulse rate, and many more things [55,56]. For instance, for the athletes involved in swimming, long jump, and heavy weightlifting, there is a specific type of costume that comes with attached or in-built IoT sensors to measure their every move and analyze later if required. Not only in sports or fitness but wearables are also being used in the fashion industry in the form of smart rings or smart jackets. The smart rings are designed in such a way that they look stylish and will measure your body calories, body temperature, steps taken through mobile application whereas the smart jackets are mainly created and used by consumers to measure the body temperature if the jacket has a heater inside it, oxygen level, SPGS/GPRS to keep a track on the person wearing it. As mentioned earlier, there are glucose-monitoring apps, similarly, companies like Dexcom and Eversense manufacture glucose-monitoring wearables. The wearable designed by Eversense is the first FDA-approved monitor whose sensors can track glucose levels up to 90 days. The Q-Collar, a wearable device to protect from brain concussions, is worn on the neck. It helps in reducing the brain movement inside the skull by increasing the blood volume through the neck. Wearable technology is developing majorly for all the external body parts as in the markets we now also have an option for Smart Contact Lenses instead of regular lenses. These smart lenses are approved by FDA and CE due to their feature of recording eye dimension changes automatically [57]. While talking about wearable technology, how can we forget about gaming technology and its connection with wearables? Almost everyone in these days is aware of virtual reality; a device that creates a virtual 360 degrees environment of anything being played on the screen is used by gamers to have a virtual reality experience of video games. Companies like Oculus, Supernatural, and Lucid Sight are well known for their VR devices in the gaming industry.

Implantables

An electronic wireless device that is partially or completely implanted in the human body was first introduced in the 1950s. It is produced and utilized to upgrade already-existing biological structures or to sustain or replace missing biological structures. In the past few decades, implantable scientific gadgets or structures had been superior through traits in technological know-how and engineering, specifically in microelectronics, biotechnology, and materials. These sensors are used in various diagnoses and treatments of patients in healthcare. Verichip, an implantable chip approved by FDA, is a radiofrequency identification device for patients that can be used for medical history, allergies, or diseases, or, in medical terms, can be used for a patient's PHR. It can also be used to check blood glucose levels and body temperature. The main aim of this is to provide treatment to those who are in an emergency using their medical data from the chip. To have access to the chip, doctors will have to scan it or use its code from the chip which is implanted mostly on the arm of humans. Recently, it has been observed that many people are suffering from heart-related problems such as blockage or low heart rates [58]. To overcome this, the researchers have developed artificial pacemakers which are implanted either internally or externally in place of the human heart for performing the functions such as pumping of blood by generating electrical signals within the pacemaker. Healthcare technology is not only developing for humans but also animals. There is a chip named, Home Again, that is implanted in pet animals to keep a track of them. This has helped many pet owners to find lost pets by tracing their steps.

1.5.2 Limitations of M-health

It is quite commonly said that tablets and smartphones cannot replace laptops and computer systems and initially they were not intended to do so either. But the recent advancement in processing power, battery life, and keyboard portability in devices like iPad and flagship smartphones has narrowed the gap between computers and tablets. But even after such a significant breakthrough, there are quite a few limitations in these technologies.

- These technologies are recently introduced and do take time to feed the input in them but advanced solutions of AI which include voice recognition features in NLP can be implemented for faster input feed. There are new features like swipe keyboard which work on intelligent pattern recognition.
- Smartphone screen sizes have been increased in the past decade from 4 to 6.5 inches but still, it is not practical to use it for all sorts of domains, and productivity gets reduced eventually. Tablets have larger screens but then they lag certain features which smartphones have. So, a solution that converges both of them has to be innovated so that clinicians do not have to buy both devices. Currently companies like Samsung and Huawei have come up with solutions for folding smartphones, but the problems related to fragility and cost still prevail [59,60].
- These devices come with different operating systems and processors which lead to problems related to interoperability. But with new open-source multiplatform frameworks, medical software which can work on all the devices smoothly can be developed.
- Patient data is of the utmost importance in the healthcare industry. As technology advances, new methods of data breach and hacking are a big threat to clinical data. Techniques like encryption and integrated biometrics can aid these problems.

- Nowadays, smartphones carry a lot of personal data like Employee IDs, Bank details, Driving License, personal photos, AADHAAR details, etc. which can be easily accessed by introducing spyware and malware in devices. Hence, strict policies and regulations should be set by the government for smartphone companies by making it mandatory for them to provide anti-viral software, encryption software, etc. In big markets like India, currently, only Apple, Google, and Samsung are providing a significant amount of security. Apple is known for its privacy policies and Google has been improving security in stock android and making it even more secure. Samsung has developed Knox security software which is now being provided in all segments of their smartphones. The major challenge in Indian markets is the Chinese brands that sell smartphones at lower prices and attract many consumers and they do this by selling their private data for personalized ads. Mobile applications related to health, being hosted on the app store and play store, should have strong security features and policies to maintain the integrity of the application.

1.6 Limits and ethical issues

When you work with any organization or company, what matters the most is one's ethics, i.e., how you behave with colleagues, staff members, higher authorities, do they follow the laws and legalities of workspace or not, etc. Similarly, in the healthcare industry, there are some limits and ethics which are to be obeyed and followed by every member of the industry. The ethics in health informatics is a combination of ethics from Medicine, Ethics, and Informatics. The main objective of ethics is to provide respect for one's privacy, intellectual property, and security. When any of these is denied in any manner, then it comes under ethical issues. These issues can occur due to various reasons such as:

- The difference in view of ethics: Ethics exists for an orderly and legitimate society, so societies need to take well-defined ethical behavior. To overcome this difference, society should have some code of conduct and set of ethics, laws to be strictly followed by everyone.
- Suppose for some research work in an organization, they are using public reports which are supposed to be used with their concern for privacy reasons and they have not taken the required steps for that. Thus, in the near future of completion of the research or even in between any individual knowing this, thing can fill a case against the organization for prohibiting their identity and privacy and will lead to an unethical issue for them. The necessary rules to be practiced while researching public data to avoid such issues.
- Data breaches in big institutes, hospitals, healthcare-related companies, or organizations can also lead to ethical issues because breaching means leakage of data containing private details of individuals, such as names, past medical history, and genetic reports and when all this goes to access in wrong hands, then will create an ethical issue.

- Healthcare professionals (HCPs) not performing their duties by bringing their profession into partiality, crossing the limitations informed by higher authorities, not being loyal to their appointed work and system, leaking data to other organizations, etc.
- When discovering new medicines for particular diseases or prostates to replace missing body parts, it should be manufactured in such a manner that it is accessible to all the needy ones. Its cost and availability should not act as an issue to anyone.
- One of the major reasons for ethical problems is expensive treatments or facilities in developing countries. In developing countries like India, medical facilities are provided at any demanded rates by the hospitals or organizations. There is no proper code of conduct for charging these facilities and thus the poor or middle-class people majority times face ill-treatments and eventually death in case of chronic or deathly diseases.
- Due to a shift of medical ethics from manual to the digital world, now most of the databases are stored and used in the form of EHRs in hospitals and organizations, and this can create problems of uninformed access or consent causing abuse of patients' rights, security, and integrity.

1.6.1 Ethics to be followed in healthcare sector

The core foundation of the healthcare industry including hospitals, clinics, medical institutes, government-affiliated or personal organizations, etc., is built on the legal principles and ethics which make their core strength for the patients, stakeholders, staff members, and also for the owners or higher authorities. The code of conduct is derived from informatic ethics since this is one of the origin factors for health informatics ethics as discussed earlier, and has to be followed by all the Healthcare Professionals (HIPs) [61,62]. The code of conduct is further classified based on the principles, ethics, and duties of these HIPs because of the role played by them in the industry at a certain point in time. Now let us discuss each of them in detail.

Fundamental ethical principles:
Whenever people are involved in some form of social responsibility, there are predefined principles for the individual. Similarly, HI also has some important principles and they are as follows:

1. Principle of autonomy: An individual has their personal choice for selecting the treatment, physician, or insurance as a patient.
2. Principle of equality and justice: All patients should be treated equally despite their professional or economical background by the physicians.
3. Principle of beneficence: All people are obliged to promote the kindness of others if the nature of their kindness is consistent with the basic and ethical values of those involved.
4. Principle of non-malfeasance: Everyone is obliged to prevent harm to others within their power without undue harm to themselves.
5. Principle of impossibility: All duties and responsibilities must be able to be carried out in light of the current situation.

6. Principle of integrity: One has to be honest with themselves and they have to accomplish the obligations to their full ability.

Informatics ethical principles:

Clinicians have lots of information regarding patients which have to be kept confidential. There are a certain set of ethical principles which do imply this purpose and they are as follows:

1. Principle of information-privacy and disposition: All persons including patients and medical staff have basic fundamental rights to privacy and have the right to know what information, data is being collected, where it is being stored, whether data is being shared or manipulated, how it is being linked.
2. Principle of openness: A person whose information has been collected should be informed about storage, use, manipulation, linkage, and collection in a timely fashion to the people whose data has been collected.
3. Principle of security: Collected data of the persons should be protected by advanced security layers to protect data against breach, unauthorized access, use.
4. Principle of access: Once the electronic records are collected, the subjects have the right to access, check completeness, accurateness, and relevance.
5. Principle of accountability: Any violation of a person's right to privacy and their ability to govern their data should be reported, and the offender should accept responsibility for the violation and any resulting repercussions.

1.6.2 Important healthcare policies

Medical institution policies and procedures ensure compliance with federal and state laws and regulations. These are important to reduce the health risks and increase the safety of everyone in the organization and their legal obligations. Policies and procedures are essential to ensure consistency and standardization throughout the organization to clarify employees with their daily activities and set clear goals for individual work tasks. The healthcare sector consists of following policies that play a crucial role in today's world.

HIPAA

The federal regulations that provide data privacy requirements on a national level to the patient's permission about the medical information are the foundation of the Health Insurance Portability and Accountability Act of 1996 (HIPAA). Due to the numerous health data breaches brought on by cyberattacks and ransomware attacks on health insurance firms and providers in recent years, the law has received more attention. The Privacy Rule and Security Rule are the two main HIPAA sectors that have been separated by the US Department of Health and Human Services (HHS) depending on how they operate.

HIPAA privacy rule

The HIPAA privacy rule, often known as the privacy standard for individually identifiable health information, established the first national standard in the United

States for the protection of a patient's private or proprietary health information (PHI). Regulations limiting the use and disclosure of sensitive PHI have been passed by HHS [63,64]. This is patient privacy by requiring doctors to provide them with an account for each entity that shares PHI for billing and management purposes while allowing relevant health information to pass through the appropriate channels. To protect is the goal. The privacy laws also ensure that patients have the right to seek their PHI from healthcare providers who are HIPAA-eligible. Organizations that qualify as HIPAA Eligible Entities must abide by the HIPAA Privacy Rules. The contracts that qualifying businesses must sign with HIPAA Business Partners and must also contain precise restrictions on how BA may use or disclose PHI. Healthcare providers, health plans, healthcare clearinghouses, and commercial partners are all covered entities under this. Regardless of the extent of service, a healthcare professional electronically transmits medical data relevant to a specific transaction. These transactions include applications, benefit eligibility requests, transfer approval requests, and other transactions. The healthcare providers, health insurance companies, or third parties are the ones who act as health planners.

HIPAA security rule

National requirements for safeguarding stored patient information are established by security standards for safeguarding electronically protected medical information, sometimes known as HIPAA security guidelines. It is based on the cybersecurity architecture developed by the National Institute of Standards and Technology (NIST). The OCR follows HIPAA security guidelines designed to strike a balance between patient safety and improvements in medical technology. To guarantee the secure transmission, storage, and receipt of PHI, the rule mandates the installation of both mechanical and technological safety mechanisms.

HITECH

The Health Information Technology Act for Economic and Clinical Health was introduced by the Obama administration in 2009 and is part of the American Recovery and Reinvestment Act. The HITECH Act was proposed to encourage and expand the adoption of healthcare information technology, especially the use of EHR by healthcare providers. The law also eliminated loopholes in the HIPAA Act by strengthening the wording of HIPAA. This ensures that business partners of HIPAA-supported companies are HIPAA compliant, and the data subject is notified when health information is leaked [65]. While broadening the extent of security and privacy protections provided by the Health Insurance Portability and Accountability Act, the HITECH Act increases the transmission of electronically protected health information (ePHI) (HIPAA). These include tighter enforcement procedures and higher liability for non-compliance. The primary HIPAA-related HITECH provisions are as follows:

- Historically, HIPAA has not been strictly enforced, but the passage of the final rules in 2013 has clarified and strengthened enforcement activities. In the event of a breach, both the company and the business partner will be subject to sanctions.

- Civil law penalties are raised for intention. Individuals cannot file proceedings against vendors for HITECH violations, but the Attorney General may file proceedings on behalf of state residents.
- The Department of Health and Human Services (HHS) is required to carry out regular audits of affected companies and business partners.

Digital fitness records are required to be embraced to strengthen healthcare, improve performance and care coordination, and make it simpler for fitness information to be exchanged across exclusive protected institutions. Many healthcare professionals are desired to switch from paper to electronic health records, but the expense of doing so was prohibitive. Incentives for hospitals and other healthcare providers have been created under the HITECH Act. Many healthcare professionals continued to utilize paper records even if the bill was not approved. The HITECH Act is a law that obliges medical institutions and their business partners to follow HIPAA privacy and security regulations. These regulations ensure that medical information is kept confidential, and its use and disclosure are limited to only those who need it. The act also requires medical institutions to provide patients with their medical records on request and implement measures to protect their information [66]. The law did not require HIPAA compliance. This is because it was already a requirement, but companies that are found to be out of compliance may be subject to heavy fines.

Telehealth is a broad term that encompasses methodologies like remote patient monitoring, reminder services, and other remote care technologies. M-health is a subset of the broader term telehealth [67]. It is a term coined in 2008 by Rockefeller Foundation in seminal conference "eHealth connects," and this was a clear representation of the evolution of E-health. In layman's terms, M-health is the utilization of wireless technology for delivering healthcare services and necessary data and information. This can be achieved through mobile phones, patient monitoring devices, personal digital assistants, smart monitoring IoT devices, tablets, and any other devices which are capable of wireless connectivity. M-Health means using mobile devices to monitor or detect biological changes in the human body, and device management units such as hospitals and clinics. Service providers collect clinical data and use it for care provision.

1.7 Healthcare informatics technologies

1.7.1 Internet of Things

IoT, the Internet of Things, is estimated to revolutionize the way monitoring data capture occurs. The main essence of IoT is that they are uniquely identifiable devices with the ability to programmatically connect to a network, collect data, and collaborate. These innovative devices are capable enough of collecting and sharing data with the cloud and other devices. It plays a major role in monitoring real-life incidents and analyzing the collected data quickly and accurately [68]. Various devices like smart bands, personal health monitors, life support devices at hospitals, smart home devices for elderly people activity monitoring and fall detection, and various other similar devices can be part of an IoT ecosystem. Healthcare Informatics and Analytics system can be provided with real-time data through this ecosystem [69–71]. To summarize, IoT is a source for data and events to be recorded for any kind of informatics and analytics.

1.7.2 Web data

Web data is the term used to describe all data made available over the World Wide Web. These online data include atom feeds, video streams, social networking information from websites like LinkedIn, Instagram, and Twitter, as well as video streaming. There are two forms of web data: structured and unstructured. From the standpoint of healthcare informatics and analytics, web data may play a significant role and serves as a crucial source of information. Due to the rise of availability of Internet, health-based websites, blogs, and discussion forums are increasingly being used for discussing symptoms, home remedies, treatments options, experiences, and so on. All of this causes a significant volume of unstructured data to be produced. This unstructured data may be examined to learn more about illnesses, infections, the most popular web searches for home treatments, and other information. If this vast quantity and diversity of data is utilized effectively, it may be possible to pinpoint specific issues that individuals are having and, via the analysis of pertinent data, come up with solutions.

1.7.3 Data stores

Data storage like transitional databases, operational databases, knowledge bases, big data stores, and others is collectively referred to as Data Stores [72]. EHR, databases of hospitals and clinics, Genome Database, drug research databases, knowledge bases, databases of insurance pharmaceuticals companies, and others are available relevant data in the case of the healthcare domain. Depending on the requirements and need for analysis, data can be selected from any of the data stores mentioned before and can be analyzed.

1.7.4 Big data analytics

Big data analytics refers to analytical processing that can handle large amounts of data that are moving quickly and in a wide range of different directions. Data transformation, cleansing, information extraction, and augmentation are key processes needed for big data analytics. Numerous probabilistic statistical models and methods support it [73]. Big data analytics heavily relies on selecting the appropriate data and formulating analytical problems to succeed [74]. Big data analytics may add additional information, insights, patterns, decision-making, and actions to the health data as complexity rises. Big data analytics offers a wide range of visualization approaches that may be used to provide insightful data communication for end users.

1.8 Case study

Impact of COVID-19 on the healthcare IT market

All medical practices had to be under immense pressure due to COVID-19 and all the healthcare facilities across the globe were overwhelmed by the large number of patients being admitted and affected by COVID. It was getting super difficult for diagnosing many visiting patients on daily basis [75,76]. Furthermore, the rise in the frequency of coronavirus illnesses has increased the need for precise diagnostic and treatment equipment in a number of nations throughout the world. In this scenario, linked care technology proved to be really beneficial. These technologies enable healthcare personnel to monitor patients in real time by employing digitally linked sensor equipment such as blood pressure monitors, oximeters, and other devices. The need for social separation between physicians and patients increased demand for healthcare IT solutions such as remote patient monitoring and telehealth systems [77]. The requirement for the precise and timely transmission of patient health records has also grown [78]. The foundation for exchanging, sharing, and retrieving electronic health information with high security was supplied by effective tools in the form of healthcare IT solutions. Even CCTV that uses artificial intelligence and video analytics helps keep an eye on whether the right measures are taken in relation to donning masks, gloves, and maintaining social distance.

1.9 Conclusion and future scope

The healthcare industry is growing hand-in-hand with the technological aspect. The healthcare industry has come a long way since the days when there was no proper clinical record and no knowledge of life-threatening illnesses and their treatments. Earlier, physicians or doctors did not have the records of patients, but now, due to electronic databases, each and every piece of data is being stored and used by researchers to find a cure for deathly diseases. However, the misuse of this technology is also observed since genes or DNA-based records/samples are being used for discovering man-made viruses. In the coming future, we might see research based on the cure of unknown disorders, medicines, and existing treatments which would be less time-consuming and cost-friendly.

References

[1] Coiera, E. (2015). *Guide to Health Informatics*. CRC Press.

[2] Dagliati, A., Malovini, A., Tibollo, V., and Bellazzi, R. (2021). Health informatics and EHR to support clinical research in the COVID-19 pandemic: an overview. *Briefings in Bioinformatics*, 22(2), 812–822.

[3] Dang, L. M., Piran, M. J., Han, D., Min, K., and Moon, H. (2019). A survey on internet of things and cloud computing for healthcare. *Electronics*, 8(7), 768.

[4] Adans-Dester, C. P., Bamberg, S., Bertacchi, F. P., *et al.* (2020). Can mHealth technology help mitigate the effects of the COVID-19 pandemic? *IEEE Open Journal of Engineering in Medicine and Biology*, 1, 243–248.

[5] Dash, S., Shakyawar, S. K., Sharma, M., and Kaushik, S. (2019). Big data in healthcare: management, analysis and future prospects. *Journal of Big Data*, 6(1), 1–25.

[6] Velasquez, D. E., Mecklai, K., Plevyak, S., Eappen, B., Koh, K. A., and Martin, A. F. (2022). Health system-based housing navigation for patients experiencing homelessness: a new care coordination framework. In *Healthcare* (vol. 10, no. 1, p. 100608). Elsevier.

[7] Dickinson, F. M., McCauley, M., Madaj, B., and van den Broek, N. (2019). Using electronic tablets for data collection for healthcare service and maternal health assessments in low resource settings: lessons learnt. *BMC Health Services Research*, 19(1), 1–7.

[8] Durga, S., Nag, R., and Daniel, E. (2019). Survey on machine learning and deep learning algorithms used in internet of things (IoT) healthcare. In *2019 3rd International Conference on Computing Methodologies and Communication (ICCMC)* (pp. 1018–1022). IEEE.

[9] Ghaleb, E. A., Dominic, P. D. D., Fati, S. M., Muneer, A., and Ali, R. F. (2021). The assessment of big data adoption readiness with a technology–organization–environment framework: a perspective towards healthcare employees. *Sustainability*, 13(15), 8379.

[10] Gibson, C. J., Dixon, B. E., and Abrams, K. (2015). Convergent evolution of health information management and health informatics. *Applied Clinical Informatics*, 6(01), 163–184.

[11] Evans, R. S. (2016). Electronic health records: then, now, and in the future. *Yearbook of Medical Informatics*, 25(S01), S48–S61.

[12] Friede, A., Blum, H. L., and McDonald, M. (1995). Public health informatics: how information-age technology can strengthen public health. *Annual Review of Public Health*, 16(1), 239–252.

[13] Gudi, N., Lakiang, T., Pattanshetty, S., Sarbadhikari, S. N., and John, O. (2021). Challenges and prospects in India's digital health journey. *Indian Journal of Public Health*, 65(2), 209.

[14] Häyrinen, K., Saranto, K., and Nykänen, P. (2008). Definition, structure, content, use and impacts of electronic health records: a review of the research literature. *International Journal of Medical Informatics*, 77(5), 291–304.

[15] Hindmarsh, J., Jenkings, K. N., and Rapley, T. (2007). Introduction to healthcare technologies in practice. *Health Informatics Journal*, 13(1), 5–8.

[16] Ali, O., Shrestha, A., Soar, J., and Wamba, S. F. (2018). Cloud computing-enabled healthcare opportunities, issues, and applications: a systematic review. *International Journal of Information Management*, 43, 146–158.

[17] Hoerbst, A. and Ammenwerth, E. (2010). Electronic health records. *Methods of Information in Medicine*, 49(04), 320–336.

[18] Hovenga, E. J. (2010). National standards in health informatics. *Health Informatics: An Overview*, 151, 133.

[19] Huang, P. S., Paulino, Y. C., So, S., Chiu, D. K., and Ho, K. K. (2022). Guest editorial: COVID-19 pandemic and health informatics: Part 2. *Library Hi Tech*, 40(2), 281–285.

[20] Keshta, I. and Odeh, A. (2021). Security and privacy of electronic health records: concerns and challenges. *Egyptian Informatics Journal*, 22(2), 177–183, https://doi.org/10.1016/j.eij.2020.07.003.

[21] Alzahrani, A. I., Al-Samarraie, H., Eldenfria, A., Dodoo, J. E., and Alalwan, N. (2022). Users' intention to continue using mHealth services: a DEMATEL approach during the COVID-19 pandemic. *Technology in Society*, 68, 101862.

[22] Ammenwerth, E. (2019). Technology acceptance models in health informatics: TAM and UTAUT. *Studies in Health Technology and Informatics*, 263, 64–71.

[23] Imhoff, M. (2002). Health informatics. In Sibbald, W. and Bion, J. (eds), *Evaluating Critical Care: Using Health Services Research to Improve Quality* (pp. 255–269). Springer.

[24] Iserson, K. V. (2020). Healthcare ethics during a pandemic. *Western Journal of Emergency Medicine*, 21(3), 477.

[25] Apathy, N. C., Holmgren, A. J., and Adler-Milstein, J. (2021). A decade post-HITECH: critical access hospitals have electronic health records but struggle to keep up with other advanced functions. *Journal of the American Medical Informatics Association*, 28(9), 1947–1954.

[26] Au-Yong-Oliveira, M., Pesqueira, A., Sousa, M. J., Dal Mas, F., and Soliman, M. (2021). The potential of big data research in healthcare for medical doctors' learning. *Journal of Medical Systems*, 45(1), 1–14.

[27] Iqbal, J., Adnan, M., Khan, Y., *et al.* (2022). Designing a healthcare-enabled software-defined wireless body area network architecture for secure medical data and efficient diagnosis. *Journal of Healthcare Engineering*, 2022, 1–19. Article ID 9210761. https://doi.org/10.1155/2022/9210761.

[28] Jones, K. H., Ford, E. M., Lea, N., *et al.* (2020). Toward the development of data governance standards for using clinical free-text data in health research: position paper. *Journal of Medical Internet Research*, 22(6), e16760. doi:10.2196/16760.

[29] Gupta, R., Thakker, U., Tanwar, S., Obaidat, M. S., and Hsiao, K. -F. (2020). BITS: a blockchain-driven intelligent scheme for telesurgery system. In *2020 International Conference on Computer, Information and Telecommunication Systems (CITS)*, Hangzhou, China, pp. 1–5, doi:10.1109/CITS49457.2020.9232662.

[30] Kalpa, S. (2012). Health IT in Indian healthcare system: a new initiative. *Research Journal of Recent Sciences*, 2277, 2502.

[31] Stoeger, K. and Schmidhuber, M. (2020). The use of data from electronic health records in times of a pandemic—a legal and ethical assessment. *Journal of Law and the Biosciences*, 7(1), lsaa041, https://doi.org/10.1093/jlb/lsaa041.

[32] Bath, P. A. (2008). Health informatics: current issues and challenges. *Journal of Information Science*, 34(4), 501–518.

[33] Bell, K. (2018). Public policy and health informatics. In *Seminars in Oncology Nursing* (vol. 34, no. 2, pp. 184–187). WB Saunders.

[34] Kadakia, K. T., Howell, M. D., and DeSalvo, K. B. (2021). Modernizing public health data systems: lessons from the health information technology for economic and clinical health (HITECH) act. *JAMA*, 326(5), 385–386.

[35] Kohli, R. and Tan, S. S. L. (2016). Electronic health records. *Mis Quarterly*, 40(3), 553–574.

[36] Kolling, M. L., Furstenau, L. B., Sott, M. K., *et al.* (2021). Data mining in healthcare: applying strategic intelligence techniques to depict 25 years of research development. *International Journal of Environmental Research and Public Health*, 18(6), 3099.

[37] Korzun, D. and Meigal, A. (2019). Multi-source data sensing in mobile personalized healthcare systems: semantic linking and data mining. In *2019 24th Conference of Open Innovations Association (FRUCT)* (pp. 187–192). IEEE.

[38] Bero L.A. (2020). Improving the quality of systematic reviews in public health: introduction to the series. *American Journal of Public Health*. 110(11), 1601–1602. doi:10.2105/AJPH.2020.305914. PMID: 33026865; PMCID: PMC7542292.

[39] Krishankumar, R., Sivagami, R., Saha, A., Rani, P., Arun, K., and Ravichandran, K. S. (2022). Cloud vendor selection for the healthcare industry using a big data-driven decision model with probabilistic linguistic information. *Applied Intelligence*, 52(12), 13497–13519.

[40] Li, W., Chai, Y., Khan, F., *et al.* (2021). A comprehensive survey on machine learning-based big data analytics for IoT-enabled smart healthcare system. *Mobile Networks and Applications*, 26(1), 234–252.

[41] Lorenzi, N. M. and Riley, R. T. (2013). *Organizational Aspects of Health Informatics: Managing Technological Change*. Springer Science & Business Media.

[42] Malik, N. and Malik, S. K. (2022). Semantic web technology-based secure system for IoT-enabled e-healthcare services. In *Futuristic Research Trends and Applications of Internet of Things* (pp. 117–137). CRC Press.

[43] Bhuvaneshwari, P., Jeevanandhan, S., Kirubhakaran, S., Naveenkumar, R., and Ramkumar, S. (2022). Healthcare informatics and analytics framework techniques. *International Journal of Research Publication and Reviews*, 3, 2695–2699.

[44] Madanian, S., Parry, D. T., Airehrour, D., and Cherrington, M. (2019). mHealth and big-data integration: promises for healthcare system in India. *BMJ Health & Care Informatics*, 26(1).

[45] Bodavala, R. (1998). Evaluation of health management information system in India need for computerized databases in HMIS. In *Ranganayakulu Bodalava Takami Fellow Int Health Harv Sch Public Health*, 665.

[46] Maher, N. A., Senders, J. T., Hulsbergen, A. F., *et al.* (2019). Passive data collection and use in healthcare: a systematic review of ethical issues. *International Journal of Medical Informatics*, 129, 242–247.

[47] Gupta, R., Shukla, A., and Tanwar, S. (2020). BATS: a blockchain and AI-empowered drone-assisted telesurgery system towards 6G. *IEEE Transactions on Network Science and Engineering*, 8(4), 2958–2967, doi:10.1109/TNSE.2020.3043262.

[48] Mbonihankuye, S., Nkunzimana, A., and Ndagijimana, A. (2019). Healthcare data security technology: HIPAA compliance. *Wireless Communications and Mobile Computing*, 2019, 1–7.

[49] Miah, S. J., Camilleri, E., and Vu, H. Q. (2022). Big Data in healthcare research: a survey study. *Journal of Computer Information Systems*, 62(3), 480–492.

[50] Okediran, O. O., Sijuade, A. A., Wahab, W. B., and Oladimeji, A. I. (2020). A framework for a cloud-based electronic health records system for developing countries. In *2020 International Conference on Electrical, Communication, and Computer Engineering (ICECCE)* (pp. 1–5). IEEE. doi:10.1109/ICECCE49384.2020.9179276.

[51] O'Carroll, P. W., Yasnoff, W. A., Ward, M. E., Ripp, L. H., and Martin, E. L. (eds). (2003). *Public Health Informatics and Information Systems*. Springer.

[52] Brender, J. (2006). *Handbook of Evaluation Methods for Health Informatics*. London: Elsevier Academic Press.

[53] Camacho-Rivera, M., Islam, J. Y., Rivera, A., and Vidot, D. C. (2020). Attitudes toward using COVID-19 mHealth tools among adults with chronic health conditions: secondary data analysis of the COVID-19 impact survey. *JMIR mHealth and uHealth*, 8(12), e24693.

[54] Oyeleye, O. A. (2021). The HIPAA privacy rule, COVID-19, and nurses' privacy rights. *Nursing*, 51(2), 11–14.

[55] Pandey, A., Roy, N., Bhawsar, R., and Mishra, R. M. (2010). Health information system in India: issues of data availability and quality. *Demography India*, 39(1), 111–28.

[56] Gatiti, P., Ndirangu, E., Mwangi, J., Mwanzu, A., and Ramadhani, T. (2021). Enhancing healthcare quality in hospitals through electronic health records: a systematic review, *Journal of Health Informatics in Developing Countries*, 15(2), 1.

[57] Preshaw, D. H., Brazil, K., McLaughlin, D., and Frolic, A. (2016). Ethical issues experienced by healthcare workers in nursing homes: literature review. *Nursing Ethics*, 23(5), 490–506.

[58] Qadeer, I., Saxena, K. B., and Arathi, P. M. (Eds.). (2021). *Universalising Healthcare in India: From Care to Coverage*. Springer, Singapore.

[59] Rahimi, B., Nadri, H., Afshar, H. L., and Timpka, T. (2018). A systematic review of the technology acceptance model in health informatics. *Applied Clinical Informatics*, 9(03), 604–634.

[60] Rudin, R. S., Friedberg, M. W., Shekelle, P., Shah, N., and Bates, D. W. (2020). Getting value from electronic health records: research needed to

improve practice. *Annals of Internal Medicine*, 172(Suppl 11), S130–S136. doi:10.7326/M19-0878. PMID: 32479182.

[61] Carson, N., Campbell-Smit, B., Walters, W., Sharp, K., and Smailes, P. (2021). Electronic health record training for nurse leaders. *Online Journal of Nursing Informatics (OJNI)*, 25 (2)

[62] Saeidnia, H. R., Mohammadzadeh, Z., and Hassanzadeh, M. (2021). Evaluation of mobile phone healthcare applications during the Covid-19 pandemic. In *Public Health and Informatics* (pp. 1100–1101). IOS Press.

[63] Saheb, T. and Saheb, M. (2019). Analyzing and visualizing knowledge structures of health informatics from 1974 to 2018: a bibliometric and social network analysis. *Healthcare Informatics Research*, 25(2), 61–72.

[64] Sapci, A. H. and Sapci, H. A. (2020). Artificial intelligence education and tools for medical and health informatics students: systematic review. *JMIR Medical Education*, 6(1), e19285.

[65] Selvaraj, S. and Sundaravaradhan, S. (2020). Challenges and opportunities in IoT healthcare systems: a systematic review. *SN Applied Sciences*, 2(1), 1–8.

[66] Sharma, S. K. and Kumar, A. (2018). Role of information technology in healthcare services: visitor's perception. *Journal of Multidisciplinary Research in Healthcare*, 4(2), 83–93.

[67] World Health Organization (2020). Ethical Standards for Research During Public Health Emergencies: Distilling Existing Guidance to Support COVID-19 R&D. World Health Organization. https://apps.who.int/iris/handle/10665/331507. License: CC BY-NC- SA 3.0 IGO.

[68] Mistry, C., Thakker, U., Gupta, R., *et al.* (2021). MedBlock: an AI-enabled and blockchain-driven medical healthcare system for COVID-19. In *ICC 2021-IEEE International Conference on Communications* (pp. 1–6). IEEE. doi:10.1109/ICC42927.2021.9500397.

[69] Sheriff, C. I., Naqishbandi, T., and Geetha, A. (2015). Healthcare informatics and analytics framework. In *2015 International Conference on Computer Communication and Informatics (ICCCI)* (pp. 1–6). IEEE.

[70] Madhavan, S., Bastarache, L., Brown, J. S., *et al.* (2021). Use of electronic health records to support a public health response to the COVID-19 pandemic in the United States: a perspective from 15 academic medical centers. *Journal of the American Medical Informatics Association*, 28(2), 393–401, https://doi.org/10.1093/jamia/ocaa287.

[71] Gupta, R., Shukla, A., and Tanwar, S. (2020). AaYusH: a smart contract-based telesurgery system for Healthcare 4.0. In *2020 IEEE International Conference on Communications Workshops (ICC Workshops)*, Dublin, Ireland, pp. 1–6, doi:10.1109/ICCWorkshops49005.2020.9145044.

[72] Tsai C.H., Eghdam A., Davoody N., Wright G., Flowerday S., and Koch S. (2020). Effects of electronic health record implementation and barriers to adoption and use: a scoping review and qualitative analysis of the content. *Life (Basel)*, 10(12), 327. Published 2020 Dec 4. doi:10.3390/life10120327.

[73] Bhavin, M., Tanwar, S., Sharma, N., Tyagi, S., and Kumar, N. (2021). Blockchain and quantum blind signature-based hybrid scheme for healthcare

5.0 applications. *Journal of Information Security and Applications*, 56, 102673.

[74] Tanwar, S., Vora, J, Kaneriya, S., *et al.* (2020). Human arthritis analysis in fog computing environment using Bayesian network classifier and thread protocol. *IEEE Consumer Electronics Magazine*, 9(1), 88–94, doi:10.1109/MCE.2019.2941456.

[75] Wadhwa, M. (2019). ICT interventions for improved health service quality and delivery in India: a literature review.

[76] Zheng, Y. L., Ding, X. R., Poon, C. C. Y., *et al.* (2014). Unobtrusive sensing and wearable devices for health informatics. *IEEE Transactions on Biomedical Engineering*, 61(5), 1538–1554.

[77] Vora, J., Tanwar, S., Tyagi, S., Kumar, N., and Rodrigues, J. J. P. C. (2017). Home-based exercise system for patients using IoT enabled smart speaker. In *2017 IEEE 19th International Conference on e-Health Networking, Applications and Services (Healthcom)* (pp. 1–6). IEEE. doi:10.1109/HealthCom.2017.8210826.

[78] Wu, D. T., Chen, A. T., Manning, J. D., *et al.* (2019). Evaluating visual analytics for health informatics applications: a systematic review from the American Medical Informatics Association Visual Analytics Working Group Task Force on Evaluation. *Journal of the American Medical Informatics Association*, 26(4), 314–323.

Chapter 2

Healthcare informatics: an overview of privacy and security

Lourdes Ruiz[1] *and Nguyen Huu Phuoc Dai*[1]

Abstract

Technology has become a ubiquitous element in healthcare systems, generating vast amounts of data from patients and healthcare providers. Consequently, healthcare information safety and privacy constitute serious concerns nowadays. This chapter presents a general overview of privacy and cybersecurity challenges via a meta-research of literature review. It discusses the development of technology informatics in healthcare, the ways health data is generated, stored, managed, and analyzed for decision making and effective patient outcomes. Concepts such as privacy, confidentiality, and security are analyzed in conjunction with two legal frameworks that regulate health information: the General Data Privacy Regulation (GDPR) in the European Union and the Health Insurance Portability and Accountability Act (HIPAA) in the United States. Moreover, challenges, threats, and upcoming trends concerning healthcare informatics are evaluated, best practices and solutions are formulated. This chapter aims to increase cybersecurity awareness for patients, healthcare staff, and organizations and presents countermeasures for future research. It also focuses on several methods to protect healthcare information, including information security risk management, data encryption, and data protection laws/regulations. COVID-19 pandemic highlighted the crucial need to enhance IT systems security. As a result, this chapter offers students, scholars, and scientists a comprehensive background concerning healthcare informatics on two relevant topics: privacy and security.

Keywords: Healthcare data management; Patient data protection; Cybersecurity in healthcare; Health information technology (HIT); Electronic health records (EHR); Healthcare data analytics; Data privacy regulations

2.1 Introduction

2.1.1 Problem definition

Healthcare systems are experiencing a continuous digital transformation. Technological advances such as telemedicine, artificial intelligence in medical

[1]Doctoral School on Safety and Security Sciences, Óbuda University, Hungary

devices, illnesses diagnosis, blockchain in health records, 3D printing, health apps in smartphones, and health tracking wearable devices are examples of how technology reshapes healthcare. As a result, a vast amount of information is generated, shared, stored, and used daily in health systems. Health information includes numerous data from different sources such as administrative, demographic, diagnosis, treatment, prescriptions, laboratory tests, and insurance.

Electronic medical records (EMR) are replacing paper charts as digitalization occurs. EMRs contain test results, diagnosis, and treatment physicians acquire within a medical center. They aid in monitoring, tracking health data over time, and identifying patients needing preventive appointments. Health information systems (HIS) collect, store, manage, retrieve, and transmit EMRs. These systems are helpful to enhance the patient's outcomes, for research purposes, and for guiding decision and policy making via big data analytics [1].

Moreover, electronic health records (EHRs) comprise the medical data acquired from all healthcare providers participating in a patient's care. These records possess sharing and data portability capabilities between different providers, health centers, and across cities and countries. Personal health records (PHRs) encompass the same information collected in EMRs. However, they are designed to be accessed and administered by the patients via smartphones or wearable devices [2].

As health information is growing, it becomes troublesome to store it. Cloud storage is a valuable solution for retrieving remote and real-time data and connecting with various health providers' databases and networks [3]. Figure 2.1

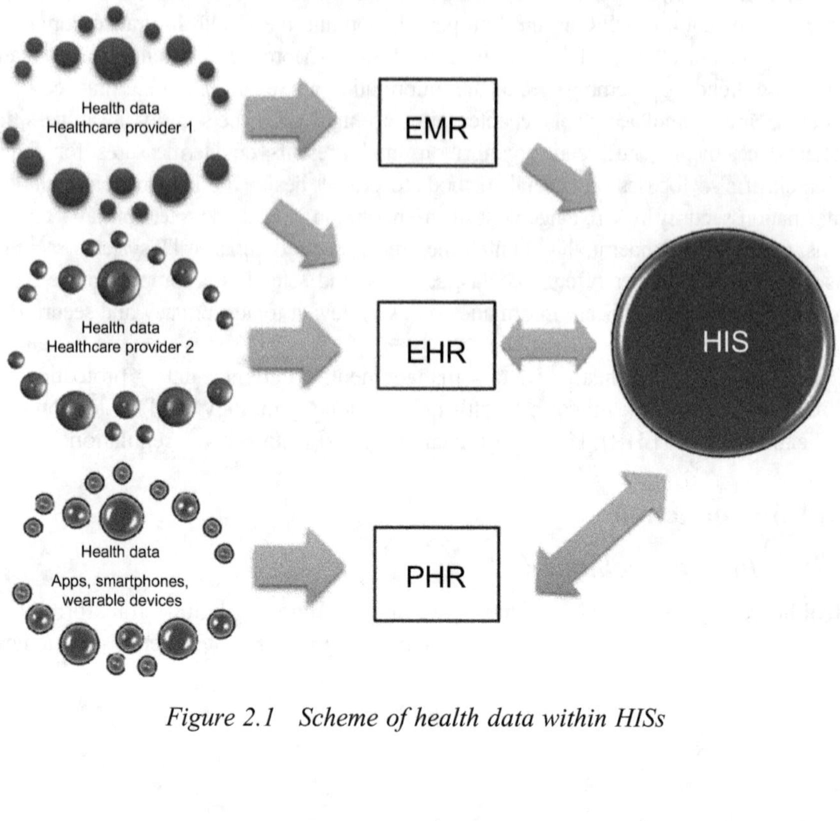

Figure 2.1 Scheme of health data within HISs

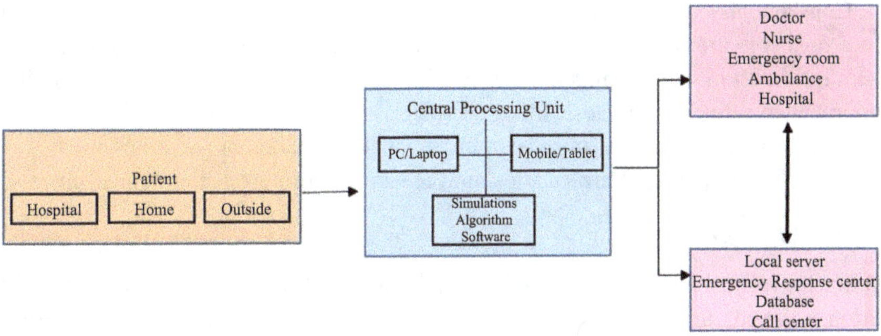

Figure 2.2 Health data communication across health systems [4]

illustrates how the different health data (EMR, EHR, PHR) interact with the HIS that can be cloud-based managed.

Healthcare informatics comprises various fields such as healthcare, information systems, databases, and information technology security. It collects, stores, manages, and analyzes health data to facilitate decision making, improves patients' care and communication throughout health actors, departments, centers, and IT staff, as shown in Figure 2.2. Moreover, medical data is administered, integrated. As a result, it is easy to access, accurate and secure.

> During the pandemic, an overload of inaccurate information, rumors, and conspiracy theories was rapidly spread via different media, known as infodemic. It causes anxiety and panic because of the tremendous amounts of COVID-19-related information. Thus, it is hard to find trustworthy sources. Healthcare informatics provide robust tools and methodologies for analyzing large amounts of data to provide scientific-based evidence to combat fake information [5].

Artificial Intelligence (AI) is a crosscutting topic in healthcare systems. AI takes advantage of the large volume of health data generated using machine learning methods. Furthermore, it employs computer-aided decision-making for disease diagnostics, surgery, rehabilitation, prediction tasks, drug management, and treatment. This technology can reduce costs and repetitive actions that stimulate critical thinking and creativity. During the COVID-19 pandemic, AI became a critical tool for health services by allowing information exchange to infer health risks, outcomes, and effective disease management [6].

The advent of technology in health systems is indeed beneficial. However, the nature of the data generated, stored, and transmitted in these systems poses substantial concerns regarding privacy and security. Moreover, healthcare systems work on the Internet, cloud-based environment, and wireless communications. Therefore, hackers can use complex methods to access these data through security holes on the Internet and these platforms. Besides, the transmission of healthcare data between healthcare service centers and networks can cause potential security issues and cyber-attacks from inside and outside the system network. These attacks affect individuals' data and damage the whole healthcare system [7].

Consequently, this chapter revises key terms such as privacy and confidentiality. These concepts are necessary to understand legal frameworks concerning security in health information which obliges the protection of patients' health data. It also offers a comprehensive review of the leading security issues happening in healthcare information systems, in order to raise awareness of their vulnerability, helps in the formulation of effective countermeasures, and advances in future technological threats.

2.1.2 Motivation

Healthcare informatics is a new multidisciplinary field that comprises health, information science, and technology. It manages how health information is stored, retrieved, and utilized. Likewise, it ensures that communication protocols between health facilities are effective within the whole healthcare chain to achieve better patient outcomes. COVID-19 pandemic transformed healthcare and accentuated the importance of the correct uses of health data. Health informatics assist emergency systems by retrieving a complete patient profile and making rapid treatment decisions. However, it augmented the health information generated, posing risks to the systems administering the data. In addition, the sensitiveness of the data and the interconnection of health systems make it a valuable target for criminals.

Hence, maintaining secure systems that protect valuable data is a paramount concern nowadays. This chapter provides a theoretical framework regarding different aspects of health data security. Thus, the reader will acquire a solid understanding of the different cyber threats, the available solutions, and future trends.

2.1.3 Contribution

This chapter is directed to healthcare IT staff, scientists, and students that use or manage data for accurate decision-making. Healthcare informatics is a field that is constantly growing, and expertise is needed in different organizations such as governmental, pharmaceuticals, biotech, and EHR companies. Thus, learning about privacy, security, and confidentiality of data is vital for people working on making sense of health data. Moreover, being aware of the constant risks affecting health information is relevant for individuals that analyze data and turn it into knowledge across the health spectrum. It aids data-driven specialists in comprehending the health systems as a whole to formulate a timely response.

This chapter will describe and evaluate concepts such as privacy, confidentiality, and security in healthcare information. Furthermore, challenges and threats in health data protection will be listed, and solutions will be formulated. Finally, upcoming trends regarding privacy, security best practices, and solutions will be enumerated.

2.1.4 Objectives and scope

This chapter's overarching goals are:

- Provide a comprehensive approach towards two complex topics: privacy and security in health data.
- Describe the current security situation in healthcare systems via legal framework analysis.

- Recognize and raise awareness of the potential threats in healthcare systems.
- Revise the future connotations and upcoming trends in healthcare informatics for the public and scientists.

The scope of this chapter comprises healthcare informatics security, starting with the definition of privacy, confidentiality, and security and how they are embedded in legal frameworks. Additionally, security threats in health systems are examined, solutions are proposed, and future trends are enumerated.

2.1.5 Organization of chapter

The rest of the chapter is organized as: Section 2.1 introduces the past, present, and future conditions of healthcare information systems (HISs). Section 2.2 conceptualizes privacy, confidentiality, and security in healthcare. Section 2.3 characterizes several legal frameworks enacted to protect healthcare data. Section 2.4 evaluates current privacy challenges faced in healthcare. Section 2.5 specifies and analyzes potential threats and cyber-attacks towards healthcare information. Section 2.6 lists countermeasures and solutions in privacy and security for health information systems. Section 2.7 summarizes the main upcoming trends regarding healthcare informatics. Section 2.8 highlights case study based discussion and section 2.9 concludes the chapter with future scope.

In addition, Figure 2.3 presents the visualization of the chapter's main topics and subtopics.

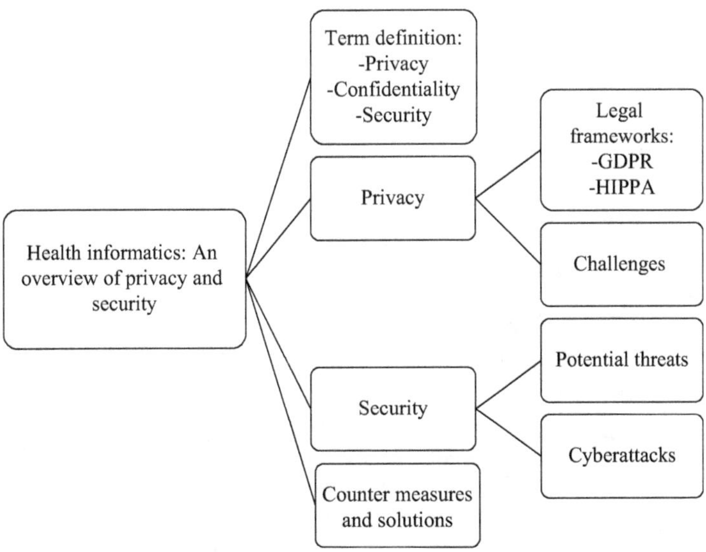

Figure 2.3 Organization of the chapter

2.2 Privacy, confidentiality, and security in healthcare information

Health data is sensitive and valuable because it includes personal information such as physical, mental health, behavior, personal relations, and even economic status. Health research is fueled by this information that aids in developing fast diagnoses, new treatments, and illness prevention. Moreover, data collection is beneficial for the patient and also for society. Hence, data protection is a primordial concern in health systems.

Health data breaches in the United States increased 55% in 2020 and 17% in 2021, affecting millions of people and causing severe economic costs for the companies involved [8]. Furthermore, hacking and IT incidents have been the leading cause of data breaches since 2017 due to healthcare digitalization and cloud storage in health systems, as displayed in Figure 2.4. The trend for 2022 is an increase in data breach incidents since the pandemic put pressure on health systems.

Health organizations need to provide effective methods to protect patients' data, timely anticipate and respond to the growing threats present in their digital systems. The concepts of privacy, confidentiality, and security serve as the baseline to understand the risks and hazards posed to health data and build effective ways to safeguard it. Section 2.2.1–2.2.3 explains privacy, confidentiality and security.

2.2.1 Privacy

Privacy is a broad concept that has been challenging to characterize in its meaning, perception, and scope. However, its execution and application comprise of three elements [10,11]:

- Decisional privacy deals with the individual's autonomy to decide about fundamental liberties concerning marriage, contraception, procreation, family relations, and raising children.

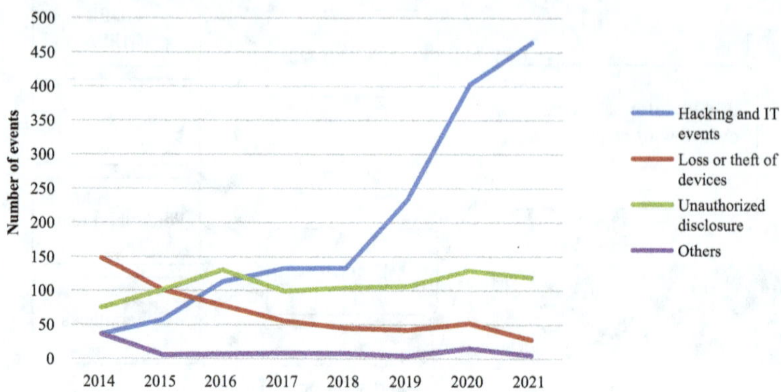

Figure 2.4 Causes of health data breaches in the United States [8,9]

- Protection of the individual in case of a reasonable expectation of privacy.
- Informational privacy encompasses the individuals' right to control the disclosure and usage of information related to them [12].

Informational privacy constitutes the main concern among health systems. Privacy enhances the communication between the doctor and the patient, which promotes better outcomes, autonomy and protects the patient from discrimination. It delimitates who and under what kind of circumstances someone has access to the patients' health data. Also, it comprises the processes of data collection, storage, usage, and concepts of purpose and proportionality. Likewise, it benefits health research. Patients are willing to provide consent to use their data or participate in a clinical study if privacy and anonymity are assured.

Ultimately, privacy enhances essential values in humans, such as decision-making, individuality, respect, dignity, and self-worth [13]. On the other hand, privacy infringement harms the person affected, such as economic, social, diminished self-image, and can also be extended to the family.

2.2.2 Confidentiality

Confidentiality in health is based on the Oath of Hippocrates, which respects the patients' privacy and autonomy. It refers to safeguarding the information acquired between a clinician and a patient and preventing it from being shared with third parties. It pertains to how health data was collected and how it will be held and utilized by the health institution that acquired it. Moreover, it addresses concerns such as further usage of the data and when the patient's permission is needed for its usage.

Confidentiality plays a vital role in the healing process because it stimulates trust and intimacy between patient and physician. As a result, the patient can disclose information that eases the illness diagnosis and treatment. Furthermore, it prevents patients from harm caused by the disclosure and unauthorized usage of health information [14].

2.2.3 Security

Security is defined as the methods, management, and technical procedures required to impede unauthorized access, modify, use, and disclose data saved in a computer system, deny service, and safeguard the system from any physical damage [15].

Health data security is necessary given the large amounts of personal data acquired and used in health systems. The potential harm for people affected during a security breach includes identity theft, economic and intrinsic damage to the individual because third parties disclose and know personal information. Moreover, it hinders health research because public trust is lost, and people are hesitant to participate in clinical trials.

2.2.4 Privacy versus security

Privacy and security are two interconnected elements, but they also share essential differences. Privacy refers to the correct collection, usage, and sharing of an

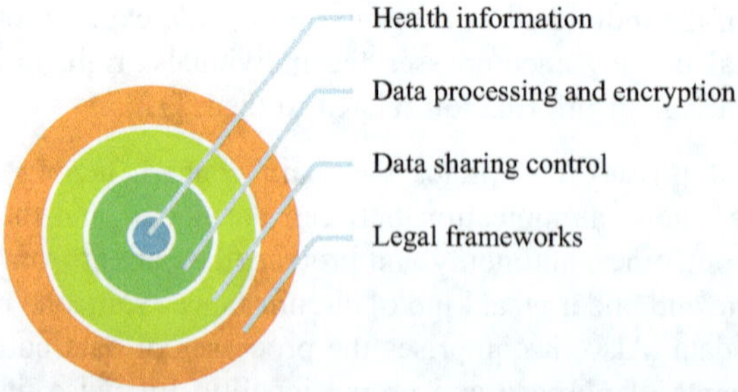

Figure 2.5 Principles assuring privacy and security in health information [17]

individual's information regarding his/her right to protect the data from disclosure to third parties and formulation of policies and regulations. On the other hand, security pertains to safeguarding the confidentiality, integrity, and availability of this information and the health institution that acquired the data. Confidentiality refers to the prevention of unauthorized access to health data. Integrity concerns to assure that the data is not altered in any way. Availability ensures that data is available for an authorized individual. Moreover, privacy deals with the capability of deciding what and where the patient's information goes, and security guarantees that these decisions are respected [16].

A security breach happens when an unauthorized party hacks the system and compromises security. It can potentially be a confidentiality breach. Nevertheless, no security procedure can prevent a privacy invasion by an individual with authorized access to the data. A privacy violation may or may not be related to a moral failure, but a confidentiality breach has a moral connotation.

Consequently, privacy and security data protection represent an ethical commitment and compliance as required in the legal frameworks described in the next section. Figure 2.5 summarizes the different protection layers used for safeguarding privacy and security of health data, starting at the moment of acquisition and finishing with legal compliance.

2.3 Legal framework

Privacy, confidentiality, and security concerns among patients are ongoing issues regarding health data acquisition, processing, and storage. Therefore, promulgation and enforcement of legal frameworks to protect health data and preserve the trust between the public, healthcare providers, and researchers are established in response to these concerns.

Health data privacy significance is recognized by assuring its protection by law. The principles of fair information practices articulated by the Organization for Economic Co-operation and Development (OECD) served as a basis for international and national laws and are described below [18]:

- **Limitation in data collection:** data acquisition should have limits. Data should be obtained by lawful procedures and with the knowledge or consent of the data subject.
- **Data quality:** data should be precise, complete, constantly updated, and be appropriate to the purpose of usage.
- **Purpose:** data collection purposes should be stated at the time of acquisition. Data usage is limited to completing such purposes and must be specified every time the purpose changes.
- **Usage limitation:** data cannot be shared, disclosed, accessed, or used for purposes other than those specified, except when consent was given by the data subject or by authority of law.
- **Safeguards:** data should be protected by proper security methods against loss, unapproved access, damage, alteration, and data breach.
- **Openness:** refers to creating a general policy regarding the procedures, actions, and methods used in the data collected. In addition, mechanisms to quickly establish the data's existence, nature, usage purposes, and effective means to identify the data controller and residence need to be instated.
- **Participation:** a person has the following rights about his/her participation:
  - To know if a data controller possesses his/ her personal information.
  - To get a copy of the data within an adequate time comprehensible for him/ her.
  - To get a reason if the petition for access to the data is refused.
  - To dispute such denial.
  - Question the data relating to him/her. If the case is favorable, the data controller must delete, rectify, complete, or correct the data.
- **Accountability:** the data controller should be responsible for abiding by the principles enumerated and explained below.

The principles explained above paved the road for developing major legal frameworks such as the General Data Protection Regulation (GDPR) in the European Union and the Health Insurance Portability and Accountability Act (HIPAA) in the United States.

2.3.1 GDPR

The GDPR was established to integrate data privacy laws among the European Union (EU) member countries. It constitutes the strictest privacy and security data law in the world. It was adopted in April 2016 and enforced in May 2018. It has a global impact since it regulates how organizations collect data about people in the EU. It imposes very high fines on companies that violate this regulation, such as a maximum of €20 million or 4% of the total global revenue. Furthermore, individuals that are affected are entitled to compensation [19].

GDPR's scope entails protecting personal data collected from a data subject by a data controller. Personal data encompasses all information that can directly or indirectly identify a person, such as [20]:

- Identity data such as name, address, gender, identification numbers
- Web information such as location, IP address, and cookies
- Health
- Genetic
- Biometric
- Racial
- Political opinions
- Religious beliefs
- Sexual orientation
- Pseudonymous

GDPR's seven principles are [21]:

- **Lawfulness, fairness, transparency:** are needed for processing personal data.
- **Purpose limitation:** data should be processed for legitimate reasons stated to the data subject at the moment of acquisition.
- **Data minimization:** data should be acquired and processed just when it is necessary and for the stated purposes.
- **Accuracy:** data should be kept accurate and updated.
- **Storage limitation**: data should be stored just for the necessary time stated in the purpose.
- **Integrity, confidentiality, security:** data should be processed in a way that guarantees integrity, confidentiality, and security.
- **Accountability:** data controllers should prove the GDPR's principles compliance.

GDPR's approach is towards protecting and guaranteeing the individuals' right to privacy. Hence, personal data can be processed just in the following instances [22]:

- Data subject gave unambiguous consent
- The data subject is part of a contract that needs data to be processed
- For complying with a legal regulation
- For saving somebody's life
- For executing a task of public interest or in an official function
- For legitimate interest

Consequently, GDPR acknowledges and enforces citizen's data privacy rights to [23]:

- Be informed
- Access
- Rectification
- Be forgotten
- Processing restriction
- Portability
- Objection
- Not be subject to a decision based on automated data processing and profiling

2.3.2 Health Insurance Portability and Accountability Act

The Health Insurance Portability and Accountability Act (HIPAA) legislation in the United States safeguards personal data and regulates health insurance coverage. It was enacted by Congress in 1996 and became effective in 2003. However, it had some additions over time, such as the Breach Notification Rule in 2009 and the Final Omnibus Rule in 2013 due to the significant changes in healthcare systems.

The HIPAA privacy rule states national standards to protect personal identifiable health information privacy. It restricts the usage and disclosure of sensitive data without the patient's authorization. Also, it establishes patients' rights to comprehend and control their health data usage. The entities subject to the privacy rule include healthcare providers, health plans, healthcare clearinghouses, and health business associates.

The Privacy Rule aims to find a balance between adequate health data protection and the usage and transfer of health information needed among healthcare institutions for treatment, care, and patient's well-being [24].

Furthermore, entities subject to the privacy rule are allowed to use and disclose health information without authorization in the following cases [25]:

- To the patient
- For treatment, payment, and healthcare operations
- The patient has the opportunity to agree or object to the disclosure
- Incidental use and disclosure
- For public interest according to the 12 national priority purposes:
  - When required by law
  - Public health activities
  - Victims of abuse, neglect
  - Health oversight activities
  - Judicial proceedings
  - Law enforcement authorities
  - Identification of a deceased person, cause of death
  - Cadaveric organ, eye, or tissue donation
  - Research restricted to certain conditions
  - Threat to health and safety
  - Essential government functions
  - Workers' compensation
- Limited data set for research, healthcare operations, and public health operations can be used and disclosed under specific data usage agreements.

The HIPAA Security Rule protects all identifiable electronic health information created, kept, transferred, and received by an entity subject to the HIPAA privacy rule. For compliance assurance, an entity must guarantee confidentiality, integrity, and availability of electronic health data. Also identify, anticipate, and protect against any threats to the data security, unauthorized usage, and disclosure of health data; certify and ensure employees' compliance. Moreover, the breach notification rule obligates an entity covered by HIPAA to report in case of a personal health information breach. Non-compliance with HIPAA is subject to fines and criminal penalties.

Different laws around the world assure health data privacy protection. Privacy frameworks share the same principles and are adapted to specific regions and industries. GDPR and HIPAA are just a couple of examples described to understand the essential elements of these types of regulations. Nevertheless, GDPR supplies a broader scope regarding personal data privacy protection and enforcement, while HIPAA is specifically for health information and health-related institutions.

2.4 Challenges in privacy

New technologies such as cloud storage, IoT, big data mining, AI, data analytics, remote medical care [26], bioinformatics, predictive modeling [27], and more are becoming pervasive in health systems. Hence, these systems face new health data privacy, confidentiality, and security challenges. Some of these issues are evaluated, and countermeasures are proposed to tackle the uprising concerns.

2.4.1 Medical electronic documents

The transition between paper and electronic medical records generates concerns about data protection and potential breaches. Health data digitalization comes with exceptional advantages. It eases the access, transmission, and reproduction of a large amount of data, storage, and integration of medical data scattered in different healthcare providers and geographic locations and offers the possibility to link different health databases. Additionally, personal data has become a valuable asset that can be used for other purposes such as marketing and promotion, surpassing the initial aim of diagnosing and healing. Nonetheless, these benefits mean that electronic health data is vulnerable to hacking unless security measures exist.

The collection, storing, and transfer of digital health information is a reality that comes with various troubles concerning privacy. Formulating and enforcing legal frameworks that concede and protect individuals' privacy rights is the first step towards privacy procurement. Awareness and empowerment of citizens as the owners of their data and catalysts of the decision-making are critical aspects for exercising privacy rights and autonomy. Furthermore, updating the legal regulations according to the technology development is essential to assure privacy protection.

2.4.2 New health data

Health data is constantly generated, and as research and technology evolve, other types of personal data are becoming available and posing concerns about privacy and confidentiality. Identifiable personal health data nowadays goes beyond names, addresses, gender, or medical data. It includes genome sequencing data, data collected from diagnostic imaging devices, bio-molecular disease markers, data acquired from mobile wearable devices, and more.

Genetic privacy refers to the individuals' rights and family protection from genetic information disclosure. An individual can be easily identifiable thanks to the unique and immutable genomic code. It can be used to discriminate against patients in the workplace and for insurance purposes. The genetic analysis detects different types of diseases and congenital problems in clinical trials. This information can be shared between different databases and researchers, posing privacy concerns. Moreover, genetic test information can be included in EHRs with portability capabilities and can be exposed to unauthorized access and data usage.

Hence, treating genetic data as special health data or 'genetic exceptionalism' has been discussed to protect genetic information against misuse, data mining, and other techniques that can risk the patient's identity [28]. Also, written consent is crucial for assuring patients' privacy and confidentiality. However, absolute privacy and confidentiality are impossible to guarantee in medical research regarding genomic data. Thus, an open-consent is proposed, in which the research subject grants unrestricted disclosure for research purposes of genotype and phenotype data. Furthermore, it emphasizes fully informed consent where its main principle is veracity concerning usage and data sharing to obtain consent. Another solution is data masking or anonymization. In this process, synthetic data is created to mimic the original data to use later for testing, training, and research purposes [29].

Health data via wellness apps, social media, and devices connected to the Internet is continuously created. Furthermore, emergent medical data obtained through online behavior and activities identify health conditions such as depression, alcoholism, or suicidal attitudes. Big data mining among tech companies such as Google, Microsoft, and Amazon is prevalent. These companies use large health data sets from EHRs to train machine learning algorithms to identify medical conditions, raising public privacy concerns. Although the positive outcomes, these activities are not regulated and can be subjected to improper use such as consumer profiling, publicity, and discrimination [30].

Individuals should be owners of their data and decide whether they want to participate or opt out from online health programs. In addition, public concern is critical to advocate new health privacy laws. For example, anti-discrimination laws regarding identifiable health data need to be enacted.

2.5 Potential threats and cyber-attacks towards healthcare information

While the development of technologies can enhance the effectiveness of healthcare delivery services, the advantages of these technologies need to meet the requirements for privacy and security concerns of the client. For example, data from medical records are transferred within the healthcare system via the Internet and wireless communications. As a result, it raises the compromised abilities of the security and privacy of users, which are described below.

2.5.1 Potential threats in healthcare information

With the support of new hi-tech devices and technologies, healthcare providers or organizations can easily manage their patients' EHR and personally identifiable information (PII). However, medical information consists of valuable patient financial data. Therefore, it can lead to hackers' attacks to gain access to these data for many purposes. Moreover, this data is stored and depends on the IT system, bringing many potential security issues from IT network systems.

There are several studies about potential cyber threats that can strongly affect HISs. For instance, Maglogiannis *et al.* [31] research categorized 25 major threats in EHRs, including hardware damages, user mistakes, software failures, and natural accidents. In addition, another analysis reported 19 primary types of threats in HISs based on user requirements, employees, and system assets [32]. On the other hand, Samy *et al.* [33] indicated that 17 types of security vulnerabilities of HISs were based on accidental or undesirable reasons/incidents involving natural, human, and environmental risks. It was also annotated that the server power problem is one of the most critical threats for HISs.

Moreover, Consolidated Technologies, Inc. (CTI) addressed eight types of threats: employees, malware, providers, device errors/loss or lack of security, unlimited access to a device in the system, and hardware problems [34]. However, in this paper, the authors separate these potential threats into two crucial types: internal and external threats. The former consists of all threats inside the HISs or healthcare organizations and can affect these systems directly or indirectly. For example, hardware devices, software, employees/patients, network, and inside threats. The latter includes many threats from outside the system, such as working environment issues, potential attacks, and natural disasters. Furthermore, the impacts and sub-categories of these threats to healthcare systems from internal and external types are described in Figure 2.6 and Table 2.1.

2.5.2 Cyber-attacks towards healthcare information

Health information systems bring many advantages in monitoring and managing patient data, especially for EHR and PII. Moreover, they contribute to developing faster, more accessible, and safer medical centers to improve patient care services via a network system. However, healthcare informatics systems are based on the IT system that operates in a cloud environment. Therefore, health systems also face various cyber-attacks in the IT system and cloud computing, such as security

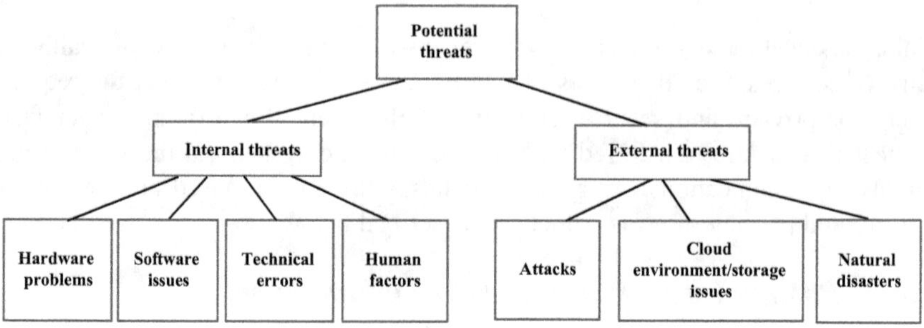

Figure 2.6 The potential type of threats in healthcare systems

Table 2.1 The potential sub-category threats and their impacts on healthcare informatics systems

Type	Sub-categories	Description	Impact	References
Internal threats	*Hardware problems*	Inadequate storage capacity, lack of maintenance, physical damage, insufficient electricity, server damage	High	[33,35]
	Software issues	Application conflict, license, failure, lack of maintenance	Medium	[33,35]
	Technical errors	Power failure, Internet corruption, network interruption, operation system problems, out-updated software/hardware/device	High	[33,35]
	Human factors	Mal-configuration, misuse data, access to data intentional or unintentional, lack of training, lack of skill or knowledge	Medium	[33,35,36]
External threats	*Attacks*	DoS attacks, terrorists, malicious software (viruses, malware, Trojans), hackers, outsiders, spoofing attacks.	High	[35,36]
	Cloud environment/ storage issues	Malicious software (malware, ransomware), data theft, third-party vulnerabilities, data breaches, DoS attacks, transmission, access control, misconfiguration, unintentional disclosure of credentials, data loss/privacy	High	[37,38]
	Natural disasters	Fire, earthquake, flood, water, lightning, storm	High	[39]

challenges in a wireless body area network (WBAN) [40], Wi-Fi security issues [41], cybersecurity threats during the Covid19 [42], risks of the IoT [43,44], threats to healthcare data [45], and cloud security issues [37,38]. Besides, Butt *et al.* [46] indicated two main types of attacks for the e-healthcare environment: routing and location-based attacks.

Concerning the importance and the severe impact on healthcare systems, four hazardous types of cyber-attacks are going to be described, as in Table 2.2.

2.5.2.1 Ransomware

One of the most popular and dramatic attacks on the HISs is ransomware attacks. This attack uses malware or malicious codes to encrypt all data, including the patients' personal identify information, health records, and request the money to decrypt the data. Many publications and reports showed that ransomware could cause massive damage to patients and medical organizations, and HISs. For example, Jack McCarthy reported more than 400 thousand ransomware attacks with damage cost about 325 million dollars in 2015 [60]. In another way, well-known ransomware, namely "WannaCry" attacked more than 150 nations and approximately 300,000 machines worldwide. Hence, it affected about 80 national healthcare service organizations in Britain in 2017 [61,62]. In addition, more than

Table 2.2 The impact of four major types of cyber-attacks on the HISs

Type	Description	Impact	References
Ransomware/ malware	- Encrypting files and requesting money for getting data back - Destroying the data, modifying the integrity of data	High	[47–50]
DoS attack	- Interrupting the network or system	High	[51–53]
Unauthorized information access	- Gaining personal health information - Increasing the adverse effects of stress on patients - Damaging reputation of healthcare institutions	High	[45,54]
Phishing attack	- Gaining valuable information (username, password, medical records) - Gaining unauthorized remote access, identity theft	High	[55–59]

600 medical centers in 2020 were under attack by ransomware, and it cost approximately 21 billion dollars [63].

Furthermore, there was a dramatic increase in attacks (mostly ransomware and DoS attacks) to healthcare organizations in Europe, Asia, Latin Americas, South East Asia, and Iberia in 2020 [64] (Figure 2.7). Remarkably, in 2020, during the COVID-19 pandemic, there was a new type of ransomware, namely Corona Virus, spreading via a fake website that allowed victims to download the fabricated setup file to get passwords and encrypt data [42].

Figure 2.7 shows a warning bell for individuals, companies, governmental offices about the severe negative impacts of ransomware on national critical infrastructure systems, as well as healthcare systems.

2.5.2.2 DoS/DDoS attack

Denial of Service (DoS) or Distributed Denial of Service (DDoS) is an attack when hackers prevent users, clients, or patients from accessing to network, system, website, or equipment in the HISs [53]. Flood attacks and crash attacks are two main categories of DoS attacks [52]. Nevertheless, DDoS attacks are more dangerous and complicated than DoS attacks because they use many botnets from various places to attack the victims on a large scale. Consequently, it is extremely hard to detect, identify, and mitigate them. Furthermore, this attack can cause severe damage and losses for medical organizations, for example, overloaded hospital network systems, corrupted network connections, interrupted operations, and the like.

Since 2014, DoS/DDoS attacks have been increased and targeted in industry and healthcare sectors in many regions [64]. Indeed, DDoS attacks increased 2% from 2017 to 2018 in healthcare services companies [65]. In 2017, 45% of medical organizations were under DDoS attacks in the United States [51]. Mainly, during the COVID-19 outbreak, there were several cyber incidents in the world related to

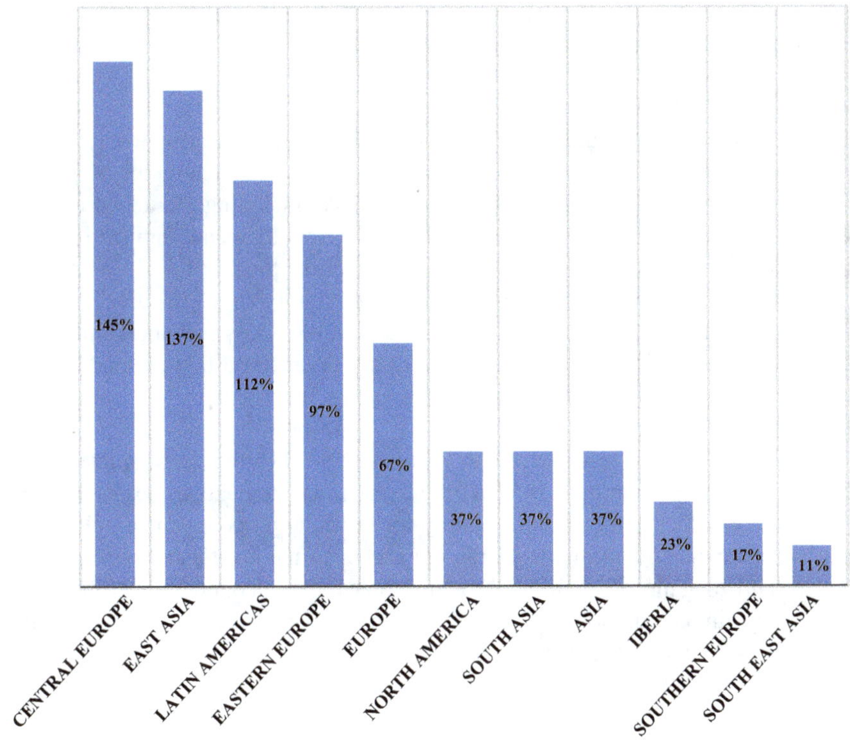

Figure 2.7 The increasing percentage of regional attacks on healthcare service companies [64]

DDoS attacks targeting HISs in many organizations. In countries such as the US Health and Human Services (HHS), World Health Organization (WHO), Hammersmith Medicines Research (MHR), and the Paris AP-HP Hospital Authority [66,67] these attacks had affected and damaged the HISs and other crucial documents attached to healthcare.

2.5.2.3 Data breach/unauthorized access

There are several ways of a data breach in electronic records. They can be used by various reasons such as curiosity, insider threats, and cyberattacks [54]. Besides, with many aims, including stealing medical treatments from patients' insurance [45], selling the protected health information, or using it for themselves, creating fake insurance demands [53], hackers can apply complicated methods to gain access to EHRs.

A data breach in healthcare organizations is significant because it directly affects victims and influences healthcare centers. For example, it may damage the victims' spirit, emotions, or psychology. An example of a data breach caused by a ransomware attack is denominated Blackbaud. It involved 3.4 million patient medical records in 2020 [68]. This incident affected nearly 80 healthcare organizations with a cost of approximately 3.6 million dollars. A data breach can affect healthcare institutions' reputations and decrease trust amongst patients, custodians, and agents during their treatment period [54,69].

2.5.2.4 Phishing attack

Phishing is a technique that allows hackers to gain valuable information from users such as usernames, passwords, medical records, or medical treatments by sending messages via emails to ask victims to click on links including malicious codes or download malware [59]. Moreover, this method refers to social engineering techniques to send messages via emails to targeted victims [56].

Phishing attacks can be classified into three main types [70] and have four purposes: finance, identity theft, unauthorized remote access, or integrating malicious malware. Besides, this kind of attack uses the method that pretends to be a legitimate website or e-mail to target victims; hence, it is tough to detect and handle this cybersecurity incident.

One of the most scathing attacks in the USA healthcare organizations is phishing attacks, accounting for 57% of cybersecurity incidents in 2019 [58]. This type of attack targets patients' information for financial purposes and employees in healthcare organizations for illegal remote access to the healthcare system. Thus, hackers aim to gain information through employees' accounts. For example, in 2019, a phishing simulation study in six US healthcare organizations reported that more than 400,000 workers were quickly targeted by approximately 3 million phishing emails [55,59]. Therefore, this attack can cause massive damage to healthcare systems and healthcare institutions.

2.6 Countermeasures/solutions

2.6.1 Information security risk management

As mentioned in previous sections, healthcare systems deal with many different security challenges of internal and external threats or attacks. They are leading to various types of damage for individuals and organizations. These damages may dramatically affect system operations, information assets, users, and healthcare centers. Hence, organizations must build an information risk management plan or information security management system (ISMS). These systems identify, evaluate, mitigate the risks/threats, and protect the availability, confidentiality, integrity of information [71]. Figure 2.8 presents a close-knit circle of four steps to protect healthcare data in the system. These steps need to be closely linked and performed continuously to ensure data safety. Healthcare institutions rely on the International Organization for Standardization (ISO) to create and establish effective ISMS. Its primary goal is to safeguard the security and privacy of sensitive information in any company by the application of standards such as ISO 45001 (formerly OHSAS 18001:2007) [72], ISO 27799:2016 [73], ISO 9000 series, especially ISO 27001:2013 [74]. Applying various ISOs to ensure the patients' security and privacy of healthcare information data is not a perfect solution for healthcare organizations. However, it requires the combination of other methods such as information

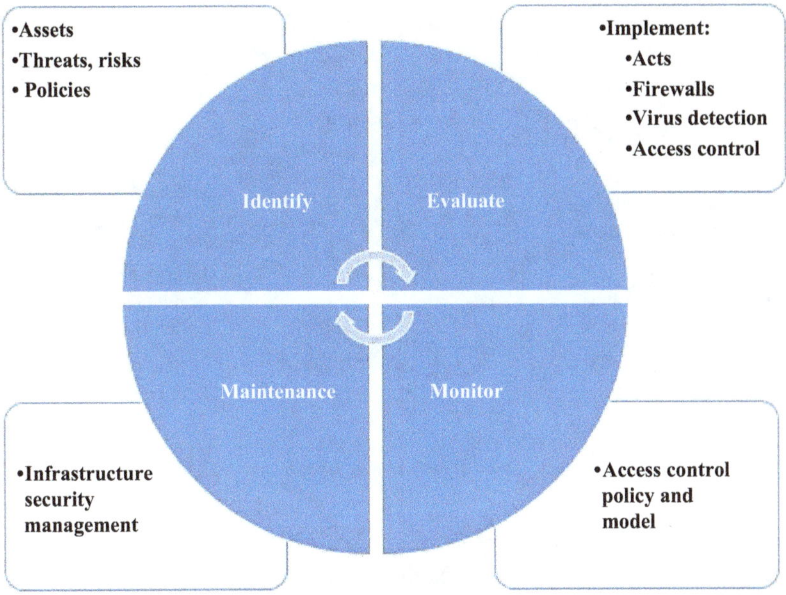

Figure 2.8 Information security risk management plan

security systems for authorized data disclosure, regulations, security policies, and procedures of processing data. Additionally, it demands building an IT staff to control the flow of data and ensure confidentiality, integrity, and availability of data.

2.6.2 New approaches for healthcare information

Healthcare data consisted of patient data, clinical data, doctor's information, and fundamental data in the past. With the boost of IT technology, healthcare is supported by a diversity of hi-tech devices. Consequently, healthcare data such as biometric information, blood pressure, electronic medical records, data from remote sensors, and social media information can be generated from these devices (Figure 2.9). All these different types of health information contribute to an extensive healthcare database [76].

The amount of healthcare data has increased quickly. Hence, information processes are becoming crucial. However, these data need to be handled to become valuable. Notably, data needs a large amount of storage and takes time to process, as well as the Internet is creating the raw data regularly that needs to be managed. As a result, big data and cloud computing are the best ways to store and process data in the healthcare aspect. Nevertheless, there are several security issues related to big data and cloud computing.

Therefore, various researchers proposed different methods to ensure the security and privacy of e-health data. For instance, a new secure framework, namely Multi Authority Attribute Based Encryptions (MA-ABE) for guaranteeing

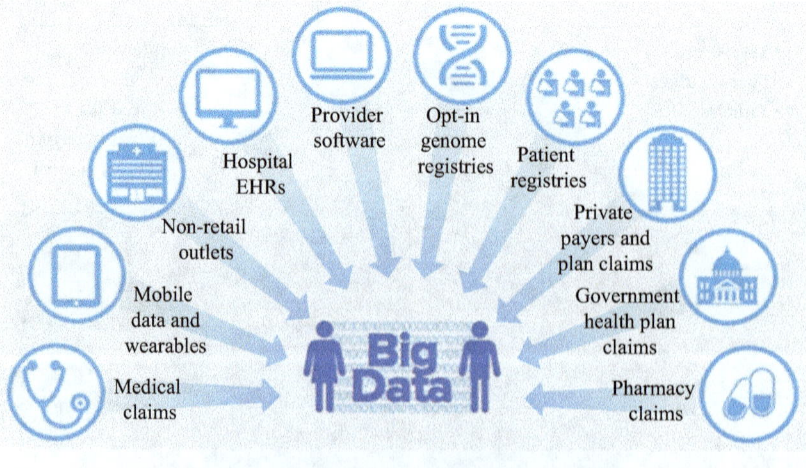

Figure 2.9 Big data in healthcare [75]

the transmission of the Protected Health Information (PHI) to health services after security check is indicated in [77]. Furthermore, another method for preserving the security and privacy of medical records and digital imaging and communication in medicine (DICOM) was proposed by [78]. Besides, due to cloud-based environment issues, Mehedi and Hossain [79] suggested a two-phase security protocol using pairing-based cryptography to counter security attacks during healthcare data transmission in the cloud.

In addition, other potential solutions, security frameworks, and various security mechanisms using cryptographic technologies to protect healthcare data in big data and cloud security issues are illustrated in [75].

2.6.3 Solutions for data security in healthcare

The concept of information security includes the following keys: confidentiality, integrity, availability, and authentication. Therefore, healthcare organizations must apply sufficient data security solutions to protect patient information, employee data, assets against cyber threats and cyberattacks from inside and outside the system. Based on a meta-analysis from relevant researches and trusted documents, several countermeasures are described in Table 2.3 with the primary purpose to preserve the security of data access, data transmission, and data storage in the healthcare system.

2.6.4 Solutions for securing data privacy in healthcare information

2.6.4.1 Methods to preserve the data privacy

Currently, there are various new methods to protect data privacy for users and organizations from threats inside and cyberattacks from outside. Two significant ways to solve healthcare information privacy issues are depicted below:

Table 2.3 Countermeasures to secure data security of healthcare information

Techniques	Description	References
Authentication	- Safeguarding the integrity of data and identity of the sender - Using digital signature mechanism, pluggable authentication modules (PAM)	[80–82] [83,84]
Encryption	- Preventing exposure to data breaches (packet sniffing, data theft) - Applying several algorithms such as DES, 3DES, RC4, RC6, RSA, AES, Rijndael	[7,85]
Cryptography API	- Applying many cryptographic models such as Generic Security Services – GSS-API (Internet Engineering Task Force), The Generic Crypto Services-GCS-API (X/Open), Cryptoki (RSA)	[47–50, 56, 58,75,78, 80–83,85–88]
Data masking	- Replacing sensitive data with unidentifiable data (K-anonymity, p-sensitive anonymity)	[89,90]
Access control	- Using role-based access control (RBAC) and attribute-based access control (ABAC)	[63,64]
Data search	- Using multi-keyword search scheme over encrypted health records and decryptable attribute-based keyword search scheme	[85,91,92]
Blockchain and cloud computing	- Building a secure, scalable, and openly coordinated platform - Including digital signatures to prevent a single point of failure - Encrypting network transactions	[80,93–97]

- De-identification
  De-identification is a conventional method to safeguard individual data privacy. This method includes k-anonymity, l-diversity, and t-closeness [98,99]. Moreover, it can help reduce and mitigate the threats from re-identification; however, this method is not suitable for preventing big data privacy from hackers because they can gain more information in the big data [100]. In order to solve this problem, several recent solutions are indicated by [99].
- HybrEx
  Cloud computing responds to the necessity of ample storage for healthcare information. Nevertheless, confidentiality and privacy regarding these data in a cloud environment pose severe concerns for healthcare centers. Hence, Ko *et al.* [101] proposed the new model, namely the HybrEx, to resolve these problems. Furthermore, this novel model supports complete and fast integrity checking models for data in both types of cloud computing, such as public and private clouds. Although this solution still has three main challenges during the

implementation process for organizations, some guidelines are suggested for healthcare centers to mitigate them in.

- Blockchain-based for e-healthcare
  Cloud computing creates a new way to store and share vital data between healthcare services, medical organizations, hospital systems, and healthcare providers. It can help reduce the cost and facilitate the process among those systems. However, patients are afraid of privacy loss from cloud service providers; they often refuse to give their identities. A potential solution to keep the patients' identity is a blind signature-based secured e-healthcare system [102]. This proposed solution helps to safeguard the confidentiality level of patients' information by using a layer of anonymity. Moreover, several studies have applied blockchain technology for e-healthcare, such as [103–105].
- Other solutions
  Besides the above mentioned solutions, there are several researches that can preserve the privacy of healthcare informatics during the process, such as [7,96,106–113]. These bring the potential solutions to maintain privacy and security for e-healthcare systems.

2.6.4.2 Regional and international regulations

Regulation is a framework, policy, or a set of rules to ensure all medical records, billings, and patients' personal information satisfy the demands of consistent standards regarding documentation, privacy, and security. These standards should be followed by doctors, hospitals, healthcare agencies, or other healthcare providers. Currently, there are many different regulations in the world to protect the privacy and security of data/information, especially in data protection law for not only citizens but also organizations as in Table 2.4.

Table 2.4 Privacy and security frameworks around the world

Region	Country	Standards
Europe	European Union	GDPR
	UK	Data Protection Act (DPA)
America	USA	HIPAA Act, Patient Safety and Quality Improvement Act (PSQIA), HITECH Act, Patient Protection and Affordable Care Act (PPACA)
	Canada	The Personal Information Protection and Electronic Documents Act (PIPEDA)
Latin America	Brazil	Brazilian General Data Protection Act, Personal Data Protection Law
Asia/Asia-Pacific	India	IT Act and IT (Amendment) Act
	South Korea	Personal Information Protection Act
	China	Cybersecurity Law, Cybersecurity multi-level Protection system 2.0

(Continues)

Table 2.4 (*Continued*)

Region	Country	Standards
	Hong Kong	The Personal Data (Privacy) Ordinance (Cap. 486 of the Laws of Hong Kong) (Ordinance) in 1996
	Taiwan	Personal Data Protection Act (PDPA)
	Japan	The Act on the Protection of Personal Information ("APPI")
	Russia	Russian Federal Law on Personal Data
	Malaysia	The Personal Data Protection Act 2010 (PDPA)
	The Philippines	The Data Privacy Act of 2012
	Vietnam	Cybersecurity law
	Australia	Privacy Act
	New Zealand	Privacy Act 2020
	Singapore	Singapore Personal Data Protection Act
Africa	Angola	Data Protection Law
	Morocco	The 09-08 Act

2.7 Upcoming trends in health informatics

Smart homes are devised for the forthcoming years. Appliances such as TVs, mirrors, or toilets will measure and collect health data and be embedded in the people's daily routines. Thus, feeding sensitive data to non-health-related corporations not regulated by privacy frameworks poses confidentiality risks.

COVID-19 outbreak pushed health systems to the limit, forcing them to adapt quickly to cope with the global emergency. The pandemic demanded government authorities to use private data such as location to control public movement and, consequently, control the spread of the disease. Moreover, it raised public health awareness, transforming how health data privacy will be envisioned in the future. It steered the efforts towards solidarity-based healthcare in which the emphasis is on the common well-being instead of the individual privacy rights to achieve health benefits. Consequently, it is envisaged that privacy regulations will be more flexible concerning medical research, allowing scientists access to large amounts of health data.

Interoperability between healthcare systems, devices, and applications is forecasted for the coming years to cooperate, exchange, and integrate health data and key actors regardless of regions and boundaries. Consumer-centric features will be embedded in health systems. Health data digitalization is an ongoing process that will contribute to data analysis. It will revolutionize healthcare in all areas, such as precision medicine and patient-based care.

2.8 Case study

Ransomware constitutes the most prevalent cyber-attack in healthcare systems nowadays. Health data is very vulnerable to being stolen since they constitute a low

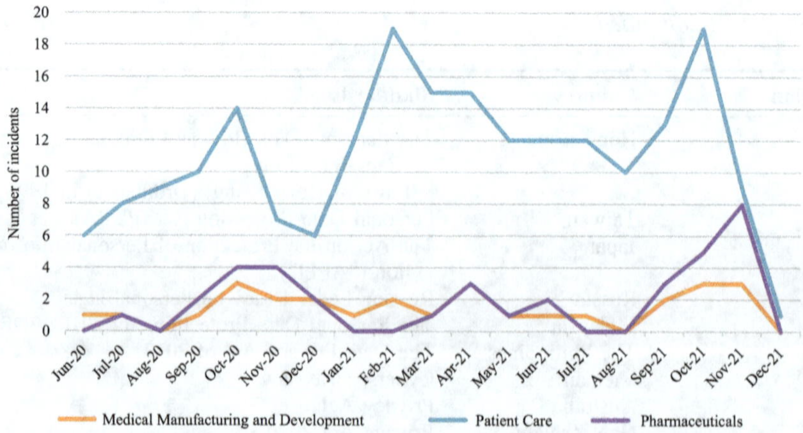

Figure 2.10 Number of cyber incidents in healthcare sectors [114]

risk but a high reward for attackers that can sell this information and there is a low chance of being caught. Additionally, ransomware increases the mortality rate, complications in medical procedures, delays in tests, transfers to other health centers, and more extended stays.

For example, in October 2021, the Johnson Memorial Health Hospital in Franklin, IN, US suffered a ransomware attack exposing patients' data and disabling the whole computer network. Likewise, in November, Hillel Yaffe Medical Center in Hadera, Israel, suffered a ransomware attack. As a result, health data was released, affecting 290,000 patients. These cases are not isolated. The COVID-19 pandemic contributed to the increment of these attacks, especially in patient care services, as seen in Figure 2.10.

Furthermore, IoT health devices are an easy target for cybercriminals and constantly grow among healthcare services. These devices have poor cyber security controls, and they are an easy access point to health systems networks due to the lack of visibility within the connected devices. Countermeasures to enhance IoT security include endpoint detection and response, visibility, and network segmentation [114,115,116].

It is vital to train cybersecurity specialists, increase funding, and raise awareness among healthcare employees regarding a holistic approach towards cybersecurity. Moreover, the governments need to deploy actions to protect healthcare systems, enforce transparency concerning the report of incidents, investigation, and prosecution.

2.9 Conclusion

Health data privacy and security are ongoing research topics. These two concepts are deeply connected. Privacy protection via legal frameworks needs to be accompanied by effective security methods to prevent breaches that affect the patient and the trust in healthcare providers and scientists. Privacy and security regulations should not be considered a hindrance to medical research. In contrast, it presents an opportunity to develop better techniques to protect the patients and strengthen their trust via honest communication of the usage, purpose, and sharing of personal data. Data protection should be conceived by design and default in every company's process to comply with the current regulations.

The advent of new technologies in healthcare such as Cloud, Big Data, IoT, AI, and data analytics pose unknown privacy and confidentiality challenges. Furthermore, due to IT and cloud-based environments, the potential security threats and cyber-attacks are a significant concern in healthcare information systems for users, patients, healthcare organizations, and governments. These challenges are required to be addressed. Countermeasures should be incorporated into existing legal frameworks to keep up with the advancement of technology. The characterization of sensitive data depends on the context, user, and social norms that can change throughout time.

For this reason, legislation that focuses on the correct data usage is critical to strengthen privacy rights. Privacy rights need to be hand in hand with emergency responsiveness. For example, big health data analysis via machine learning can be practical for faster vaccines, diagnosis, and disease treatment in crises. However, it implies the loss of privacy rights. A balance between the health benefits and risks needs to be weighted considering data ethics and actions that assure data usage for medical research and prevent misuse.

This chapter conveyed a comprehensive overview of healthcare data privacy and security issues via relevant publications, research, and documents. Furthermore, current solutions to secure healthcare information were presented, such as information security risk management system and several methods for data security, privacy-preserving including cryptography, blockchain, triad CIA concepts, de-identification, hybrEX, regional and international regulations. This chapter aims to raise security awareness for users, patients, healthcare centers, and the government and proposes several trends and solutions for future research.

References

[1] C. Brook, "What is a Health Information System?" *Digital Guardian*, 2020. https://digitalguardian.com/blog/what-health-information-system (accessed May 18, 2021).

[2] S. Tanwar, S. Tyagi, and N. Kumar, *Security and Privacy of Electronic Healthcare Records: Concepts, Paradigms and Solutions*. Institution of Engineering and Technology, 2020.

[3] C. Esposito, A. De Santis, G. Tortora, H. Chang, and K. K. R. Choo, "Blockchain: a panacea for healthcare cloud-based data security and privacy?" *IEEE Cloud Comput.*, vol. 5, no. 1, pp. 31–37, 2018, doi: 10.1109/MCC.2018.011791712.

[4] J. Xu and C. Liu, "Infodemic vs. pandemic factors associated to public anxiety in the early stage of the COVID-19 outbreak: a cross-sectional study in China," *Front. Public Heal.*, vol. 9, p. 1286, 2021, doi: 10.3389/FPUBH.2021.723648/BIBTEX.

[5] M. M. Baig and H. Gholamhosseini, "Smart health monitoring systems: an overview of design and modeling," *J. Med. Syst.*, vol. 37, no. 2, 2013, doi: 10.1007/S10916-012-9898-Z.

[6] S. Secinaro, D. Calandra, A. Secinaro, V. Muthurangu, and P. Biancone, "The role of artificial intelligence in healthcare: a structured literature review," *BMC Med. Inform. Decis. Mak.*, vol. 21, no. 1, pp. 1–23, 2021, doi: 10.1186/S12911-021-01488-9/FIGURES/12.

[7] P. K. D. Pramanik, G. Pareek, and A. Nayyar, *Security and Privacy in Remote Healthcare: Issues, Solutions, and Standards.* Elsevier Inc., 2019.

[8] Bitglass, "Healthcare breach report 2021 hacking and IT incidents on the rise," 2021.

[9] U.S. Department of Health & Human Services, "Breach portal: notice to the secretary of HHS breach of unsecured protected health incidents," 2022. https://ocrportal.hhs.gov/ocr/breach/breach_report.jsf (accessed Jan 7, 2022).

[10] S. J. Nass, L. A. Levit, and L. O. Gostin, "The value and importance of health information privacy," in *Beyond the HIPAA Privacy Rule: Enhancing Privacy, Improving Health Through Research.* National Academies Press (US), 2009.

[11] M. Gupta and R. Sharman, *Handbook of Research on Social and Organizational Liabilities in Information Security.* IGI Global, Hershey, PA, 2008.

[12] A. Westin, "Privacy and freedom," *Adm. Law Rev.*, vol. 22, no. 1, pp. 101–106, 1967.

[13] A. Westin, "How the public views privacy and health research," 2008.

[14] Institute of Medicine (US) Committee on Regional Health Data Networks, M. S. Donaldson, and K. N. Lohr, *Health Data in the Information Age.* Washington, DC: National Academies Press, 1994.

[15] R. Turn and W. Ware, *Privacy and Security Issues in Information Systems.* RAND Corporation, 2017.

[16] K. Abouelmehdi, A. Beni-Hessane, and H. Khaloufi, "Big healthcare data: preserving security and privacy," *J. Big Data*, vol. 5, no. 1, pp. 1–18, 2018, doi: 10.1186/s40537-017-0110-7.

[17] Y. Lu and R. O. Sinnott, "Security and privacy solutions for smart healthcare systems," in *Innovation in Health Informatics: A Smart Healthcare Primer*, pp. 189–216. Academic Press, 2020, doi: 10.1016/B978-0-12-819043-2.00008-3.

[18] OECD, OECD Guidelines on the Protection of Privacy and Transborder Flows of Personal Data, Organization for Economic Co-operation and Development, 2013. https://www.oecd.org/digital/ieconomy/oecdguidelinesontheprotectionofprivacyandtransborderflowsofpersonaldata.htm (accessed May 27, 2021).

[19] M. Burgess, "What is GDPR? The summary guide to GDPR compliance in the UK," *Wired*, 2020. https://www.wired.co.uk/article/what-is-gdpr-uk-eu-legislation-compliance-summary-fines-2018 (accessed Jun 8, 2021).

[20] M. Nadeuau, "What is the GDPR, its requirements and facts?" *CSO United States*, 2020. https://www.csoonline.com/article/3202771/general-data-protection-regulation-gdpr-requirements-deadlines-and-facts.html (accessed Jun 4, 2021).

[21] B. Wolford, "What is GDPR, the EU's new data protection law?" *GDPR.EU*, 2021. https://gdpr.eu/what-is-gdpr/?cn-reloaded=1 (accessed Jun 4, 2021).

[22] Intersoft consulting, "Art. 6 GDPR – Lawfulness of processing," 2021. https://gdpr-info.eu/art-6-gdpr/ (accessed Jun 8, 2021).

[23] "EUR-Lex – 32016R0679," *EUR-Lex*, 2021. https://eur-lex.europa.eu/eli/reg/2016/679/oj (accessed Jun 8, 2021).

[24] CDC, "Health insurance portability and accountability act of 1996 (HIPAA)," *Centers for Disease Control and Prevention*, 2018. https://www.cdc.gov/phlp/publications/topic/hipaa.html (accessed Jun 9, 2021).

[25] HHS, "Summary of the HIPAA Privacy Rule," *The U.S. Department of Health and Human Services (HHS)*. https://www.hhs.gov/hipaa/for-professionals/privacy/laws-regulations/index.html (accessed Jun 9, 2021).

[26] C. Mistry, U. Thakker, R. Gupta, *et al.*, "MedBlock: an AI-enabled and blockchain-driven medical healthcare system for COVID-19," in *ICC 2021 – IEEE International Conference on Communications*, Montreal, QC, Canada, 2021, pp. 1–6, doi:10.1109/ICC42927.2021.9500397.

[27] S. Tanwar, J. Vora, S. Kaneriya, *et al.*, "Human arthritis analysis in fog computing environment using Bayesian network classifier and thread protocol," *IEEE Consumer Electronics Magazine*, vol. 9, no. 1, pp. 88–94, 2020, doi:10.1109/MCE.2019.2941456.

[28] A. L. McGuire, R. Fisher, P. Cusenza, *et al.*, "Confidentiality, privacy, and security of genetic and genomic test information in electronic health records: points to consider," *Genet. Med.*, vol. 10, no. 7. pp. 495–499, 2008, doi: 10.1097/GIM.0b013e31817a8aaa.

[29] J. E. Lunshof, R. Chadwick, D. B. Vorhaus, and G. M. Church, "From genetic privacy to open consent," *Nat. Rev. Genet.*, vol. 9, no. 5, pp. 406–411, 2008, doi: 10.1038/nrg2360.

[30] J. Gershman, "Health data after Covid-19: more laws, less privacy," *Wall Street Journal*, 2020. https://www.wsj.com/articles/protecting-health-data-after-covid-19-more-laws-less-privacy-11599750100 (accessed Oct 1, 2020).

[31] I. Maglogiannis, E. Zafiropoulos, A. Platis, and C. Lambrinoudakis, "Risk analysis of a patient monitoring system using Bayesian Network modeling," *J. Biomed. Inform.*, vol. 39, no. 6, pp. 637–647, 2006, doi: 10.1016/j.jbi.2005.10.003.

[32] I. Maglogiannis and E. Zafiropoulos, "Modeling risk in distributed healthcare information systems," in *Annual International Conference of the IEEE Engineering in Medicine and Biology – Proceedings*, pp. 5447–5450, 2006, doi: 10.1109/IEMBS.2006.260351.

[33] G. Narayana Samy, R. Ahmad, and Z. Ismail, "Security threats categories in healthcare information systems," *Health Informatics J.*, vol. 16, no. 3, pp. 201–209, 2010, doi: 10.1177/1460458210377468.

[34] CTI, "Security Threats in HealthCare Systems," 2019.

[35] A. Fatima and R. Colomo-Palacios, "Security aspects in healthcare information systems: a systematic mapping," *Proc. Comput. Sci.*, vol. 138, pp. 12–19, 2018, doi: 10.1016/j.procs.2018.10.003.

[36] HHS Office of the Assistant Secretary for Preparedness, "Health industry cybersecurity practices: managing threats and protecting patients," 2019.

[37] E. Mehraeen, M. Ghazisaeedi, J. Farzi, and S. Mirshekari, "Security challenges in healthcare cloud computing: a systematic review," *Glob. J. Health Sci.*, vol. 9, no. 3, p. 157, 2016, doi: 10.5539/gjhs.v9n3p157.

[38] Y. Al-Issa, M. A. Ottom, and A. Tamrawi, "eHealth cloud security challenges: a survey," *J. Healthc. Eng.*, vol. 2019, 2019, doi: 10.1155/2019/7516035.

[39] G. N. Samy, R. Ahmad, and Z. Ismail, "Threats to health information security," in *5th International Conference on Information Assurance and Security, IAS 2009*, vol. 2, pp. 540–543, 2009, doi: 10.1109/IAS.2009.312.

[40] L. Mucchi, S. Jayousi, A. Martinelli, S. Caputo, and P. Marcocci, "An overview of security threats, solutions and challenges in WBANs for healthcare," in *International Symposium on Medical Information & Communication Technology (ISMICT)*, pp. 1–6, 2019, doi: 10.1109/ISMICT.2019.8743798.

[41] H. P. D. Nguyen and D. D. Nguyen, "Drone application in smart cities: the general overview of security vulnerabilities and countermeasures for data communication," *Stud. Syst. Decis. Control*, vol. 332, pp. 185–210, 2021, doi: 10.1007/978-3-030-63339-4_7.

[42] N. A. Khan, S. N. Brohi, and N. Zaman, "Ten deadly cyber security threats amid COVID-19 pandemic," *TechRxiv Powered by IEEE*, pp. 1–6, 2020, doi: 10.36227/techrxiv.12278792.v1.

[43] N. S. Abouzakhar, A. Jones, and O. Angelopoulou, "Internet of things security: a review of risks and threats to healthcare sector," in *Proceedings – 2017 IEEE International Conference on Internet of Things (iThings) and IEEE Green Computing and Communications (GreenCom) and IEEE Cyber, Physical and Social Computing (CPSCom) and IEEE Smart Data (SmartData), iThings/GreenCom/CPSCom/SmartData 2017*, pp. 373–378, 2018, doi: 10.1109/iThings-GreenCom-CPSCom-SmartData.2017.62.

[44] S. S. Bhuyan, U. Y. Kabir, J. M. Escareno, *et al.*, "Transforming healthcare cybersecurity from reactive to proactive: current status and future recommendations, " *J. Med. Syst.*, vol. 44, no. 5, 2020, doi: 10.1007/s10916-019-1507-y.

[45] S. Sujitparapitaya, A. Shirani, and M. Roldan, "Issues in information systems," *Issues Inf. Syst.*, vol. 13, no. 2, pp. 112–122, 2012.

[46] S. A. Butt, J. L. Diaz-Martinez, T. Jamal, A. Ali, E. De-La-Hoz-Franco, and M. Shoaib, "IoT smart health security threats," In *Proceedings – 2019 19th*

International Conference on Computational Science and Its Applications (ICCSA 2019), pp. 26–31, 2019, doi: 10.1109/ICCSA.2019.000-8.

[47] Perspectives in Health Information Management, "Ransomware in healthcare facilities: a harbinger of the future?," 2018.

[48] L. Dyrda, "The 5 most significant cyberattacks in healthcare for 2020", *Becker's Hospital Review*, 2020.

[49] University of Illinois at Chicago, *Cybersecurity: How Can It Be Improved in Health Care?*

[50] V. Kumar and C. Walker, "Cyber-attacks: rising threat to healthcare," 2017.

[51] J. Davis, "Denial-of-service attacks on healthcare poised to explode," *Healthcare IT News*, May 16, 2017. https://www.healthcareitnews.com/news/denial-service-attacks-healthcare-poised-explode.

[52] J. West, "What is a denial of service attack and why is healthcare targeted?" *Paubox*, March 27, 2023. [Online]. Available: https://www.paubox.com/blog/denial-of-service-attack [Accessed: 10-Apr-2023]

[53] M. Mrcela and I. Vuletic, "Healthcare, privacy, big data and cybercrime: which one is the weakest link?" *Ann. Heal. Law*, vol. 27, no. 2, p. 257, 2018.

[54] Information and Privacy Commissioner of Ontario, *Detecting and Deterring Unauthorized Access to Personal Health Information*, pp. 1–27, 2015.

[55] W. J. Gordon, A. Wright, R. Aiyagari, *et al.*, "Assessment of employee susceptibility to phishing attacks at US health care institutions," *JAMA Netw. Open*, vol. 2, no. 3, pp. 3–11, 2019, doi: 10.1001/jamanetworkopen.2019.0393.

[56] D. Nordgren, "Phishing attacks targeting hospitals: a study over phishing knowledge at Blekingesjukhuset," Thesis, Blekinge Tekniska Högskola, Institutionen för datalogi och datorsystemteknik, 2018. http://urn.kb.se/resolve?urn=urn:nbn:se:bth-16681.

[57] W. J. Gordon, A. Wright, R. J. Glynn, *et al.*, "Evaluation of a mandatory phishing training program for high-risk employees at a US healthcare system," *J. Am. Med. Informatics Assoc.*, vol. 26, no. 6, pp. 547–552, 2019, doi: 10.1093/jamia/ocz005.

[58] A. Vaidya, "Phishing attacks most common cybersecurity incident at US healthcare organizations," *MedCity News*, 2020.

[59] W. Priestman, T. Anstis, I. G. Sebire, S. Sridharan, and N. J. Sebire, "Phishing in healthcare organisations: threats, mitigation and approaches," *BMJ Heal. Care Informatics*, vol. 26, no. 1, pp. 1–6, 2019, doi: 10.1136/bmjhci-2019-100031.

[60] J. McCarthy, "Ransomware to wreak havoc in 2016, ICIT study says," *Healthcare IT News*, Mar 21, 2016.

[61] S. Ghafur, S. Kristensen, K. Honeyford, G. Martin, A. Darzi, and P. Aylin, "A retrospective impact analysis of the WannaCry cyberattack on the NHS," *npj Digit. Med.*, vol. 2, no. 1, pp. 1–7, 2019, doi: 10.1038/s41746-019-0161-6.

[62] L. Heather, "Report: 40% of healthcare organizations hit by WannaCry in past 6 months," 2019. [Online]. Available: https://www.fiercehealthcare.com/

tech/lingering-impacts-from-wannacry-40-healthcare-organizations-suffered-from-attack-past-6-months [Accessed: 10-Apr-2023].

[63] H. Mitchel, "Ransomware attacks on healthcare organizations cost nearly $21B last year, study finds," *Becker's Hospital Review*, 2021.

[64] Check Point, *Attacks Targeting Healthcare Organizations Spike Globally as COVID-19 Cases Rise Again*, 2021.

[65] F. Donovan, "Healthcare DDoS attacks on organizations edged up in 2018," 2019. [Online]. Available: https://hitinfrastructure.com/news/healthcare-ddos-attacks-on-organizations-edged-up-in-2018 [Accessed: 10-Apr-2023]

[66] E. Chickowski, "7 Cyberattacks against pandemic-stressed healthcare orgs," 2020. [Online]. Available: https://securityboulevard.com/2020/04/7-cyber-attacks-against-pandemic-stressed-healthcare-orgs/ [Accessed: 10-Apr-2023].

[67] A. R. Choudhury, "Cyber-attacks on healthcare indicate criminals never let go of opportunity," 2020. [Online] Available: https://govinsider.asia/intl-en/article/cyber-attacks-on-healthcare-indicate-criminals-never-let-go-of-opportunity [Accessed: 10-Apr-2023]

[68] F. Langston, "Recent spike in healthcare breach reports due to Blackbaud ransomware attack," 2020. [Online]. Available: https://www.criticalinsight.com/resources/news/article/recent-spike-in-healthcare-breach-reports-due-to-blackbaud-ransomware-attack [Accessed: 10-Apr-2023].

[69] V. A. Patel, P. Bhattacharya, S. Tanwar, *et al.*, "Adoption of federated learning for healthcare informatics: emerging applications and future directions," *IEEE Access*, vol. 10, pp. 90792–90826, 2022, doi:10.1109/ACCESS.2022.3201876.

[70] "Public-Private Analytic Exchange Program (AEP) Deliverables," Homeland Security, November 10, 2022. https://www.dhs.gov/aep-deliverables.

[71] K. A. Mahmood, "Information security management of healthcare system: a case study of Blekinge region healthcare," Master Thesis, Blekinge Institute of Technology, Sweden; 2010.

[72] NQA, *ISO 45001: Occupational Health and Safety Standard*, 2021.

[73] ISO, *ISO 27799:2016 Health Informatics—Information Security Management in Health Using ISO/IEC 27002*, 2021.

[74] NetGrowth, *ISO 27001:2013 Introduction*, 2021.

[75] M. Siddique, M. A. Mirza, M. Ahmad, J. Chaudhry, and R. Islam, "A survey of big data security solutions in healthcare," *Lecture Notes of the Institute for Computer Sciences, Social Informatics and Telecommunications Engineering, LNICST*, vol. 255, pp. 391–406, 2018, doi: 10.1007/978-3-030-01704-0_21.

[76] A. Widmer, R. Schaer, D. Markonis, and H. Müller, "Gesture interaction for content-based medical image retrieval." in *Proceedings of International Conference on Multimedia Retrieval*, pp. 503–506, 2014.

[77] A. E. Youssef, "A framework for secure healthcare systems based on big data analytics in mobile cloud computing environments," *Int. J. Ambient Syst. Appl.*, vol. 2, no. 2, pp. 1–11, 2014, doi: 10.5121/ijasa.2014.2201.

[78] C. Gunamalai and S. Sivasubramanian, "A novel method of security and privacy for personal medical record and DICOM images in cloud computing," *ARPN J. Eng. Appl. Sci.*, vol. 10, no. 10, pp. 4635–4638, 2015.

[79] M. Masud and M. S. Hossain, "Secure data-exchange protocol in a cloud-based collaborative health care environment," *Multimed. Tools Appl.*, vol. 77, no. 9, pp. 11121–11135, 2018, doi: 10.1007/s11042-017-5294-5.

[80] A. Al-Noman Patwary, A. Fu, R. K. Naha, *et al.*, "Authentication, access control, privacy, threats and trust management towards securing fog computing environments: a review," pp. 1–34, 2020, arXiv preprint arXiv:2003.00395.

[81] Y. Y. Al-Salqan, "Security and confidentiality in healthcare informatics," in *Proceedings of the Workshop on Enabling Technologies: Infrastructure for Collaborative Enterprises, WET ICE*, pp. 371–375, 1998, doi: 10.1145/270152.270180.

[82] Y. L. Liao and C. Tsai, "Analysis of computer crime characteristics in Taiwan," *LNCS*, vol. 3917, 2006.

[83] K. Abouelmehdi, A. Beni-Hssane, H. Khaloufi, and M. Saadi, "Big data security and privacy in healthcare: a review," *Procedia Comput. Sci.*, vol. 113, pp. 73–80, 2017, doi: 10.1016/j.procs.2017.08.292.

[84] M. Bhavin, S. Tanwar, N. Sharma, S. Tyagi, and N. Kumar., "Blockchain and quantum blind signature-based hybrid scheme for healthcare 5.0 applications," *J. Inf. Secur. Appl.*, 56, 102673, 2021.

[85] S. Bajrić, "Data security and privacy issues in healthcare," *Appl. Med. Informatics*, vol. 42, no. 1, pp. 19–27, 2020.

[86] J. Linn, "Generic security service application program interface," *J. Chem. Inf. Model.*, vol. 53, no. 9, pp. 1689–1699, 2013.

[87] NSA Cross Organization CAPI Team, "Security service API," pp. 1–25, 1995.

[88] B. Kaliski, "Cryptoki: a cryptographic token interface," version 1.0, 1995.

[89] L. Sweeney, "Achieving k-anonymity privacy protection using generalization and suppression," *Int. J. Uncertainty, Fuzziness Knowledge-Based Syst.*, vol. 10, no. 5, pp. 571–588, 2002, doi: 10.1142/S021848850200165X.

[90] T. M. Truta and B. Vinay, "Privacy protection: P-sensitive k – anonymity property," in *ICDEW '06: Proceedings of the 22nd International Conference on Data Engineering Workshops (ICDEW'06)*, 2006, doi: 10.1109/ICDEW.2006.116.

[91] Y. Miao, J. Ma, X. Liu, F. Wei, Z. Liu, and X. A. Wang, "m^2-ABKS: attribute-based multi-keyword search over encrypted personal health records in multi-owner setting," *Journal of Medical Systems*, vol. 40, 246, 2016.

[92] L. Guo, Z. Li, W. C. Yau, and S. Y. Tan, "A decryptable attribute-based keyword search scheme on eHealth cloud in internet of things platforms," *IEEE Access*, vol. 8, pp. 26107–26118, 2020, doi: 10.1109/ACCESS.2020.2971088.

[93] L. Huang and H. H. Lee, "A medical data privacy protection scheme based on blockchain and cloud computing," *Wirel. Commun. Mob. Comput.*, vol. 2020, 2020, doi: 10.1155/2020/8859961.

[94] G. N. Nguyen, N. H. Le Viet, G. P. Joshi, and B. Shrestha, "Intelligent tunicate swarm-optimization-algorithm-based lightweight security mechanism in internet of health things," *Computers, Materials & Continua*, vol. 66, pp. 551–562, 2020.

[95] K. P. Kibiwott, Y. Zhao, J. Kogo, and F. Zhang, "Verifiable fully outsourced attribute-based signcryption system for IoT eHealth big data in cloud computing," *Math. Biosci. Eng.*, vol. 16, no. 5, pp. 3561–3594, 2019, doi: 10.3934/mbe.2019178.

[96] P. Bhattacharya, S. Tanwar, U. Bodkhe, S. Tyagi, and N. Kumar, "BinDaaS: blockchain-based deep-learning as-a-service in healthcare 4.0 applications," *IEEE Trans. Netw. Sci. Eng.*, vol. 8, no. 2, pp. 1242–1255, 2021, doi: 10.1109/TNSE.2019.2961932.

[97] J. J. Hathaliya, R. Gupta, S. Tanwar, and P. Sharma, "A smart contract-based secure data sharing scheme in healthcare 5.0," in *2021 IEEE Globecom Workshops (GC Wkshps), Madrid, Spain*, pp. 1–6, 2021, doi:10.1109/GCWkshps52748.2021.9681956.

[98] N. Li, "t-Closeness: privacy beyond k-anonymity and diversity," in *Proceeding of the IEEE 23rd International Conference on Data Engineering (ICDE 07)*, no. 2, pp. 106–115, 2011.

[99] P. Jain, M. Gyanchandani, and N. Khare, "Big data privacy: a technological perspective and review," *J. Big Data*, vol. 3, no. 1, 2016, doi: 10.1186/s40537-016-0059-y.

[100] S. Tanwar, S. Tyagi, and N. Kumar (eds.), *Security and Privacy of Electronic Healthcare Records: Concepts, Paradigms and Solutions*. Institution of Engineering and Technology, 2019.

[101] S. Y. Ko, K. Jeon, and R. Morales, "The Hybrex model for confidentiality and privacy in cloud computing," in *3rd USENIX Workshop on Hot Topics in Cloud. Computing, HotCloud 2011*, 2011.

[102] J. Vora, P. Devmurari, S. Tanwar, S. Tyagi, N. Kumar, and M. S. Obaidat, "Blind signatures based secured E-healthcare system," in *CITS 2018 – 2018 International Conference on Computer, Information, and Telecommunication Systems*, pp. 1–5, 2018, doi: 10.1109/CITS.2018.8440186.

[103] R. Gupta, S. Tanwar, S. Tyagi, N. Kumar, M. S. Obaidat, and B. Sadoun, "HaBiTs: blockchain-based telesurgery framework for healthcare 4.0," in *CITS 2019 – Proceeding 2019 International Conference on Computer, Information, and Telecommunication Systems*, pp. 1–5, 2019, doi: 10.1109/CITS.2019.8862127.

[104] A. Kumar, R. Krishnamurthi, A. Nayyar, K. Sharma, V. Grover, and E. Hossain, "A novel smart healthcare design, simulation, and implementation using healthcare 4.0 processes," *IEEE Access*, vol. 8, pp. 118433–118471, 2020, doi: 10.1109/ACCESS.2020.3004790.

[105] R. Gupta, A. Shukla, and S. Tanwar, "AaYusH: a smart contract-based telesurgery system for healthcare 4.0," in *2020 IEEE International Conference on Communications Workshops (ICC Workshops)*, Dublin, Ireland, 2020, pp. 1–6, doi:10.1109/ICCWorkshops49005.2020.9145044.

[106] W. Zeng, P. Chen, and Y. Yi, "Private aggregation scheme based on erasable data-hiding in wireless sensor networks," in *Proceedings – 2016 IEEE 4th International Conference on Future Internet of Things and Cloud (FiCloud), 2016*, vol. 10, no. 9, pp. 76–83, 2016, doi:10.1109/FiCloud.2016.19.

[107] P. K. D. Pramanik, A. Nayyar, and G. Pareek, *WBAN: Driving e-Healthcare Beyond Telemedicine to Remote Health Monitoring: Architecture and Protocols.* Elsevier Inc., 2019.

[108] J. L. Fernández-Alemán, I. C. Señor, P. Ángel O. Lozoya, and A. Toval, "Security and privacy in electronic health records: a systematic literature review," *J. Biomed. Inform.*, vol. 46, no. 3, pp. 541–562, 2013, doi: 10.1016/j.jbi.2012.12.003.

[109] S. Ali, Y. Hafeez, N. Zaman, *et al.*, "Towards pattern-based change verification framework for cloud-enabled healthcare component-based," *IEEE Access*, vol. 8, pp. 148007–148020, 2020, doi: 10.1109/ACCESS.2020.3014671.

[110] J. J. Hathaliya and S. Tanwar, "An exhaustive survey on security and privacy issues in healthcare 4.0," *Comput. Commun.*, vol. 153, pp. 311–335, 2020, doi: 10.1016/j.comcom.2020.02.018.

[111] B. D. Deebak, F. Al-Turjman, and A. Nayyar, "Chaotic-map based authenticated security framework with privacy preservation for remote point-of-care," *Multimed. Tools Appl.*, vol. 80, no. 11, pp. 17103–17128, 2021, doi: 10.1007/s11042-020-10134-x.

[112] R. Gupta, S. Tanwar, S. Tyagi, and N. Kumar, "Tactile-internet-based telesurgery system for healthcare 4.0: an architecture, research challenges, and future directions," *IEEE Netw.*, vol. 33, no. 6, pp. 22–29, 2019, doi: 10.1109/MNET.001.1900063.

[113] M. Abouhawwash and A. M. Alessio, "Multi-objective evolutionary algorithm for PET image reconstruction: concept," *IEEE Trans. Med. Imaging*, vol. 40, no. 8, pp. 2142–2151, 2021, doi: 10.1109/TMI.2021.3073243.

[114] "Cyber Incident Tracer #HEALTH," *Cyber Peace Institute*, 2021. https://cit.cyberpeaceinstitute.org/explore (accessed Jan 18, 2022).

[115] H. S. T. Committee, "HL7 role-based access control (RBAC) role engineering process," *Security*, 2007.

[116] A. Mohan and D. M. Blough, "An attribute-based authorization policy framework with dynamic conflict resolution," in *ACM International Conference Proceeding Series*, pp. 37–50, 2010, doi: 10.1145/1750389.1750395.

Chapter 3

Health informatics and its contribution to health sectors

Ginu George[1], V. Saravanakrishnan[1] and Ankit Agarwal[2]

Abstract

In most developed countries, healthcare sectors take more than 10% of the GDP, and it is one of the most significant and most rapidly growing sectors globally. With such growth of the healthcare department, data management becomes challenging; a robust platform helps to address these challenges. Health Informatics (HI) is an upcoming development, an interdisciplinary field in healthcare sectors; it combines the Internet of Things (IoT) and Artificial Intelligence (AI) in the healthcare software, which helps boost the overall operational efficiency of the healthcare departments. These AI algorithms integrated into IoT devices help acquire, store, retrieve, and use health and medical-related data. Patient data are enormous in healthcare sectors, and it is required for various purposes by hospital administrators, insurance agents, doctors, nurses, and other health departments. Accessing and managing these datasets often becomes challenging; HI is one of those innovations that has helped address these challenges to a large extent. The chapter discusses informatics, related definitions, HI, and its relation with other disciplines. The chapter also provides an educational overview of the evolution of HI, different HI technologies, benefits and challenges of HI to its various stakeholders. It ends with some thoughts on HI's future growth.

Keywords: Artificial intelligence; Health informatics; Health sector; Internet of Things

3.1 Introduction to health informatics

3.1.1 Meaning of informatics

The word informatics was first coined by Steinbuch [1] as *informatik* and later was developed by Drefyus [2] in French as *Informatique*, which refers to extensive data

[1]Department of Commerce, Christ (Deemed to be University), India
[2]Adelaide Business School, The University of Adelaide, Australia

of information arranged systematically. It helps better manage information generating, storing, processing, and presenting. Informatics has been interchangeably used for computers, computer sciences, information technology, and information system [3,4]. Some of the definitions that support these concepts are the following: Bernstam *et al.* [5], p. 106 defined informatics as "a science of information, where information is defined as data with meaning." According to [6], informatics is defined as "a field of science that investigates principles and technologies for defining semantics, creating value, and giving order to the world, by processing information." According to [7], informatics is "a discipline that combines into one-field information technologies, computer science, and business administration." They state that informatics involves:

> the practice of information systems engineering and information processing. It studies the structure, behavior, and interactions of natural and artificial systems that collect, generate, store, process, transmit and present information. Informatics combines aspects of software engineering, human-computer interaction, and the study of organizations and information technology; one can say it studies computers and people.

In general, it refers to applying information and technology in day-to-day processes. Yatsko and Suslow [7] state that the central concept of informatics is transforming information with the help of computation and communication by various organisms and artifacts.

3.1.2 *Meaning of health informatics*

Informatic is seen in all aspects of human life, be it in media, business, science, health, communities, and many more; informatics has dramatically enhanced the business functions [8]. One such field that is evolving, enhanced and relatively new is health informatics (HI) [9]. HI can be referred to as having a better insight into the meaning, relationship, and different factors of health care information. According to Saba and McCormick ([10], p. 232), HI is "the integration of healthcare sciences, computer science, information science, and cognitive science to assist in the management of healthcare information." McLane *et al.* [9] referred HI as the "use of analysis, design, implementation, and evaluation of information systems to improve processes and outcomes, driven by computer and cognitive sciences, to support improved health outcomes."

HI is widely accepted and considered an individual discipline that discusses other fields such as information science, computer science, social science, behavioral science, and healthcare. HI overlaps with other disciplines and has often been synonymously used with other names such as medical informatics, clinical informatics, pharmacy informatics, public health informatics, biomedical informatics, and bioinformatics [11,12]. Figure 3.1 depicts how different disciplines are used as synonymous terms.

According to [13], medical informatics refers to studying and applying techniques that help in improving the management of patient data, population data, and information related to patience and community health. The focus of medical

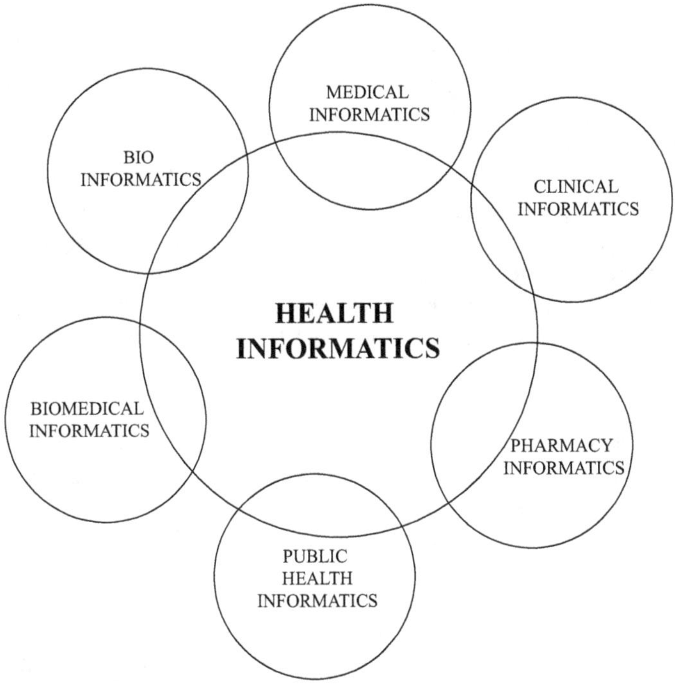

Figure 3.1 HI model

informatics is more on information management rather than technology. Wilson ([14], p. 12) defines information management as "management of an organisation's information resources and involves the management of information technology [IT]." At the same time, IT is one component of the infrastructure that is utilized by businesses for collecting and transforming data.

Clinical informatics refers to the use of advanced technologies that enable a better understanding of health and health problems, further boosting the conduct of research and developing effective interventions on public health and the provision of health and social care [15]. Pharmacy informatics extensively deal with pharmacy-related health data for varied purposes. According to [16], p. 201, pharmacy informatics is defined as "the use and integration of data, information, knowledge, technology, and automation in the medication-use process to improve health outcomes." The various factors covered are pharmacy management concerning barcoding of medicines, prescribing, and dispensing of drugs. It also involves people across the pharmacy, such as pharmacists, technicians from pharmacies, doctors, nurses, information technology personnel, and other professionals from healthcare [17].

Public health informatics is a systematic and logical application of data, computer science, and information and communication technologies (ICTs) on public health, research, and related learning; it prevents and promotes health [18,19].

Biomedical informatics (BMI) is an interdisciplinary field that focuses on understanding the effective use of information, facts, and data related to biomedical for solving problems. It helps in making decisions that further improve public health. According to Shortliffe *et al.* (2000, cited in [22], p. 105) defined, "biomedical informatics as the scientific field that deals with biomedical information, data and knowledge – their storage, retrieval and optimal use for problem-solving and decision making." Bioinformatics refers to implementing tools and techniques in different areas such as biology, medicine, math, physics, and computer science to analyze, capture, and interpret biological data [20]. This discipline helps primarily manage biology and medicine data and includes computer program tools such as BLAST and Ensemble, particularly analyzing human genome projects.

3.1.3 Evolution

HI is considered an evolving discipline that had existed since the 1950s, which was the beginning time when computers were used in healthcare [21,22]. According to Nelson [23], informatics has extended in the 1960s by experimenting with new technologies in nursing and medicine education. One of the recent papers by Masic [24] stated that in a conference held in Prague on the history of medical informatics, George Mihalas discussed the evolution of HI elaborately. According to Masic and Chronaki [25], the development of HI is classified into five stages; early-stage, childhood/youth, consolidation of HI, maturity stage, and full integration; the stages are explained below.

- *Early-stage of HI was up to 1975*: During this period, the concept started evolving, discussed, and most researchers and scientists were more interested in researching these areas related to informatics [26]. More attempts were made in signal analysis, HI in the medical laboratory, and other areas such as database, modeling, and biostatistics.
- *The next stage referred to as childhood/youth of HI (1975–1990)*: This stage focused on strengthening the concept of HI. Many governments developed and set up organizations nationally and internationally that exclusively focused on building and contributing to HI [27]. The study also mentioned how governments focused on HI conferences where speakers and researchers from all parts of the world discussed HI and its role. More attempts were made to bring extensive systematization in the areas relating to HI. During this period of advancements, there were other developments, such as introducing new methodologies, health information systems, management of patient records with computers, and other advanced decision support expert systems.
- *Unifying of HI (1990–2000)*: HI is referred to with different names and more than one meaning assigned to the HI concept. Although in the past, the term HI has been interchangeably used as medical informatics and biomedical informatics [28], according to Protti *et al.* [29], the debate on whether HI can be considered a single and scientific discipline had already started. It first happened in a panel discussion in an IMA working conference at Heidelberg. During this discussion, the experts from field medicine concluded that HI is a

standalone discipline with its methodology and systems [29]. HI was initially referred to as medical informatics, coined by Francois Grémy and Peter Reichertz in Europe [30]. Later, the concept became more structured and was redefined as health informatics and biomedical informatics [28]. Along with considering HI as an individual discipline, the other developments that took off during this phase included universities introducing Master programs in HI and other courses such as Healthcare Management, Health Economics, IT for Rural Healthcare, and more [28]. Implementation of imaging, telemedicine, and hospital information systems occurred during this phase, and extensive funding had been allocated for e-health research [24]. However, the authors did not reveal the amount of funding involved.

- *Developed stage of HI (2000–2010)*: This phase brought a more in-depth understanding of HI potentials and how it can address the various challenges faced in the present healthcare system. Other developments during this phase were the reduction of health disparity among underserved populations who are not unable to incur the medical expenses, computerized health records helping to reduce cost and errors, and more technologically integrated health care systems [25].
- When most businesses have integrated technology into their day-to-day business activities and functions, one could argue that in just a matter of time, the implementation of technology takes place in all areas of medical and health [31] and the classic example such as thermal scanning, COVID patient tracking, and other smart devices [32]. With the massive volume of health and medical data such as patient records, laboratory test reports, admission related reports to be managed, it is necessary to automate administration-related data. Health informatics is one such remarkable innovation and development that has taken place in the medical field. HI is connected to the early stages of implementing computers in medicine. With the development of ICTs in the last 20 years, the evolution of HI has taken different forms and shapes, which is discussed earlier. The Internet has profoundly influenced all parts of health informatics [33] and has contributed to many new technological innovations such as artificial intelligence (AI), Internet of Things (IoT), telemedicine, telehealth, mobile health, big data brought into medical and health [34].

Without a doubt, the development of HI has made data sharing and handling much faster and easier. For example, earlier in the development of HI, medical reports such as lab reports, X-ray reports, and other test reports were generally provided in the printed form. Similarly, consultation reports and discharge summaries were all printed, and patients were handed hard copies. But post the introduction and adoption of HI, we have seen significant changes in handling these reports and documents; most of them are electronically collected and transferred between the patient and different health entities [35]. In the last few years, another term that has caught attention is "eHealth" [36]. While HI focuses more on medical data and information [37], eHealth concentrates on using health and information technology in medical and health care [38].

According to Eysenbach ([38], p. 2), eHealth refers to "an emerging field in the intersection of medical informatics, public health, and business, referring to health services and information delivered or enhanced through the Internet and related technologies." Different authors have made other references to eHealth, such as telemedicine and telemonitoring, electronically managed health records, and pervasive and user-centric healthcare [34,39–42]. The benefits of eHealth are tremendous, such as better quality in the patient and health care, empowerment of patients, cost-effectiveness, and health promotions [43,44].

According to Norris [45], patient empowerment has been one of the driving forces for introducing HI; it is the government's responsibility to ensure health strategies focus on empowering patients. The study also pointed out some driving forces to achieve patient empowerment: an easy and transparent flow of communication with hospitals and general physicians, providing home treatment for aged and physically challenged patients, and web-based medical sites where patients can seek all kinds of health patient-related information. In this context, this chapter addresses the following objectives:

- To understand the role of IoT and AI in HI
- To discuss in-depth on the evolving technologies in HI
- To understand the various benefits of HI

3.1.4 Organization of chapter

This chapters consist of six sections which is purely based on the review of literature, the following are the section which are discussed in this chapter are as follows:

Section 3.2 gives an overview on IoT and its role in healthcare, AI and its role in healthcare, and how AI and IoT is playing a role in health informatics. Section 3.3 enlightens detailed discussion on the evolving technologies in HI and emerging technologies in HI. Section 3.4 provides an explanation on the various benefits of health informatics to its different stakeholders. Section 3.5 deals with the various challenges of health informatics which is impeding the implementation of HI. Section 3.6 discusses on the future of health informatics in the hospital sectors and the change that can be expected in the future. Section 3.7 concludes the chapter with future scope.

3.2 Role of IoT and AI in HI

3.2.1 Internet of Things

IoT is described as the network of physical objects enclosed with the Internet, sensors, and software for data exchange over other systems through the internet. It is a concept that integrates various sensors and electronic devices into one extensive interconnected system [13]. IoT is a network of interrelated devices that can collect and transmit data without requiring human interaction. It can be defined as objects with built-in sensors that can monitor and communicate with each other. An IoT ecosystem consists of web-connected devices that collect and act on data collected by their environment, such as processors, sensors, and communication hardware. These devices connect to an IoT gateway or another edge device to receive and store the data they collect. As more companies realize the value of the IoT, they are increasingly adopting it to improve their operations and gain a competitive advantage [46]. The IoT is a technology that enables healthcare facilities to monitor and quarantine all the infected individuals during the COVID-19 pandemic.

Before the IoT, patients had limited access to doctors. There was no way to monitor their health conditions remotely [47]. Thus, according to the study, the emergence of the IoT-enabled patients to monitor their health conditions remotely. This has allowed them to interact with their doctors more efficiently and effectively. It has also helped reduce healthcare costs by allowing doctors to provide better care [47].

The IoT is transforming healthcare by transforming how devices and people interact. IoT applications transform healthcare's delivery [48]. Wearable devices, such as fitness bands and glucometers, are becoming integral parts of the healthcare system. They can provide patients with personalized attention and monitor their vital signs. Duarte and João-Paulo [49] discussed that wearable devices, such as fitness bands and glucometers, are becoming integral parts of the healthcare system. According to him, they can provide patients with personalized attention and monitor their vital signs. Rajashekar and Manish [48] also discussed that healthcare professionals can now monitor patients' health and provide the best possible treatment plan based on their individual needs using IoT. They can also connect with them and provide immediate medical attention if needed.

3.2.1.1 IoT in healthcare

According to Kumar *et al.* [50], the number of elderly people population above 50 years had created pressure on the healthcare industry. For instance, the rehabilitation of stroke victims requires a long-term commitment. The study also stated that medical rehabilitation is a branch of therapy introduced to improve the quality of life by helping individuals overcome mental and physical restraints. Through the IoT, medical facilities can be enhanced by intelligentizing their systems to help prevent manual errors and provide better treatment [50–52].

Feki *et al.* [53] described that the advent of an IoT-based system allows medical rehabilitation services to the patients at remote locations. Patients can be treated through smart rehabilitation, which is a process that uses the medical resources within a community. It is also expected that IoT-based intelligence will soon be a priceless tool in the healthcare industry ([54,55]).

Jara *et al.* [56] mentioned that numerous achievements are made in healthcare, such as monitoring and control, interoperability, and security. Despite the significant progress, the challenges of implementing an intelligent and secure IoT-based healthcare system remain in the form of data resource management, security, access control, and stakeholder collaboration. Due to the rise in the aged population dependent on the healthcare system, the demand for healthcare services has become more predominant. An IoT-based smart rehabilitation system is helping address this issue by connecting all the available resources such as radio-frequency identification (RFID), wireless Internet, Bluetooth, and camera, performing various tasks such as remote surgeries, monitoring, and diagnosis [57]. The system connects e-healthcare facilities such as hospitals, rehabilitation centers, and medical practitioners. It uses a centralized database to store and analyze all the data related to the system's operation [58]. The various components of the system are connected to the Internet using RFID technology [59]. A dedicated resource center is created to meet the varying requirements of the individual patients [60].

3.2.1.2 COVID-19 and IoT developments

IoT can track the status of a patient's vital statistics, such as blood pressure and heart rate. The implementation of this technology can improve the efficiency of healthcare workers by reducing their workload. It can also help minimize expenses and mistakes like faulty detection of COVID-19 during the outbreak [61]. According to Inn [62], with the onset of the COVID-19 epidemic, civilians have become aware of its symptoms. Developing an informed network can quickly identify which individuals are infected and can be treated [63]. Systematic reporting about the symptoms and recovery helps authorities plan on the quarantine period. IoT also provides possible evidence to predict the pandemic caused by COVID-19 [64]. With the help of the IoT, patients can monitor their vital signs and take advantage of the multiple features of their health. It can also help track their conditions without leaving their homes [65]. With the help of this technology, healthcare insurance companies can detect fraud claims and improve the efficiency of their operations [66].

3.2.2 AI

AI is a branch of computer science focused on solving cognitive problems commonly associated with humans. It goes beyond science fiction to encompass pattern recognition and learning. Professor Pedro Domingos of the University of Texas describes the various tribes of machine learning as follows: symbolists, evolutionary biologists, Bayesians, and analogizers [67,68].

Modern advances in statistical computing have led to new fields such as Bayesians and machine learning. These topics are related to the discipline of AI [68]. On the other hand, supervised learning techniques are used for training systems designed to perform specific tasks [69]. Unsupervised training is achieved without the need for the desired output. With the increasing amount of data businesses collect, AI is becoming more intelligent and learning faster [70]. It is driving the development of deep learning and machine learning solutions that take advantage of the massive amounts of data mining through various means [71].

3.2.2.1 History

AI was first described in the 1950s, and many of its limitations were first built. Many of the disadvantages are addressed in the advent of deep learning [72]. With the development of AI algorithms and self-learning, we can now analyze complex algorithms and improve the accuracy and efficiency of our clinical practice [73].

AI is a broad term that refers to developing algorithms that perform tasks similar to humans [74].

According to Rajaraman [75], John McCarthy, the father of AI, coined AI in 1955. McCarthy [76] defines AI as the construction of intelligent machines akin to Alan Turing, an English mathematician, computer scientist, logician, cryptanalyst, philosopher, and theoretical biologist [77]. According to Muggleton [78], Alan Turing was the first to use the concept of AI after proving that machines could mimic human behavior. Thus, computational power has gained momentum, allowing real-time calculations [79]. There are many terms for this field, such as machine learning, deep learning, and vision computer. It involves learning how to identify patterns from the environment. Currently, AI focuses on developing deep learning, an algorithm that learns and makes decisions independently [80]. AI has also become integrated into our daily lives in various forms, such as virtual assistants (e.g. Siri, Alexa, Google Assistant) and mass transportation. It has also begun to be utilized in medicine to improve patient care [81]. Machine learning focuses on improving the efficiency of medical examinations through the use of magnetic resonance imaging (MRI) machines, radiological images, and electronic medical records [82]. Computer vision is a process utilized by AI to understand an image or video better [73]. Also, this field has been in the news over the years, with gastroenterology and endoscopy's most notable contribution [73].

3.2.2.2 AI in healthcare

Figure 3.2 highlights diverse areas of impact of AI in healthcare. Ardakani *et al.* [83] studied the convolutional neural networks, also known as CNN, a deep learning technology used to diagnose sickness using medical imaging such as MRI, X-ray, and CT scans. Several tools for diagnosing COVID-19 have been created in

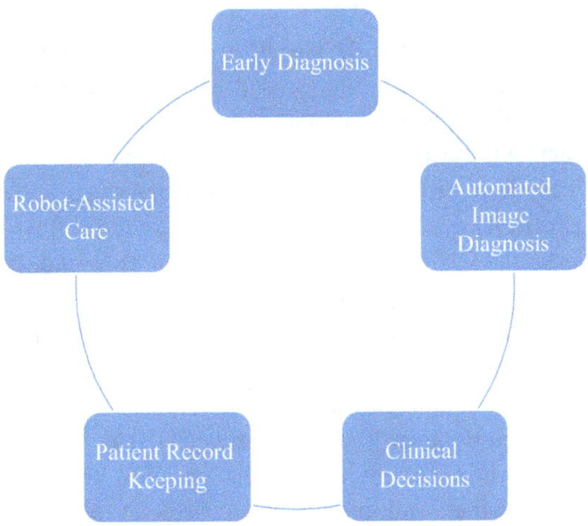

Figure 3.2 Areas of impact for AI in healthcare

the last couple of months due to deep learning and machine learning technologies [84]. A team from Carnegie Mellon University has created an app that can detect if an individual is infected. The app is not a primary diagnosis tool but a screening tool for people at high risk of getting COVID-19 [85].

Imran *et al.* [85] proposed a computational system called AI4COVID-19 to diagnose COVID-19 based on an individual's coughing sounds. The system can detect different types of coughs, such as symptoms of COVID-19. For example, it can also distinguish between healthy and disease-free coughs. The algorithm is trained using a public dataset known as ESC-50. Imran *et al.* [86] used a data set that included 3,597 environmental sounds and 1,838 sounds made by humans. The recordings were then computed and fed into a neural network. The system then fed the corresponding image to the COVID-19 diagnostic system to confirm whether it learned the different coughing sounds and isolated those that point to cough in COVID-19 patients. The system provides a valid result if the three outputs are the same [86].

3.2.3 IoT and AI in HI

Modern AI technology can help improve the efficiency of the medical field by providing a solution to the most critical situation that the experts have been facing while analyzing the data. For instance, medical practitioners have been dealing with the issue of accessing patient data [87]. Apart from helping diagnose and treating patients, AI can also predict future diseases and provide helpful suggestions to prevent them [88]. According to Lorig *et al.* [89], AI can look back at a patient's past medical records and suggest ways to prevent a specific illness. Despite the IoT exposure, speed and accuracy are still not ideal. But AI can still learn from the patterns, it creates and make sense of the data it collects [90]. AI depends on enhancement by self-mechanism. AI is used to facilitate intelligence for IoT [91].

3.2.4 IoT and AI during COVID

The pandemic caused by the COVID has profoundly affected the world, putting technology in the spotlight. During the pandemic, AI and IoT helped maintain the social distance between people with contactless touch in healthcare facilities [92]. The synergy between AI and IoT can untether the prospective future in the health care industry. It has transformed the healthcare ecosystem and helped improve overall operational efficiency. Some of the benefits are the following: streamlining healthcare providers' work, focusing more on patient care and medicaments, and automating administration-related tasks in the outpatient department [93].

AI and IoT also enable health care providers to provide emergency care and monitor patients' conditions remotely. Virtual health assistants are one such innovation that helps geriatric and chronic illness patients monitor patients' blood glucose, oxygen, and temperature at their homes [94]. AI also enables healthcare providers to give real-time patient conditions updates and other benefits such as scheduling appointments and paying their bills at their convenience [95]. Monitoring and improving patient satisfaction can help hospitals lower costs and increase their bottom line [96].

Shakeel *et al.* [97] argued that with the help of deep learning technology, radionics could detect abnormalities in an image that could diagnose breast cancer and lung cancer. Designing the treatment plans for cancer patients can be very time-consuming and complex due to the extensive type of cancer with different stages. When cancer tissues are manually detected [98], AI-powered treatment plans can help cancer patients get the latest information about their condition within minutes. Although surgical robots are used in hospitals to perform procedures, they have also provided emergency supplies to the public, such as meals, medications, lab specimens, and so on [99,100].

3.2.5 IoT and AI in diabetes

Dankwa-Mullan *et al.* [100] stated that over 425 million individuals worldwide have diabetes, accounting for 12% of the world's health expenditures. In their study, authors claimed to have reviewed close to 450 articles on distributed diabetes and AI. Authors discussed various imaginative approaches that aim to change how diabetes care is provided. New AI-controlled technologies such as diabetes management frameworks, insulin siphons, and cell phone applications are already being developed, such as mySugr and Wellthy diabetes. AI-based applications could help people with diabetes care achieve better blood glucose control, decrease hypoglycemics scenes, and improve overall health [102].

3.2.6 IoT and AI in cardiovascular

Hoermann *et al.* [103] studied the relationship between high-risk factors such as smoking and diabetes and found the lack of treatment to be the leading cause of death in India. Baranowski and Lytle [104] explored how the man-caused insight frameworks are utilized in medication development to study novel variants and phenotypes in patient existing conditions. AI have been used for assessing and estimating the prevalence of cardiovascular contamination [105]. The advancements in AI technology are expected to play a significant role in developing precision medicine for heart disease [106].

Chatterjee *et al.* [107] studied the various aspects of an IoT-based healthcare platform and the various components of a decision support system to identify the various risk groups and diseases that can be avoided using the IoT platform. The system identified the different risk groups based on a certain kind of cardiovascular disease such as abnormal heart rhythms, coronary artery disease, and heart attack [108]. A healthcare platform that uses the IoT can serve as a common platform for all healthcare providers, patients, and governments [109].

3.2.7 IoT and AI in drug developments

The current process of developing new drugs can be very time-consuming and costly. Through AI, there is a chance to improve the process. AI-powered drug discovery is becoming more prevalent at standard cost and record time [110]. According to Ottesen *et al.* [111], new medical technology uses computer simulations to interact between different drugs and human physiology; the technology is already being used in regulatory evaluation and product development. Hummod is a technology developed by VPH (Virtual Physiological Human Institute), a research institute that uses virtual human models to study heart disease and osteoporosis [112].

Liang *et al.* [113] argued that AI can analyze the resistance patterns to anticancer drugs in cancer cells. AI can also help doctors make informed decisions about treating patients with different types of cancer [114]. McDougall [114] found that AI can predict the effectiveness of chemotherapy drugs and limit the use of radiation therapy.

3.3 Evolving technologies in HI

3.3.1 Introduction to evolving technologies

The World Health Organization (WHO) defines health technology as applying organized knowledge and skills in medicines, medical devices, vaccines, procedures, and systems developed to solve a health problem and improve quality of life [115]. The Organization for Economic Co-operation and Development (OECD) defines health technology and innovations as applying knowledge to solve practical clinical and health problems, including products, procedures, and practice styles that alter healthcare delivery [116].

Health technology refers to activities that aim to improve the efficiency and safety of healthcare [117]. Early adoption is essential for technology-driven tools as it can help minimize the risk of errors and provide a safer and more comfortable environment for users. Other factors such as ease of use and affordability are also considered to select the most appropriate technology for the health system. A few factors that influence the adoption of the new technology are the time of its impact, size of its impact, state of its development, and its ease of use.

3.3.1.1 AI

According to American Marketing Association [118], there is nothing quite as exciting as the potential of AI in healthcare marketing. The use of AI in healthcare is expected to grow at a robust rate of 40% through 2022, significantly higher than the industry's 2% annual growth [119].

According to Kripalani *et al.* [120], automated tasks include remembering to take medication, helping patients keep track of their drugs and identifying people at high risk of experiencing medical emergencies. Also, his study further studied that AI can also recommend the best dose for each patient based on their unique body chemistry. AI can be utilized in healthcare marketing differently from other industries [121]. As AI becomes more prevalent in healthcare marketing, marketers must keep up with the latest developments [122].

Vaishya *et al.* [61] mentioned that AI is a technology that can help improve healthcare by analyzing and designing treatments for different conditions. It can also diagnose and develop drugs for less common diseases. According to Meskó and Gorog [123], Atomwise Inc. produces AI for drug discovery applications to treat Ebola. Authors also mentioned that the company launched in 2015 to find safe and effective medicines by reprogramming existing medication.

DeepMind, a Google subsidiary, developed an AI chatbot to analyze breast cancer. Chung and Park [124] argued that the system could identify cancer using pre-cleared data sets, and several companies are working on AI-related projects in healthcare. They asserted that from developing new drugs to disrupting the medical image, these companies push the envelope regarding how tech can improve healthcare.

3.3.1.2 Virtual reality (VR) and augmented reality (AR)

VR is transforming the lives of doctors and patients alike. It allows them to interact with each other while in a hospital or traveling across the globe [125]. Also, he mentioned that VR is also used in various surgical procedures and to train qualified surgeons. Pedersen *et al.* [126] study revealed that surgeons' overall performance improved by about 30% after using VR for a couple of months. Further, he said VR headsets are used for women undergoing labor pain to visualize a relaxing landscape. A pilot study conducted in 2019 showed that VR helped decrease patients' anxiety and pain levels [127].

According to Juanes-Méndez *et al.* [128], AR and VR techniques started their research in medicine more than two decades ago. Authors mentioned that initially, their use was restricted to a small and targeted audience, and VR has been developing its independent path, which has led to the need for more sophisticated hardware and software [128]. Due to its ability to represent the world in a virtual format, it has become more prevalent in society. This study described that AR has advantages over other technologies like video games and movies. Its goal is to provide real-world information in a virtual environment. The study also discussed that AR is very natural and straightforward to use, and its purpose is to provide the user with all the necessary details without distractions or interruptions. Aside from these, various other devices such as stereoscopic vision goggles, gloves, and fixed outdoor systems use AR and VR [128]. The study also discussed that one of the main advantages of AR is that it eliminates the need for constant interruptions. Instead, it allows the user to perform simple and natural activities. The rise of augmented reality has revolutionized the efficiency and cost of surgery. By combining data processing and real-time diagnostic images, surgeons can now plan for surgical procedures with greater accuracy and efficiency [129].

Nicolau *et al.* [130] discussed that augmented reality has changed the way surgeons operate. The study also mentioned that MRI and digital imaging technologies allow surgeons to plan and perform procedures more efficiently. AR is a type of technology that uses two-dimensional images and other data to create a three-dimensional model of a patient [131]. The study further discussed that it is becoming the main component of healthcare technology as it allows users to see and hear information in real-time. It could help medical students prepare for real-life operations by giving them hands-on experience with anatomy. Students can now gain hands-on experience with human anatomy through AR, which is usually impossible with digital body representations. SyncThink, a company founded by Magic Leap, is working with other companies to develop AR-based therapeutic platforms for healthcare. Although there are no commercial products yet, AR-based platforms are expected to be introduced in the healthcare market in the next couple of years.

3.3.1.3 Sensors and IoT

Appelboom *et al.* [132] highlighted that with the rise of wearable technology, health trackers, and sensors, people can now remotely monitor and control their health conditions. In the study, authors mentioned that people can now monitor their health and live a better life with these devices. These include the Fitbit Ionic, which tracks sleep and helps track various activities, and the Polar H10 smartwatch. Sano *et al.* [133] studied sleep quality, stress level, and mental health using personality traits using wearable sensors and mobile phones. These wearable sensors can also connect with their doctor and provide them with helpful information, and with these devices, patients can now take control of their health.

Geocks *et al.* [134] discussed that wearable devices are already transforming biomedicine by enabling continuous, accurate monitoring of health and widely using healthcare practitioners. Authors also mentioned that these devices are also being used to develop automated intervention and prevention programs and can also help to predict an individual's health event and provide feedback on their health status. Dunn *et al.* [135] explored various advancements in wearable sensing technology in the past couple of years which include the ability to monitor a patient's vital signs, sleep patterns, and ICU alarms [135].

3.3.1.4 Medical tricorder

Muthu *et al.* [136] discussed that one of the essential things a healthcare professional should have a device that can analyze and diagnose every disease. It has led to the rapid growth of healthcare technologies and the emergence of devices that can measure various parameters [136]. Further, the researcher mentioned that one such device is the Viatom CheckMe Pro, which can monitor physiological conditions such as heart rate, blood pressure, and oxygen saturation. Other companies are also developing similar devices that can monitor multiple parameters. BioIntellisense's BioSticker is an FDA-cleared device that can monitor a wide range of physiological parameters, such as heart rate, temperature, and body position [137].

Gavriel *et al.* [138] studied the possibility of using smartphones to monitor and record the heart activity of subjects. Companies like Cloud DX and Xprize have developed a wearable application called Medical Tricorder to capture a person's heart movements quickly. Wusk and Gabler [139] explored that the app can extract the heart rate using the Ballisto Cardio Graph signal or BCG, which is a type of sensor that measures the mechanical aspects of a person's body. Further, the research mentioned that the signal could provide valuable information about the mechanical integrity of the heart; it is typically powerful enough for diagnostic purposes. By analyzing the data collected by the app, it could detect abnormalities in the heart at an early stage, which could be very helpful for patients which in-turn opened up the possibility of turning smartphones into healthcare devices.

3.3.1.5 Nanotechnology

According to Henzen *et al.* [140], in 2018, researchers at MIT launched an electronic capsule that can relay information and drugs to a smartphone, and smart patches are becoming more commonplace in the market. One company from France, Grapheal, showed off its wearable patch at CES 2020 [141]. Further, their study discussed that more medical devices with minimal interference could be developed due to advancements in nanotechnology.

Chakravarthy *et al.* [142] studied the development of nanomachines and various enhanced and improved applications, including military applications, healthcare, and industry. Further in the study, it was discussed that the internet is used to enhance nanotechnology, and healthcare monitoring involves developing a system based on loT's capabilities. It will provide a better and more accessible healthcare system [142]. According to a study by Pilarski *et al.* [143], nanoscale medicine is rapidly becoming the future of medicine. Combining therapeutic agents with nanotechnology has led to effective diagnostic and therapeutic drugs [144]. Further in the study, it is discussed that despite the advantages of nanotechnology, its use still requires extensive studies to confirm its effectiveness. The researcher aims to study the concept of a nanotherapeutic system and explore nanomaterials' various characteristics [145]. From the scholarly research, it is evident that nanotechnology is proliferating. The use of nanotechnology in healthcare is of significant importance.

3.3.1.6 Robotics

According to Taylor *et al.* [146], medical robots are becoming more prevalent in medicine. Some of these include surgical robots and exoskeletons. In 2019, the first exoskeleton-assisted surgery was performed, which inspired other developers to create similar devices. These robots were used for various applications, such as helping people with spinal cord injuries [147]. Being companions with robots can help people with mental health issues and improve their quality of life. Some examples of such are the Pepper and Jibo robots [148].

A medical robot is a machine designed to perform various tasks, such as assisting surgeons and nurses [149]. The study mentioned that it came into existence during the 1960s to help people with various medical conditions. Over the years, robots have evolved to support individuals with special needs, such as those with cognitive disorders and physical injuries [150]. Thus, it is evident from the research that robots are used in the healthcare industry for surgery, assisting surgeons and nurses.

3.3.1.7 3D printing

3D printing (also known as 3DP) is a process that consists of modifying an old inject printer to produce 3D images. 3D printing is used across the healthcare industry, with companies such as 3D Systems and Biogen planning to capitalize on this technology [151]. Further in the research, it was discussed that in November 2019, researchers from the New York Polytechnic Institute created a method of 3D-printing living skin and blood vessels. This procedure was beneficial for treating burn victims [151]. Currently, 3D-printed drugs are in circulation in the US Researchers are also developing 3D-printed pills that allow patients to stick to their medication plan [152].

Thus, the pharmaceutical industry is on the verge of a revolution, as 3D printing is expected to transform the way drugs are designed and manufactured. 3D printing's potential to transform the way medicines are designed and manufactured is explored. 3D printing is expected to become the most significant disruptive technology in the global pharmaceutical industry. Its adoption could raise various ethical, legal, and regulatory considerations [153]. Due to its rapid expansion, it has become a widely used tool for various industrial applications. 3DP is used in various printing applications in medicine, dentistry, and implants. 3D printing is extensively used to produce customized medical equipment such as artificial limbs and implants in medicine [154]. Contrary to popular belief, 3D printing does not involve drugs or active sites, instead, designed to work seamlessly with a patient's needs [155].

3.3.1.8 Distributed ledger/blockchain

Blockchain is already used in healthcare to transfer patient records and billing information. Its multi-threaded nature lets it securely exchange and maintain the same information across multiple parties [156]. According to a Harvard Business Review study, owning all of your data could help solve various issues, such as access to and security of your records [157]. They also explained that blockchain allows people to control their data, almost unalterable. This concept helps to collect and distribute data about a global pandemic.

3.3.1.9 Cloud computing

The evolution of technology has allowed healthcare solutions to be seamlessly integrated with various other systems and enabled a more innovative health care cloud. It is expected to create new opportunities and challenges in the healthcare industry. The Software as a Service (SaaS) model is used to manage the rapidly changing needs of healthcare professionals. Chauhan and Kumar [158] discussed the advantages and challenges of using cloud-based healthcare systems and how to assist healthcare practitioners to become more productive.

According to Omar *et al.* [159], it was discussed in an Amazon blog that as a part of its cloud transformation, the Rush University Medical Center worked with the Chicago Department of Public Health to create a public health analytics hub. In the blog, the researcher discussed the hub combines various hospital data sets to provide a comprehensive view of the care given to COVID-19 patients. The hub is a modernized version of the public health infrastructure that meets the varying needs of different organizations. There are various ways to submit data, such as manual and semi-automated methods. Amazon Web Services (AWS) provided the necessary infrastructure support to respond to the pandemic. AWS allowed us to manage our infrastructure efficiently through server less computing capabilities [159].

3.3.2 Impact of emerging technologies in HI

The emerging technologies in HI significantly contribute to the healthcare sector. Based on a report, "Accenture Digital Health Technology Vision," prepared by Accenture Consulting in 2019, revealed that health care executives are already experimenting with new technologies in their respective domains.

During the pandemic, HI technologies became more critical and essential [160,161]. Ye *et al.* [161] developed a technology framework from HI perspectives such as mobile, web-based service, 5G telemedicine, intelligent-based diagnosis, big data analysis such as digital-based contract tracing with the help of QR codes or epidemic prediction which was specifically used in China during the COVID-19 pandemic. Many patients had also utilized remote healthcare monitoring during the pandemic times when most of the hospitals were facing overflow in admission [162], and in the United States, an increase of 154% was seen using telehealth during the last week of March 2020 compared to the same time in the previous year 2019 [163].

Though the presence and uses of HI-based technologies were more realized during the pandemic times, innovative technologies in the healthcare sector were present much earlier, and older patients and other stakeholders were using them for booking appointments, online consultation, telemedicine, online access to medical test, and other patient records [164]. The acceptance of remote healthcare became more prevalent as people became accustomed to its convenience through different services such as home monitoring, retail clinics to web visits [165]. Patients have become the cornerstone for healthcare sectors. Some of the most notable trends in healthcare in 2021 are a collaborative ecosystem, customization for patients, personalized care, virtual care, and AI and automation [166].

3.4 Benefits of HI

HI has shown a tremendous advancement in dealing with medical resources, equipment, and techniques to optimize collecting, archiving, retrieving, and communicating to the required stakeholders. The main objective of HI is to improve decision-making, problem-solving, and ensuring the best healthcare services. HI has been an evolving concept, and now it is becoming a solution to many medical and health entities.

3.4.1 General benefits of HI

Today, the most significant challenge most healthcare industry faces is managing and handling patients and patient-related data, which can be tackled by advanced informatics in healthcare [167]. Though the staples of healthcare remain the same, the focus is on handling patients better and ensuring that medical professionals can provide the best service with lesser chaos and errors. With the inventions of HI in healthcare, medical practitioners and nurses were able to concentrate more on their core services like treating patients with utmost care and better diagnosis, which otherwise was challenging [168].

Without a doubt, HI has contributed to patient care and management. Other driving forces for the implementation of HI across hospitals are to increase the overall efficiency, better healthcare facilities, lesser cost, more accessible access to healthcare, faster and prompt communication, and providing standardized service across the population. The below paragraph discusses more in detail the different uses of HI, its benefits to other stakeholders, and how it has helped transform the healthcare system.

- *Cost reduction:* According to an article by Savvas [169], it was mentioned that healthcare data will be growing exponentially compared to other industries by 2025. Manually handling these prodigious data is always challenging. The possibility of error was high, but with the introduction to HI, managing these health data has become more accessible and cost-effective [170,171]. Some of the technologies, such as Clinical Decision Support System (CDSS), Computer Provider Order Entry Systems (CPOE), are programs that support in providing decisions based on patient records and, in turn, helps in increasing the professional practice [172]. The study also mentioned that these technologies helped identify medical errors such as repetition or mismatch based on the patient data, thus reducing the high costs incurred by the healthcare industries.
- *Efficient data management:* This sector deals with extensive data across the globe, and, for a long time, we have seen healthcare units handling these documents in the form of paper, which involves much paperwork. Perhaps this is still a prevailing problem in many hospitals and related services where they continue to follow a paper-based system. With HI, it is helping medical professionals and nurses to store all patient-related data online. Electronic health record (EHR) is a digital form of handling these data. They primarily gather all data about the patient, such as medical history, test and laboratory reports, personal data, and other related information [173]. Another benefit of electronically managed data is easy accessibility from anywhere and anytime. Since EHR is integrated with different departments such as the hospital administration department, pharmacy department, laboratory, diagnostic centers, and operation theatres, duplicate records for single patients concerning each department can be eliminated [171,174]. The study also stated that the process could be made faster, and patients could save time collecting the relevant documents from each department.
- *Patient-centered care*: HI helps patients have easier access to their health records. This systematically managed data allows the patients to seek the proper medical advice at the right time with the right expert as and when required [175]. The paper also stated that HI provides certain critical information about the patient to their family members as and when needed. Doctors and other medical practitioners can also use these data to execute appropriate medical care.

- *Increased outcomes*: HI has helped automate the administration-related work involving paper [176]. With automation, the time and money involved as reduced to a great extent for nurses, medical practitioners, physicians, and other related stakeholders [170,176]. They can focus much more on their core functions leading to improved quality services and reduced error levels significantly [177]. In their paper, the authors also mentioned that this development also facilitates the healthcare system to provide the correct diagnosis and thus provide the proper care and treatment for the patients.

3.4.2 Benefits of HI to different stakeholders

Healthcare sectors comprise different stakeholders such as medical practitioners, physicians, nurses, patients, pharmacists, caregivers, insurance companies, ministries of health, government, and other agencies. It is essential to discuss how HI has helped the different stakeholders in the healthcare sector, and the below points discuss more elaborately these:

- *Patients:* They are an essential part of the healthcare sector. One of the studies by Snyder *et al.* [175] explained that the HI developments had led to better information management on various healthcare facilities, providing patients with multiple insurance plans and available policies. The other innovative health and fitness apps facilitate sending reminder messages on health. Other benefits are online consultation, appointment and access to lab test results, telemedicine, and telemonitoring. Ensuring their safety is the main aim of all hospitals, and it is also a national priority. Thus, HI has made a tremendous impact on patients' care.
- *Physicians and nurses*: HI has helped physicians and nurses to manage patient records electronically through EHRs electronic Medication Administration Records (eMAR), and this has helped in the reduction of paperwork [178,179]. COPE has also allowed physicians and nurses to electronically transfer medical treatment details for patients to the required departments. Through COPE, information is shared digitally to the concerned departments like the hospital administration department, pharmacy, and diagnostic centers in completing the required request by nurses or doctors [180]. The study also explained how COPE has taken over many disadvantages faced while maintaining handwritten documents and led to reduced time, more efficiency, and effectiveness in communication. These electronically managed programs help provide timely care and services to the patients.
- *Hospital administration department:* Hospitals handle thousands of patients every single day, handling and managing them becomes a strenuous task for the healthcare administrators. There are different kinds of patients who come to hospitals every day, from outpatient, emergency, diagnostic, and admission. The administration department is involved with patient enrollment, appointment scheduling and confirming, billing, and bed management, all electronically managed. The automated Health Management System (HMS) helps automate tasks. The need for resources like workforce, money, and time is

drastically reduced, resulting in lower operational costs leading to operational effectiveness [181].

- *Insurance companies:* Earlier, when a patient needed to claim for insurance, it involved a lengthy procedure of collecting and gathering all the required data from the hospital and transferring it to the insurance agents, but with digitalization, the insurance companies can get easy access to the patient documents from hospitals. Electronic submission of patient reports and bills are submitted, and electronic claim transmission is done by the insurance agents directly to the hospital [182]. Since all the patient data are collected and managed digitally, there is access to extensive data that can be analyzed for various purposes every second. These insurance companies use these data to design and plan different insurance schemes catering to the public's needs [183].
- *Pharmacist:* Pharmacy informatics, a branch of HI, helps in medicine handling, reviewing the utilization of different drugs, barcoding while dispensing the drugs, and creating alert systems to improve the dispensing of medicines. Pharmacists are connected with various EHR, eMAR, CPOE, and other digital documents. Thus, it helps pharmacists ensure efficiency and safety in entering the prescription data, dispensing drugs as per the prescription and compounding, and monitoring outcomes [184].

3.5 Challenges of HI

The ultimate aim of any development is to bring a positive change and help every stakeholder. HI involves collecting, accumulating, analyzing, and presenting health data to the healthcare sectors and stakeholders. Health informatics imposes informatics concepts, policies, and processes to the actual situation to achieve better health outcomes. Various challenges which impending the implementation of HI are discussed below:

- *Data vulnerability*: Since all the patient-related data are digitally maintained, there is a high chance of data leak or hacking [185]. These kinds of software or programs need to be updated regularly; it is the responsibility of the concerned officials to check for required updates and ensure that the patient data is not leaked. But often, people who handle this software are not skilled enough; hence, these kinds of risks are high.
- *Lack of skilled personnel*: Product managers are experienced with healthcare knowledge [186]. Thus, according to the study, it becomes more challenging for such personnel to drive their technology teams to build HI-based solid software and programs. Another challenge is that medical practitioners use the end product, nurses, pharmacists, and lab technicians who are not skilled, thus causing ineffective software usage.
- *No restrictions*: Hospitals have given all hospital personnel access to data through their systems, phones, and tablets, and these can cause a data privacy breach [45]. Thus, hospitals have to strictly restrict the access of this software only through hospital systems and implement rules and regulations related to

data encryption. There is a need to bring in a system that ensures authorized personnel can access data based on their role in that hospital. It helps ensure that the patient's data are not shared with anyone at any time without appropriate authorization, and thus, confidentiality is maintained.

- *Insufficient data input*: Informatics works effectively only when the right and sufficient data input and integration is executed. But in the case of HI, it remains one of the well-known obstacles yet unsolved. HI involves data from different departments; however, gathering all the required data and integrating that at an enterprise level is a real challenge. One primary reason for this challenge is the data complexity, which is high in the case of health sectors [187].
- *Traditional healthcare environment*: Most healthcare sectors, including government-run hospitals, still work in the conventional setting, where the process, policies, framework, technologies, and people are traditional. Hence, it becomes very challenging for them to embrace these new innovative and new models into the system. It is also critical to have a good and reliable connection among various technologies that helps in the successful implementation of HI. High reliance on these technologies reduces interference, increasing data security concerns.
- *Insufficient funding*: According to Giuse and Kuhn [187], healthcare is failing to adopt HI ultimately because HI's benefits are not quantifiable. The survey report conducted by HIMSS in 2019 stated that the return on investment and the advantages from these technologies are not measurable in actual terms. Even though the hospitals believe that there is a need for the data to be managed and systemized, they are not convinced about HI's strategic uses and benefits. The relationship between information management and financial investment on HI has not been firmly established, which acts as an obstacle from implementing HI, supported in earlier studies [188,189].
- *Diverse device and integration*: HI involve wearable technologies in healthcare such as Fitbit, smartwatches, and other small electronic devices that help measure the oxygen rate, temperature, blood pressure, sugar level, and heart activity [190]. According to Ometov *et al.* [191], interoperability between these wearable devices is a big challenge. These devices need to be integrated and exchanged between many connections, including exchanging information across various interfaces. Thus, there is a need for devices to be compatible. It is essential to understand that the combination of interface requirements is high when the communication takes place between different systems, and that is the case concerning these wearable devices. Thus, due to various devices, interoperability of such devices is a big concern [191].
- *Lack of law and ethics*: The absence of solid law and ethics governing data protection and confidentiality can lead to a breach of privacy [192]. The health data managed by hospitals and their related sectors are shared with various research agencies to conduct studies related to disease and in case of such conditions. It is a matter of concern that such data shared with these research agencies are utilized only for research-related purposes and not mishandled, misused, or shared with any other parties.

- *Volume, variety, velocity, and veracity (4V)*: The 4V in HI is another challenging factor [193]. It is necessary to provide the correct data to implement HI successfully. The study discussed how 4V are challenging in implementing HI. The volume of required medical data is high, collected from different departments and various other sources. Hence, it becomes challenging and daunting for the concerned stakeholders to handle and manage these vast data gathered from varied sources. Another challenge is that the velocity of these data is high, and gathering such data by ensuring accuracy and validity is a concern. The healthcare sectors primarily manage partial data, clattering, and anomalies that threaten to make wrong patient treatment decisions.

3.6 Future of HI

The recent innovation of technology in HI has enlightened all healthcare stakeholders about its future scope. There is a need to visualize the future of HI and focus on developing it to make it more accessible, accessible and user friendly for the future. HI is a very dynamic field and will remain to change; therefore, the precise direction for HI's future growth is inherently uncertain. However, there are some key elements, and essential breakthroughs that need to be focused on that would positively support the future development of HI, which are discussed in the below section.

- *Demand for HI-related courses*: According to Rhodes [194], the Bureau of Labor Statistics projects that the number of people working in health informatics grows by 15% between 2014 and 2024. This field of study is expected to gain more importance and become the fastest-growing occupation in the United States. Due to the increasing demand for jobs related to HI and the need for skilled personnel, educational institutions are developing programs related to HI. In an interview with US News in 2014 [195], Charles Friedman, who is a Chair of Learning Health Sciences at Michigan University Medical school and a former director of Health Informatics program at the University of Michigan, stated, "HI field is exploding. Access to the health information on the web is taking off at a meteoric pace. It's creating enormous employment opportunities" [196]. The healthcare industry is migrating more towards a digitalized environment. Thus, more positions are becoming necessary to support the various initiatives to improve patient care. To effectively perform their duties, HI professionals usually enroll in graduate programs that prepare them to use multiple electronic tools and methods to perform their tasks effectively.
- *Interoperability*: Currently, many EHR systems do not effectively communicate with one another. According to Sullivan [37], interoperability allows systems to seamlessly exchange and integrate data coordinated across various geographic and organizational boundaries.
    - Health data exchange standards enable secure and appropriate access to all required data to provide a complete and accurate view of the patient's care. Verma [197] stated that the Center for Medicare and Medicaid

Services (CMS) introduced interoperability and patient access rule, aiming to improve the efficiency and effectiveness of healthcare data exchange. Through the proposed policies, the CMS aims to break down barriers to interoperability and give patients greater access to their health information [197]. This attracts more healthcare sectors and stakeholders to adopt HI technologies.

- *Health data analytics*: The rise of digital technology has opened new healthcare organizations to collect and use data to improve patient care. The ability to use Big Data in healthcare is an area of expertise that requires deep knowledge in various facets of health informatics. Today, diagnosing diseases and providing timely treatment are critical factors that need health informatics. One of these is an electrocardiogram (ECG) non-invasive tool that enables computer-aided diagnosis (CAD). Gupta *et al.* [198], aimed to develop an efficient method for processing the signals using an autoregressive model. The proposed algorithm studied two parameters, atrial tachycardia and premature atrial contractions. The results indicated that the proposed method performed better than the principal component analysis and K-nearest neighbor models. The algorithm achieved a sensitivity of 99.95%, a specificity of 99.97%, an accuracy of 99.93%, and a mean time difference of 0.557 ms. The results indicated that the proposed method could detect premature heart diseases, trending data analytics, and visualization and customized dashboard simplifies the analytics. Healthcare analytics also helps reduce the various challenges such as mining data and applying computational modeling.
- *Aid to older society*: According to Bath [199], most developing countries are seeing an increase in older people, and the number continues to increase till 2050. It indicates that there can be an increase in reliance on health sectors. The only effective way to tackle this increasing demand is by using various e-health initiatives. Better HI implementation plays a significant role, especially for countries where the aged population is higher than the working-age population, where dependence on humans for healthcare services becomes challenging. It is indicative that when a country has more of an aging population, its diseases also increase; this, in turn, leads to more demand for healthcare services, home care services, and support. HI, developments such as telemedicine, telehealth services, e-health services, interactive digital services, and online consultation support these demands. These services help empower the patients and caretakers much better [200]. Hence, healthcare sectors can develop user-friendly, easy-to-access ICT-based services that significantly support aging.
- *Equal access to all sections of the society*: The main objective of HI is to ameliorate disparity among different areas of the community [201]. Developments like HI helps in providing equal services and required information to all the individuals and groups; however, in reality, a considerable section of society is neglected and ignored. Thus, the disparity still exists among different groups like people from developing countries, homeless people, people with disabilities, and older people [202]. Riley [202] also states that the primary reason for this is that they either lack knowledge in using these e-health services or are not accessible

to this section of people. It is an effort to be taken by the concerned authorities such as healthcare sectors and governments to overcome this digital divide. Initiatives such as setting up proper healthcare services in rural and less affluent places, ensuring access to the internet, and providing necessary training to these people can help reduce these kinds of vulnerabilities.

3.7 Conclusion and future scope

This chapter intended to focus on the evolving technologies in health informatics and their contribution to health sectors. Information-intensive services are integral to the healthcare industry. The use of technology has enabled healthcare organizations to improve their efficiency and knowledge base. As technology advances, so does the role of healthcare professionals. They play a crucial role in implementing and managing the changes that come with it. Health informatics specialists and clinicians work directly with various departments and individuals. They must also have a deep understanding of the institution's vision and values.

Health and medicine are critical fields that require special attention compared to other areas such as education and commerce. Within the information science discipline, health informatics is an emerging area. Still, due to the sensitive nature of the data that is collected about people's health, it is not uncommon for companies to store and process this information in a way that is not secure. Also, the amount of data collected during a person's care is enormous. Second, various people play essential roles in the care of a person. Some include doctors, nurses, social workers, and informal careers. The developments in health informatics must address these individuals' needs and not issues related to other disciplines.

Creating a healthy environment for all children is vital for sustainable development. As the world faces a global health crisis, COVID-19 contributes to millions of people's deaths. Before the pandemic, much progress was made in improving the health of the world's population. However, more work is needed to eradicate various diseases and address the various health issues that can still lead to death. According to WHO, inadequate medical and nursing personnel are some of the factors that contribute to the lack of health care globally. Over 40% of all countries have fewer than the medical doctors per 10,000 people, and over 55% have less than 40 nursing and midwifery personnel per 10,000 people. This situation put the developed country, doctors, and research community to create an environment that everyone can access the health care facility by digitalizing.

The current chapter primarily discusses on the role of HI in hospital sectors, benefits, and various challenges based on the review of previous research. From the past literature, it has been evident that there is a dearth of primary data-based research, hence, future studies if focused more on primary research, it would add great contribution to the existing literature. In the future, researchers can focus on doing a comparative study on the implementation of HI across different countries. Research can also be focused on doing a primary-based study with hospital administrators in understanding more in depth on the benefits and challenges of HI implementation in hospitals. Academicians and researchers can conduct a study on relationship between AI, NLP, and ML in health informatics. Further studies can also be made on the changes that have happened in the hospital sectors in terms to services, patient care, disease diagnosis, and treatment with reference to pre and post implementation of health informatics.

References

[1] Steinbuch, K. (1957). Informatics: automatic information processing. *SEG-Nachrichten (technical messages of the standard Elektrik Gruppe)*, 4, 171.

[2] Dreyfus P. (1962) L'informatique. *Gestion*, 5, 240–241.

[3] Cawkell, T. (1997). Communications 98; a review. Part 2. *Aslib Proceedings*, 49(10), 255–262. https://doi.org/10.1108/eb051474

[4] Wellisch, H. (1972). From information science to informatics: A terminological investigation. *Journal of Librarianship*, 4(3), 157–187. https://doi.org/10.1177/096100067200400302

[5] Bernstam, E. V., Smith, J. W., and Johnson, T. R. (2010). What is biomedical informatics? *Journal of Biomedical Informatics*, 43(1), 104–110. https://doi.org/10.1016/j.jbi.2009.08.006.

[6] Hagiya, M. (2015). Defining Informatics across Bun-kei and Ri-kei. *Journal of Information Processing*, 23(4), 525–530. https://doi.org/10.2197/ipsjjip.23.525.

[7] Yatsko, A. and Suslow, W. (2016). Introduction to informatics. In: *Insight into Theoretical and Applied Informatics: Introduction to Information Technologies and Computer Science*. Warsaw, Poland: De Gruyter Open Poland, pp. 1–29. https://doi.org/10.1515/9783110469882-002.

[8] Joshi, S., Kulkarni, N., and Athavale, R. (2013). Business informatics a key driver for successful strategic planning. In: *2013 4th International Conference on Intelligent Systems, Modelling and Simulation*, 2013, pp. 305–309, doi:10.1109/ISMS.2013.111.

[9] McLane, T. M., Hoyt, R., Hodge, C., Weinfurter, E., Reardon, E. E., and Monsen, K. A. (2021). What industry wants: an empirical analysis of health informatics job postings. *Applied Clinical Informatics*, 12(02), 285–292. https://doi.org/10.1055/s-0041-1726423.

[10] Saba, V. K. and McCormick, K. A. (2015). *Essentials of Nursing Informatics* (6th ed.). New York, NY: McGraw Hill Professional.

[11] Dalrymple, P. W. (2011). Data, information, knowledge: the emerging field of health informatics. *Bulletin of the American Society for Information Science and Technology*, 37(5), 41–44. https://doi.org/10.1002/bult.2011.1720370512.

[12] Hersh, W. (2009). *Information Retrieval: A Health and Biomedical Perspective*. New York, NY: Springer.

[13] Wyatt, J. C. (2002). Basic concepts in medical informatics. *Journal of Epidemiology & Community Health*, 56(11), 808–812. https://doi.org/10.1136/jech.56.11.808.

[14] Wilson, T. (1989). Towards an information management curriculum. *Journal of Information Science*, 15(4–5), 203–209. https://doi.org/10.1177/016555158901500403.

[15] Simpao, A. F., Ahumada, L. M., Gálvez, J. A., and Rehman, M. A. (2014). A review of analytics and clinical informatics in health care. *Journal of Medical Systems*, 38(4), 1–7. https://doi.org/10.1007/s10916-014-0045-x.

[16] American Society of Health-System Pharmacists. (2007). ASHP statement on the pharmacist's role in informatics. *American Journal of Health-System Pharmacy*, 64(2), 200–203. https://doi.org/10.2146/ajhp060364.

[17] White, C. L. and Hohmeier, K. C. (2015). Pharmacy informatics. *Journal of Pharmacy Technology*, 31(6), 247–252. https://doi.org/10.1177/8755122515605517.

[18] Friede, A., Blum, H. L., and McDonald, M. (1995). Public health informatics: how information-age technology can strengthen public health. *Annual Review of Public Health*, 16(1), 239–252. https://doi.org/10.1146/annurev.pu.16.050195.001323.

[19] Aziz, H. A. (2017). A review of the role of public health informatics in healthcare. *Journal of Taibah University Medical Sciences*, 12(1), 78–81. https://doi.org/10.1016/j.jtumed.2016.08.011

[20] Bayat, A. (2002). Science, medicine, and the future: bioinformatics. *BMJ*, 324(7344), 1018–1022. https://doi.org/10.1136/bmj.324.7344.1018.

[21] Masic, I., Ridjanovic, Z., Pandza, H., Masic, Z., Koprivica, J., and Sivic, S. (2010). *Medical Informatics*, 2nd ed. Avicena.

[22] Shortlife, E. H., Pereault, L. E., Wiederhold, G., and Fagan, L. M. (2000). *Computer Applications in Health Care and Biomedicine*. New York, NY: Springer-Verlag.

[23] Nelson, R. (2016). *Theoretical Foundations of Health Informatics* (10th ed.). Health Informatics-E-Book: An Interprofessional Approach. Amsterdam, Netherlands: Elsevier.

[24] Masic, I. (2020). The history of medical informatics development-an overview. *International Journal on Biomedicine and Healthcare*, 8(1), 37–52. https://doi.org/10.5455/ijbh.2020.8.37-52

[25] Masic, I. and Chronaki, C. (2021). EFMI inside – the official newsletter of the European federation for medical informatics. *Acta Informatica Medica*, 29(2), 139. https://doi.org/10.5455/aim.2021.29.139-170.

[26] Collen, M. F. (1994). The origins of informatics. *Journal of the American Medical Informatics Association*, 1(2), 91–107. https://doi.org/10.1136/jamia.1994.95236152.

[27] Mihalas, G., Zvarova, J., Kulikowski, C., *et al.* (2014). History of medical informatics in Europe – a short review by different approach. *Acta Informatica Medica*, 22(1), 6. https://doi.org/10.5455/aim.2014.22.6-10.

[28] Mantas, J. (2016). Biomedical and health informatics education – the IMIA years. *Yearbook of Medical Informatics*, 25(S 01), S92–S102. https://doi.org/10.15265/iy-2016-032.

[29] Protti, D. J., Van Bemmel, J. H., Gunzenhäuser, R., *et al.* (1994). Can health/ medical informatics be regarded as a separate discipline? *Methods of Information in Medicine*, 33(03), 318–326. https://doi.org/10.1055/s-0038-1635017.

[30] Hasman, A., Mantas, J., and Zarubina, T. (2014). An abridged history of medical informatics education in Europe. *Acta Informatica Medica*, 22(1), 25. https://doi.org/10.5455/aim.2014.22.25-36

[31] Lu, G., Yan, Y., Ren, L., Saponaro, P., Sebe, N., and Kambhamettu, C. (2016). Where am I in the dark: Exploring active transfer learning on the use of indoor localization based on thermal imaging. *Neurocomputing*, 173, 83–92. https://doi.org/10.1016/j.neucom.2015.07.106

[32] Khalid, M. H., Murtaza, M., and Habbal, M. (2020). *Study of Security and Privacy Issues in Internet of Things*. IEEE. https://doi.org/10.1109/citisia50690.2020.9371828.

[33] Laugesen, J., Hassanein, K., and Yuan, Y. (2015). The impact of internet health information on patient compliance: a research model and an empirical study. *Journal of Medical Internet Research*, 17(6), e143. https://doi.org/10.2196/jmir.4333.

[34] Ye, J. (2020). The role of health technology and informatics in a global public health emergency: practices and implications from the COVID-19 pandemic. *JMIR Medical Informatics*, 8(7), e19866. https://doi.org/10.2196/19866.

[35] Sheikh, A., Cornford, T., Barber, N., *et al.* (2011). Implementation and adoption of nationwide electronic health records in secondary care in England: final qualitative results from prospective national evaluation in "early adopter" hospitals. *BMJ*, 343(171), d6054–d6054. https://doi.org/10.1136/bmj.d6054.

[36] Schreier, G., Hayn, D., and Ammenwerth, E. (2010). Health informatics meets eHealth. *Methods of Information in Medicine*, 49(03), 269–270. https://doi.org/10.1055/s-0038-1625339.

[37] Sullivan, F. (2001), What is health informatics *Journal of Health Services Research & Policy*, 6(4), 251–254.

[38] Eysenbach, G. (2001). What is e-Health? *Journal of Medical Internet Research*, 3(2), e20. https://doi.org/10.2196/jmir.3.2.e20

[39] Kun, L. G. (2001). Telehealth and the global health network in the 21st century. From homecare to public health informatics. *Computer Methods and Programs in Biomedicine*, 64(3), 155–167. https://doi.org/10.1016/s0169-2607(00)00135-8.

[40] Ruland, C. M., Brynhi, H., Andersen, R., and Bryhni, T. (2008). Developing a shared electronic health record for patients and clinicians. In: Andersen S. K., Klein M. C. A., Schluz S., Aarts J., and Mazzoleni M. C. (Eds), *eHealth Beyond the Horizon – Get IT There. Studies in Health Technology and Informatics*, 136, 57–62.

[41] Sood, S., Mbarika, V., Jugoo, S., *et al.* (2007). What is telemedicine? A collection of 104 peer-reviewed perspectives and theoretical underpinnings. *Telemedicine and E-Health*, 13(5), 573–590. https://doi.org/10.1089/tmj.2006.0073.

[42] Arnrich, B., Bardram, J., Tröster, G., and Mayora, O. (2010). Pervasive healthcare. *Methods of Information in Medicine*, 49(1), 67–73. https://doi.org/10.3414/me09-02-0044

[43] Lintonen, T. P., Konu, A. I., and Seedhouse, D. (2008). Information technology in health promotion. *Health Education Research*, 23(3), 560–566. https://doi.org/10.1093/her/cym001.

[44] Stroetmann, K. A., Jones, T., Dobrev, A., and Stroetmann, V. N. (eds.). (2006, *September). eHealth is Worth it: The Economic Benefits of Implemented eHealth Solutions at Ten European Sites*. European Commission. https://doi.org/10.13140/2.1.1551.9367.

[45] Norris, A. C. (2002). Current trends and challenges in health informatics. *Health Informatics Journal*, 8(4), 205–213. https://doi.org/10.1177/146045820200800407.

[46] Gillis, A. S. (2021, August 13). *What is Internet of Things (IoT)?* IoT Agenda. https://internetofthingsagenda.techtarget.com/definition/Internet-of-Things-IoT.

[47] Huk, A. S., and Bondarenko, M. E. (2021). The role of internet of things in healthcare. Open Electronic Archive of Kharkov National University of Radio Electronics. https://openarchive.nure.ua/bitstream/document/18199/1/Huk_Bondarenko.pdf.

[48] Rajashekhar, K. and Manish, J. (n.d.). *IoT in Healthcare Industry IoT Applications in Healthcare – Wipro*. Wipro. Retrieved February 9, 2022, from https://www.wipro.com/business-process/what-can-iot-do-for-healthcare-/.

[49] Duarte, D. and Paulo Silva Cunha, J. (2018). Wearable health devices—vital sign monitoring, systems and technologies. *Sensors*, 18(8), 2414. PubMed Central (PMC). https://www.ncbi.nlm.nih.gov/pmc/articles/PMC6111409/.

[50] Kumar, P., Kumar, Y., and Tawhid, M. A. (2021). *Machine Learning, Big Data, and IoT for Medical Informatics*, pp. xxi–xxv. https://doi.org/10.1016/b978-0-12-821777-1.00009-4.

[51] Gao, Y., Tan, K., Sun, J., Jiang, T., and Zou, X. (2019). Application of mixed reality technology in visualization of medical operations. *Chinese Medical Sciences Journal*, 69, Article no. 003564. https://doi.org/10.24920/003564.

[52] Hole, G., Hole, A. S., and McFalone-Shaw, I. (2021). Digitalization in pharmaceutical industry: What to focus on under the digital implementation process? *International Journal of Pharmaceutics: X*, 3, 100095. https://doi.org/10.1016/j.ijpx.2021.100095.

[53] Feki, M. A., Kawsar, F., Boussard, M., and Trappeniers, L. (2013). The internet of things: the next technological revolution. *Computer*, 46(2), 24–25. https://doi.org/10.1109/mc.2013.63.

[54] Farahani, B., Firouzi, F., and Luecking, M. (2021). The convergence of IoT and distributed Ledger technologies (DLT): Opportunities, challenges, and solutions. *Journal of Network and Computer Applications*, 177, 102936. https://doi.org/10.1016/j.jnca.2020.102936

[55] Li, X., Lu, R., Liang, X., Shen, X., Chen, J., and Lin, X. (2011). Smart community: an Internet of things application. *IEEE Communications Magazine*, 49(11), 68–75. https://doi.org/10.1109/mcom.2011.6069711.

[56] Jara, A. J., Alcolea, A. F., Zamora, M. A., Skarmeta, A. F., and Alsaedy, M. (2010). Drugs interaction checker based on IoT. In: *2010 Internet of Things (IOT)*. https://doi.org/10.1109/iot.2010.5678458.

[57] Tarouco, L. M., Bertholdo, L. M., Granville, L. Z., *et al.* (2012). Internet of things in healthcare: interoperability and security issues. In: *2012 IEEE International Conference on Communications (ICC)*, pp. 6121–6125. https://doi.org/10.1109/icc.2012.6364830.

[58] Castañeda, H., Holmes, S. M., Madrigal, D. S., Young, M. D., Beyeler, N., and Quesada, J. (2015). Immigration as a social determinant of health. *Annual Review of Public Health*, 36(1), 375–392. https://doi.org/10.1146/annurev-publhealth-032013-182419

[59] Nath, B., Reynolds, F., and Want, R. (2006). RFID technology and applications. *IEEE Pervasive Computing*, 5(1), 22–24. https://doi.org/10.1109/mprv.2006.13.

[60] Floerkemeier, C., Roduner, C., and Lampe, M. (2007). RFID application development with the Accada middleware platform. *IEEE Systems Journal*, 1(2), 82–94. https://doi.org/10.1109/jsyst.2007.909778.

[61] Vaishya, R., Javaid, M., Khan, I. H., and Haleem, A. (2020). Artificial intelligence (AI) applications for COVID-19 pandemic. *Diabetes & Metabolic Syndrome: Clinical Research & Reviews*, 14(4), 337–339. https://doi.org/10.1016/j.dsx.2020.04.012.

[62] Inn, T. L. (2020). Smart city technologies take on COVID-19. *World Health*, 841. https://doi.org/10.1109/MC.2013.63.

[63] Bandello, F. (2020). Cataract surgery with combined versus deferred intravitreal dexamethasone implant for diabetic macular edema: Long-term outcomes from a real-world setting. *Acta Diabetologica*, 57(10), 1193–1201. https://doi.org/10.1007/s00592-020-01509-5

[64] Sharma, S. K. and Ahmed, S. S. (2021). IoT-based analysis for controlling & spreading prediction of COVID-19 in Saudi Arabia. *Soft Computing*, 25(18), 12551–12563. https://doi.org/10.1007/s00500-021-06024-5

[65] Domingo, M. C. (2012). An overview of the Internet of Things for people with disabilities. *Journal of Network and Computer Applications*, 35(2), 584–596. https://doi.org/10.1016/j.jnca.2011.10.015

[66] Singh, R. P., Javaid, M., Haleem, A., and Suman, R. (2020). Internet of things (IoT) applications to fight against COVID-19 pandemic. *Diabetes & Metabolic Syndrome: Clinical Research & Reviews*, 14(4), 521–524. https://doi.org/10.1016/j.dsx.2020.04.041.

[67] Jung, I. (2005). ICT-pedagogy integration in teacher training: Application cases worldwide. *Journal of Educational Technology & Society*, 8(2), 94–101. https://www.jstor.org/stable/10.2307/jeductechsoci.8.2.94

[68] Kasap, H. (2021). Multi-class brain normality and abnormality diagnosis using modified faster R-CNN. *International Journal of Medical Informatics*, 155, 104576. https://doi.org/10.1016/j.ijmedinf.2021.104576

[69] Zhu, X. and Goldberg, A. B. (2009). Introduction to semi-supervised learning. *Synthesis Lectures on Artificial Intelligence and Machine Learning*, 3 (1), 1–130. https://doi.org/10.2200/s00196ed1v01y200906aim006.

[70] Czerwinski, M., Hernandez, J., and McDuff, D. (2021). Building an AI that feels: AI systems with emotional intelligence could learn faster and be more helpful. *IEEE Spectrum*, 58(5), 32–38. https://doi.org/10.1109/mspec. 2021.9423818.

[71] Zohuri, B. and Mossavar Rahmani, F. (2020). Artificial intelligence driven resiliency with machine learning and deep learning components. *Japan Journal of Research*, 1(1), 1–8. https://doi.org/10.33425/2690-8077.1002.

[72] Back, T., Hammel, U., and Schwefel, H. P. (1997). Evolutionary computation: comments on the history and current state. *IEEE Transactions on Evolutionary Computation*, 1(1), 3–17. https://doi.org/10.1109/4235.585888.

[73] Kaul, V., Enslin, S., and Gross, S. A. (2020). History of artificial intelligence in medicine. *Gastrointestinal Endoscopy*, 92(4), 807–812. https://doi.org/10.1016/j.gie.2020.06.040.

[74] He, J., Baxter, S. L., Xu, J., Xu, J., Zhou, X., and Zhang, K. (2019). The practical implementation of artificial intelligence technologies in medicine. *Nature Medicine*, 25(1), 30–36. https://doi.org/10.1038/s41591-018-0307-0.

[75] Rajaraman, V. (2014). John McCarthy –Father of artificial intelligence. *Resonance*, 19, 198–207.

[76] McCarthy, J. (1974). Artificial intelligence: a paper symposium. *Artificial Intelligence*, 5(3), 317–322. https://doi.org/10.1016/0004-3702(74)90016-2.

[77] Khemani, D. (2020). Artificial intelligence. *Resonance*, 25(1), 33–41. https://doi.org/10.1007/s12045-019-0920-3.

[78] Muggleton, S. (2014). Alan Turing and the development of Artificial Intelligence. *AI Communications*, 27(1), 3–10. https://doi.org/10.3233/aic-130579.

[79] Garvey, A. and Lesser, V. (1994). A survey of research in deliberative real-time artificial intelligence. *Real-Time Systems*, 6(3), 317–347. https://doi.org/10.1007/bf01088630.

[80] Dellermann, D., Ebel, P., Söllner, M., and Leimeister, J. M. (2019). Hybrid intelligence. *Business & Information Systems Engineering*, 61(5), 637–643. https://doi.org/10.1007/s12599-019-00595-2.

[81] Guo, Z. and Li, X. (2018). Medical image segmentation based on multi-modal convolutional neural network: Study on image fusion schemes. In: *2018 IEEE 15th International Symposium on Biomedical Imaging (ISBI 2018)*. Washington, DC, USA, pp. 903–907. https://doi.org/10.1109/isbi.2018.8363717

[82] Pesapane, F., Codari, M., and Sardanelli, F. (2018). Artificial intelligence in medical imaging: threat or opportunity? Radiologists again at the forefront of innovation in medicine. *European Radiology Experimental*, 2(1), Article number: 35. https://doi.org/10.1186/s41747-018-0061-6.

[83] Ardakani, A. A., Kanafi, A. R., Acharya, U. R., Khadem, N., and Mohammadi, A. (2020). Application of deep learning technique to manage COVID-19 in routine clinical practice using CT images: results of 10 convolutional neural networks. *Computers in Biology and Medicine*, 121, 103795. https://doi.org/10.1016/j.compbiomed.2020.103795.

[84] Naude, W. (2020). Artificial intelligence vs COVID-19: limitations, constraints and pitfalls. *AI & Society*, 35(3), 761–765. https://doi.org/10.1007/s00146-020-00978-0.

[85] Ghimire, A., Thapa, S., Jha, A. K., Kumar, A., Kumar, A., and Adhikari, S. (2020). AI and IoT solutions for tackling COVID-19 pandemic. In: *2020 4th International Conference on Electronics, Communication and Aerospace Technology (ICECA)*. https://doi.org/10.1109/iceca49313.2020.9297454.

[86] Imran, A., Posokhova, I., Qureshi, H. N., *et al.* (2020). AI4COVID-19: AI enabled preliminary diagnosis for COVID-19 from cough samples via an app. *Informatics in Medicine Unlocked*, 20, 100378.

[87] Patel, V. L., Shortliffe, E. H., Stefanelli, M., *et al.* (2009). The coming of age of artificial intelligence in medicine. *Artificial Intelligence in Medicine*, 46 (1), 5–17. https://doi.org/10.1016/j.artmed.2008.07.017.

[88] Jiang, F., Jiang, Y., Zhi, H., *et al.* (2017). Artificial intelligence in healthcare: past, present and future. *Stroke and Vascular Neurology*, 2(4), 230–243. https://doi.org/10.1136/svn-2017-000101.

[89] Lorig, K. R., Sobel, D. S., Stewart, A. L., *et al.* (1999). Evidence suggesting that a chronic disease self-management program can improve health status while reducing hospitalization. *Medical Care*, 37(1), 5–14. https://doi.org/10.1097/00005650-199901000-00003.

[90] Mushtaq, M., Bricq, J., Bhatti, M. K., *et al.* (2020). WHISPER: a tool for run-time detection of side-channel attacks. *IEEE Access*, 8, 83871–83900. https://doi.org/10.1109/access.2020.2988370.

[91] Kuzlu, M., Fair, C., and Guler, O. (2021). Role of artificial intelligence in the internet of things (IoT) cybersecurity. *Discover Internet of Things*, 1(1), 1–14. https://doi.org/10.1007/s43926-020-00001-4.

[92] Ndiaye, M., Oyewobi, S. S., Abu-Mahfouz, A. M., Hancke, G. P., Kurien, A. M., and Djouani, K. (2020). IoT in the wake of COVID-19: a survey on contributions, challenges and evolution. *IEEE Access*, 8, 186821–186839. https://doi.org/10.1109/access.2020.3030090.

[93] Manzo, J., Sinnett, M. J., Sosnowski, F., Begliomini, R., Green, J., and Pane, F. (2005). Case study: challenges, successes and lessons learned from implementing computerized physician order entry (CPOE) at two distinct health systems: implications of CPOE on the pharmacy and the medication-use process. *Hospital Pharmacy*, 40(5), 420–429. https://doi.org/10.1177/001857870504000509.

[94] Chan, M., Campo, E., Estève, D., and Fourniols, J. Y. (2009). Smart homes—current features and future perspectives. *Maturitas*, 64(2), 90–97. https://doi.org/10.1016/j.maturitas.2009.07.014.

[95] Tai-Seale, M., Downing, N. L., Jones, V. G., *et al.* (2019). Technology-enabled consumer engagement: promising practices at four health care delivery organizations. *Health Affairs*, 38(3), 383–390. https://doi.org/10.1377/hlthaff.2018.05027Tarouco.

[96] Andaleeb, S. S. (2001). Service quality perceptions and patient satisfaction: a study of hospitals in a developing country. *Social Science & Medicine*, 52 (9), 1359–1370. https://doi.org/10.1016/s0277-9536(00)00235-5.

[97] Shakeel, P. M., Burhanuddin, M., and Desa, M. I. (2019). Lung cancer detection from CT image using improved profuse clustering and deep

learning instantaneously trained neural networks. *Measurement*, 145, 702–712. https://doi.org/10.1016/j.measurement.2019.05.027.

[98] Patel, D., Shah, Y., Thakkar, N., Shah, K., and Shah, M. (2019). Implementation of Artificial Intelligence techniques for cancer detection. *Augmented Human Research*, 5(1), 7. https://doi.org/10.1007/s41133-019-0024-3.

[99] Bloss, R. (2011). Mobile hospital robots cure numerous logistic needs. *Industrial Robot: An International Journal*, 38(6), 567–571. https://doi.org/10.1108/01439911111179075

[100] Gilchrist, A. (2016). *Industry 4.0: The industrial Internet of Things*. Berkeley, CA: Apress.

[101] Dankwa-Mullan, I., Rivo, M., Sepulveda, M., Park, Y., Snowdon, J., and Rhee, K. (2019). Transforming diabetes care through artificial intelligence: the future is here. *Population Health Management*, 22(3), 229–242. https://doi.org/10.1089/pop.2018.0129.

[102] Lambrinou, E., Hansen, T. B., and Beulens, J. W. (2019). Lifestyle factors, self-management and patient empowerment in diabetes care. *European Journal of Preventive Cardiology*, 26(2_suppl), 55–63. https://doi.org/10.1177/2047487319885455.

[103] Hoermann, S., McCabe, K. L., Milne, D. N., and Calvo, R. A. (2017). Application of synchronous text-based dialogue systems in mental health interventions: systematic review. *Journal of Medical Internet Research*, 19 (8), e267. https://doi.org/10.2196/jmir.7023.

[104] Baranowski, T. and Lytle, L. (2015). Should the IDEFICS outcomes have been expected? *Obesity Reviews*, 16, 162–172. https://doi.org/10.1111/obr.12359.

[105] Seyhan, A. A. and Carini, C. (2019). Are innovation and new technologies in precision medicine paving a new era in patients centric care? *Journal of Translational Medicine*, 17(1), 1–28. https://doi.org/10.1186/s12967-019-1864-9

[106] Ahmad, R. W., Salah, K., Jayaraman, R., Yaqoob, I., Ellahham, S., and Omar, M. (2020). Blockchain and COVID-19 pandemic: applications and challenges. *IEEE TechRxiv*, 1–19. https://doi.org/10.36227/techrxiv.12936572.v1.

[107] Chatterjee, P., Cymberknop, L. J., and Armentano, R. L. (2017). IoT-based decision support system for intelligent healthcare—applied to cardiovascular diseases. In: *2017 7th International Conference on Communication Systems and Network Technologies (CSNT)*, pp. 362–366. https://doi.org/10.1109/csnt.2017.8418567.

[108] Jabeen, F., Maqsood, M., Ghazanfar, M. A., *et al.* (2019). An IoT based efficient hybrid recommender system for cardiovascular disease. *Peer-to-Peer Networking and Applications*, 12(5), 1263–1276. https://doi.org/10.1007/s12083-019-00733-3.

[109] Nambiar, R., Bhardwaj, R., Sethi, A., and Vargheese, R. (2013). A look at challenges and opportunities of big data analytics in healthcare. In: *2013 IEEE International Conference on Big Data*. https://doi.org/10.1109/bigdata.2013.6691753.

[110] Mak, K. and Pichika, M. R. (2019). Artificial intelligence in drug development: present status and future prospects. *Drug Discovery Today*, 24(3), 773–780. https://doi.org/10.1016/j.drudis.2018.11.014.

[111] Ottesen, J. T., Olufsen, M. S., and Larsen, J. K. (2004). *Applied Mathematical Models in Human Physiology*. Philadelphia, PA:SIAM.

[112] Hester, R. L., Brown, A. J., Husband, L., *et al.* (2011). HumMod: a modeling environment for the simulation of integrative human physiology. *Frontiers in Physiology*, 2, 12. https://doi.org/10.3389/fphys.2011.00012.

[113] Liang, G., Fan, W., Luo, H., and Zhu, X. (2020). The emerging roles of artificial intelligence in cancer drug development and precision therapy. *Biomedicine & Pharmacotherapy*, 128, 110255. https://doi.org/10.1016/j.biopha.2020.110255.

[114] McDougall, R. J. (2018). Computer knows best? The need for value-flexibility in medical AI. *Journal of Medical Ethics*, 45(3), 156–160. https://doi.org/10.1136/medethics-2018-105118.

[115] WHO World Health Organization (2011, August 5). Health Technology Assessment of Medical Devices. https://www.who.int/publications/i/item/9789241501361.

[116] OECD. (2017). *OECD Skills Outlook 2017: Skills and Global Value Chains*. Paris: OECD Publishing.

[117] Velasco Garrido, M., Gerhardus, A., Røttingen, J., and Busse, R. (2010). Developing health technology assessment to address health care system needs. *Health Policy*, 94(3), 196–202. https://doi.org/10.1016/j.healthpol.2009.10.002.

[118] American Marketing Association (2017). *The Past, Present and Future of AI in Marketing*. https://www.ama.org/marketing-news/the-past-present-and-future-of-ai-in-marketing/.

[119] Bansal, S., Bruno, P., Denecker, O., Goparaju, M., and Niederkorn, M. (2018, October). *Global Payments 2018: A Dynamic Industry Continues to Break New Ground*. McKinsey & Company | Global management consulting. https://www.mckinsey.com/~/media/mckinsey/industries/financial%20services/our%20insights/global%20payments%20expansive%20growth%20targeted%20opportunities/global-payments-map-2018.pdf.

[120] Kripalani, S., Robertson, R., Love-Ghaffari, M. H., *et al.* (2007). Development of an illustrated medication schedule as a low-literacy patient education tool. *Patient Education and Counseling*, 66(3), 368–377. https://doi.org/10.1016/j.pec.2007.01.020.

[121] Wang, J., Hu, X., and Zhu, D. (2008). Applications of data mining in the healthcare industry. *Encyclopedia of Healthcare Information Systems*, 68–73. https://doi.org/10.4018/978-1-59904-889-5.ch010.

[122] Lim, W. M. and Ting, D. H. (2012). Healthcare marketing: contemporary salient issues and future research directions. *International Journal of Healthcare Management*, 5(1), 3–11. https://doi.org/10.1179/204797012x13293146890048.

[123] Meskó, B. and Gorog, M. (2020). A short guide for medical professionals in the era of artificial intelligence. *NPJ Digital Medicine*, 3(1), 126. https://doi.org/10.1038/s41746-020-00333-z
[124] Chung, K. and Park, R. C. (2018). Chatbot-based heathcare service with a knowledge base for cloud computing. *Cluster Computing*, 22(S1), 1925–1937. https://doi.org/10.1007/s10586-018-2334-5.
[125] Pottle, J. (2019). Virtual reality and the transformation of medical education. *Future Healthcare Journal*, 6(3), 181–185. https://doi.org/10.7861/fhj.2019-0036.
[126] Pedersen, P., Palm, H., Ringsted, C., and Konge, L. (2014). Virtual-reality simulation to assess performance in hip fracture surgery. *Acta Orthopaedica*, 85(4), 403–407. https://doi.org/10.3109/17453674.2014.917502.
[127] Hoxhallari, E., Behr, I. J., Bradshaw, J. S., *et al.* (2019). Virtual reality improves the patient experience during wide-awake local anesthesia no tourniquet hand surgery. *Plastic and Reconstructive Surgery*, 144(2), 408–414. https://doi.org/10.1097/prs.0000000000005831 https://www.mckinsey.com/~/media/mckinsey/industries/financial%20services/our%20insights/global%20payments%20expansive%20growth%20targeted%20opportunities/global-payments-map-2018.pdf.
[128] Juanes-Méndez, J. A., Marcos-Pablos, S., Velasco-Marcos, M. A., and Izard, S. G. (2022). Medical learning through simulations with immersive virtual reality and augmented reality devices. In: *Technological Adoption and Trends in Health Sciences Teaching, Learning, and Practice*. Hershey, PA: IGI Global, pp. 265–282. https://doi.org/10.4018/978-1-7998-8871-0.ch012.
[129] Munzer, B. W., Khan, M. M., Shipman, B., and Mahajan, P. (2019). Augmented reality in emergency medicine: a scoping review. *Journal of Medical Internet Research*, 21(4), e12368. https://doi.org/10.2196/12368.
[130] Nicolau, S., Soler, L., Mutter, D., and Marescaux, J. (2011). Augmented reality in laparoscopic surgical oncology. *Surgical Oncology*, 20(3), 189–201. https://doi.org/10.1016/j.suronc.2011.07.002.
[131] Jones, C. M., Roth, B., Mercado, A. M., *et al.* (2018). The Americleft project: comparison of ratings using two-dimensional versus three-dimensional images for evaluation of nasolabial appearance in patients with unilateral cleft lip and palate. *Journal of Craniofacial Surgery*, 29(1), 105–108. https://doi.org/10.1097/scs.0000000000004102.
[132] Appelboom, G., Camacho, E., Abraham, M. E., *et al.* (2014). Smart wearable body sensors for patient self-assessment and monitoring. *Archives of Public Health*, 72(1), 28. https://doi.org/10.1186/2049-3258-72-28.
[133] Sano, A., Phillips, A. J., Yu, A. Z., *et al.* (2015). Recognizing academic performance, sleep quality, stress level, and mental health using personality traits, wearable sensors and mobile phones. In: *2015 IEEE 12th International Conference on Wearable and Implantable Body Sensor Networks (BSN)*. https://doi.org/10.1109/bsn.2015.7299420.

[134] Goecks, J., Jalili, V., Heiser, L. M., and Gray, J. W. (2020). How machine learning will transform biomedicine. *Cell*, 181(1), 92–101. https://doi.org/10.1016/j.cell.2020.03.022

[135] Dunn, J., Runge, R., and Snyder, M. (2018). Wearables and the medical revolution. *Personalized Medicine*, 15(5), 429–448. https://doi.org/10.2217/pme-2018-0044.

[136] Muthu, B., Sivaparthipan, C. B., Manogaran, G., *et al.* (2020). IOT based wearable sensor for diseases prediction and symptom analysis in healthcare sector. *Peer-to-Peer Networking and Applications*, 13(6), 2123–2134. https://doi.org/10.1007/s12083-019-00823-2.

[137] Ahir, S., Telavane, D., and Thomas, R. (2020). The impact of artificial intelligence, blockchain, big data and evolving technologies in coronavirus disease – 2019 (COVID-19) curtailment. In: *2020 International Conference on Smart Electronics and Communication (ICOSEC)*, pp. 113–120. https://doi.org/10.1109/icosec49089.2020.9215294.

[138] Gavriel, C., Parker, K. H., and Faisal, A. A. (2015). Smartphone as an ultra-low cost medical tricorder for real-time cardiological measurements via ballistocardiography. In: *2015 IEEE 12th International Conference on Wearable and Implantable Body Sensor Networks (BSN)*. https://doi.org/10.1109/bsn.2015.7299425.

[139] Wusk, G. and Gabler, H. (2018). Non-invasive detection of respiration and heart rate with a vehicle seat sensor. *Sensors*, 18(5), 1463. https://doi.org/10.3390/s18051463.

[140] Henzen, A., Van de Kamer, J., Nakamura, T., *et al.* (2004). Development of active-matrix electronic-ink displays for handheld devices. *Journal of the Society for Information Display*, 12(1), *17*. https://doi.org/10.1889/1.1824234.

[141] Jaya Lakshmi, N. and Jabalia, N. (2021). Body sensor networks as emerging trends of technology in health care system: challenges and future. In: *Internet of Things*, pp. 129–157. https://doi.org/10.1007/978-3-030-66633-0_6.

[142] Chakravarthy, V. J., Mydeen, P. H., and Seenivasan, M. (2022). Study on the internet of nanotechnology (IoNT) in a healthcare monitoring system. In: *Handbook of Research on Nano-Drug Delivery and Tissue Engineering*. London: CRC Press, pp. 151–156. https://doi.org/10.1201/9781003161196-6.

[143] Pilarski, L. M., Mehta, M. D., Caulfield, T., Kaler, K. V., and Backhouse, C. J. (2004). Microsystems and nanoscience for biomedical applications: a view to the future. *Bulletin of Science, Technology & Society*, 24(1), 40–45. https://doi.org/10.1177/0270467604263120.

[144] Sun, T., Shrike Zhang, Y., Bo, P., Hyun, D. C., Yang, M., and Xia, Y. (2021). Engineered nanoparticles for drug delivery in cancer therapy. *Nanomaterials and Neoplasms*, 22, 31–142. https://doi.org/10.1201/9780429027819-2.

[145] Bindra, H. S. and Singh, B. (2022). Nanomedicine and healthcare systems. In: *Innovative Approaches for Nanobiotechnology in Healthcare Systems*.

Hershey, PA: IGI Global, pp. 1–32. https://doi.org/10.4018/978-1-7998-8251-0.ch001.

[146] Taylor, R. H., Menciassi, A., Fichtinger, G., Fiorini, P., and Dario, P. (2016). Medical robotics and computer-integrated surgery. In: *Springer Handbook of Robotics*, New York, NY: Springer, pp. 1657–1684. https://doi.org/10.1007/978-3-319-32552-1_63.

[147] Wu, A. R. (2021). Human biomechanics perspective on robotics for gait assistance: challenges and potential solutions. *Proceedings of the Royal Society B: Biological Sciences*, 288(1956), 20211197. https://doi.org/10.1098/rspb.2021.1197.

[148] Henschel, A., Laban, G., and Cross, E. S. (2021). What makes a robot social? A review of social robots from science fiction to a home or hospital near you. *Current Robotics Reports*, 2(1), 9–19. https://doi.org/10.1007/s43154-020-00035-0.

[149] Pepito, J. A. and Locsin, R. (2019). Can nurses remain relevant in a technologically advanced future? *International Journal of Nursing Sciences*, 6 (1), 106–110. https://doi.org/10.1016/j.ijnss.2018.09.013.

[150] Kolpashchikov, D., Gerget, O., and Meshcheryakov, R. (2022). Robotics in healthcare. In: *Handbook of Artificial Intelligence in Healthcare, Vol 2: Practicalities and Prospects*. New York, NY: Springer, pp. 281–306. https://doi.org/10.1007/978-3-030-83620-7_12.

[151] Algorri, M., Abernathy, M. J., Cauchon, N. S., Christian, T. R., Lamm, C. F., and Moore, C. M. (2021). Re-envisioning pharmaceutical manufacturing: increasing agility for global patient access. *Journal of Pharmaceutical Sciences*. 111(3), 593–607. https://doi.org/10.1016/j.xphs.2021.08.032.

[152] Trenfield, S. J., Awad, A., Madla, C. M., *et al.* (2019). Shaping the future: recent advances of 3D printing in drug delivery and healthcare. *Expert Opinion on Drug Delivery*, 16(10), 1081–1094. https://doi.org/10.1080/17425247.2019.1660318.

[153] Awad, A., Trenfield, S. J., Goyanes, A., Gaisford, S., and Basit, A. W. (2018). Reshaping drug development using 3D printing. *Drug Discovery Today*, 23(8), 1547–1555. https://doi.org/10.1016/j.drudis.2018.05.025.

[154] Rengier, F., Mehndiratta, A., Von Tengg-Kobligk, H., *et al.* (2010). 3D printing based on imaging data: review of medical applications. *International Journal of Computer Assisted Radiology and Surgery*, 5(4), 335–341. https://doi.org/10.1007/s11548-010-0476-x.

[155] Dodziuk, H. (2016). Applications of 3D printing in healthcare. *Polish Journal of Cardio-Thoracic Surgery*, 3, 283–293. https://doi.org/10.5114/kitp.2016.62625.

[156] Conti, M., Sandeep Kumar, E., Lal, C., and Ruj, S. (2018). A survey on security and privacy issues of bitcoin. *IEEE Communications Surveys & Tutorials*, 20(4), 3416–3452. https://doi.org/10.1109/comst.2018.2842460.

[157] DalleMule, L. and Davenport, T. H. (2017, May 1). The 2 types of data strategies every company needs. *Harvard Business Review*. https://hbr.org/2017/05/whats-your-data-strategy.

[158] Chauhan, R. and Kumar, A. (2013). Cloud computing for improved healthcare: techniques, potential and challenges. In: *2013 E-Health and Bioengineering Conference (EHB)*, pp. 1–4. https://doi.org/10.1109/ehb.2013.6707234.

[159] Omar L., Aneesh C., Anil S., Bala H., Casey F., and Jimmy D. (2021, April 30). Rush University Medical Center creates COVID-19 analytics hub on AWS. *Amazon Web Services*. https://aws.amazon.com/blogs/publicsector/rush-medical-center-aws-analytics-hub/.

[160] Dixon, B. E., Grannis, S. J., McAndrews, C., *et al.* (2021). Leveraging data visualization and a statewide health information exchange to support COVID-19 surveillance and response: application of public health informatics. *Journal of the American Medical Informatics Association*, 28(7), 1363–1373. https://doi.org/10.1093/jamia/ocab004.

[161] Ye, Q., Zhou, J., and Wu, H. (2020). Using information technology to manage the COVID-19 pandemic: development of a technical framework based on practical experience in China. *JMIR Medical Informatics*, 8(6), e19515. https://doi.org/10.2196/19515.

[162] Vindrola-Padros, C., Sidhu, M. S., Georghiou, T., *et al.* (2021). The implementation of remote home monitoring models during the COVID-19 pandemic in England. *EClinical Medicine*, 34, 100799. https://doi.org/10.1016/j.eclinm.2021.100799.

[163] Koonin, L. M., Hoots, B., Tsang, C. A., *et al.* (2020). Trends in the use of telehealth during the emergence of the COVID-19 pandemic—United States, January–March 2020. *MMWR. Morbidity and Mortality Weekly Report*, 69(43), 1595–1599. https://doi.org/10.15585/mmwr.mm6943a3.

[164] Koch, S. and Hägglund, M. (2009). Health informatics and the delivery of care to older people. *Maturitas*, 63(3), 195–199. https://doi.org/10.1016/j.maturitas.2009.03.023.

[165] Finn, N. B. and Bria, W. F. (2009). *Digital Communication in Medical Practice,* vol. 1. Springer Publishing, pp. 161–171 https://doi.org/10.1007/978-1-84882-355-6.

[166] Martin. G., (2022, January 14). *Top 10 Emerging Trends in Health Care for 2021: The New Normal*. AHA Trustee Services. https://trustees.aha.org/top-10-emerging-trends-health-care-2021-new-normal.

[167] Otokiti, A. (2019). Using informatics to improve healthcare quality. *International Journal of Health Care Quality Assurance*, 32(2), 425–430. https://doi.org/10.1108/ijhcqa-03-2018-0062.

[168] Imhoff, M. L., Zhang, P., Wolfe, R. E., and Bounoua, L. (2010). Remote sensing of the urban heat island effect across biomes in the continental USA. *Remote Sensing of Environment*, 114(3), 504–513. https://doi.org/10.1016/j.rse.2009.10.008

[169] Savvas, A. T. (2018, November 29). *Healthcare to Create More Data Than Any Other Industry Says IDC*. BroadGroup. Retrieved February 4, 2022, from https://www.broad-group.com/data/news/documents/b1m2y1j9y5fvb8/health-care-to-create-more-data-than-any-other-industry-says-idc.

[170] Hillestad, R., Bigelow, J., Fonkych, K., *et al.* (2005). Health information technology: can HIT lower costs and improve quality? *RB-9136*. https://doi.org/10.7249/rb9136.

[171] Vidogah, G., Moreton, R., and Sallah, D. (2009). Reducing health cost: health informatics and knowledge management as a business and communication tool. In: Papadopoulos, G. A., Wojtkowski, G., Zupancic, J., Wrycza, S., and Wojtkowski, W. (eds.), *Information Systems Development Towards a Service Provision Society,* 1st ed. US: Springer, pp. 715–722. https://doi.org/10.1007/b137171_74.

[172] Kawamoto, K. and Lobach, D. F. (2003). Clinical decision support provided within physician order entry systems: a systematic review of features effective for changing clinician behavior. In: *AMIA Annual Symposium Proceedings. AMIA Symposium, 2003*, pp. 361–365.

[173] Kabakuş, A. T. and Kara, R. (2016). The importance of informatics for health care industry. *Acta Medica Anatolia*, 4(3), 124–125. https://doi.org/10.5505/actamedica.2016.81300.

[174] Dinya, E. and Tóth, T. (2013). Health informatics: eHealth and telemedicine. *Presentation, Semmelweis University*. Retrieved from https://lh-hsrc.pnu.edu.sa/wp-content /uploads/2018/07/ED_ehealth_20140203.pdf.

[175] Snyder, C. F., Wu, A. W., Miller, R. S., Jensen, R. E., Bantug, E. T., and Wolff, A. C. (2011). The role of informatics in promoting patient-centered care. *The Cancer Journal*, 17(4), 211–218. https://doi.org/10.1097/ppo.0b013e318225ff89.

[176] Siau, K., (2003). Health care informatics. *IEEE Transactions on Information Technology in Biomedicine*, 7(1), 1–7. https://doi.org/10.1109/titb.2002.805449.

[177] Alotaibi YK. and Federico F. (2017). The impact of health information technology on patient safety. *Saudi Medical Journal*, 38(12),1173–1180. 10.15537/smj.2017.12.20631.

[178] Carayon, P., Cartmill, R., Blosky, M. A., *et al.* (2011). ICU nurses' acceptance of electronic health records. *Journal of the American Medical Informatics Association*, 18(6), 812–819. https://doi.org/10.1136/amiajnl-2010-000018.

[179] Guo, J., Iribarren, S., Kapsandoy, S., Perri, S., and Staggers, N., (2011). Usability evaluation of an electronic medication administration record (eMAR) application. *Applied Clinical Informatics*, 02(02), 202–224. https://doi.org/10.4338/aci-2011-01-ra-0004.

[180] Fuller, A. E., Guirguis, L. M., Sadowski, C. A., and Makowsky, M. J. (2018). Electronic medication administration records in long-term care facilities: a scoping review. *Journal of the American Geriatrics Society*, 66 (7), 1428–1436. https://doi.org/10.1111/jgs.15384.

[181] Madera, A. and Sandri, S. (2006). Health management system. *All Days*. https://doi.org/10.2118/98922-ms.

[182] Balasubramanian, S., Shukla, V., Sethi, J. S., Islam, N., and Saloum, R. (2021). A readiness assessment framework for blockchain adoption: a healthcare case study. *Technological Forecasting and Social Change*, 165, 120536. https://doi.org/10.1016/j.techfore.2020.120536.

[183] Lowery, C. (2020). What is digital health and what do I need to know about it? *Obstetrics and Gynecology Clinics of North America*, 47(2), 215–225. https://doi.org/10.1016/j.ogc.2020.02.011.

[184] Martin, A., Brummond, P., Vlasimsky, T., *et al.* (2017). The evolving frontier of digital health: opportunities for pharmacists on the horizon. *Hospital Pharmacy*, 53(1), 7–11. https://doi.org/10.1177/0018578717738221.

[185] Seh, A. H., Zarour, M., Alenezi, M., *et al.* (2020). Healthcare data breaches: insights and implications. *Healthcare (Basel, Switzerland)*, 8(2), 133. https://doi.org/10.3390/healthcare8020133.

[186] Anwar, F. and Shamim, A. (2011). Barriers in adoption of health information technology in developing societies. *International Journal of Advanced Computer Science and Applications*, 2(8), 40–46. https://doi.org/10.14569/ijacsa.2011.020808.

[187] Giuse, D. and Kuhn, K. (2001). From hospital information systems to health information systems – problems, challenges, perspectives. *Yearbook of Medical Informatics*, 10(01), 63–76. https://doi.org/10.1055/s-0038-1638108.

[188] Ogushi, Y., Okada, Y., Kimura, M., Kumamoto, I., Sekita, Y., and Haruki, Y. (1999). Status and perspective of hospital information systems in Japan. *Methods of Information in Medicine*, 38(03), 200–206. https://doi.org/10.1055/s-0038-1634185.

[189] Stead, W. W. and Lorenzi, N. M. (1999). Health informatics: linking investment to value. *Journal of the American Medical Informatics Association*, 6(5), 341–348. https://doi.org/10.1136/jamia.1999.0060341.

[190] Burnham, J. P., Lu, C., Yaeger, L. H., Bailey, T. C., and Kollef, M. H. (2018). Using wearable technology to predict health outcomes: a literature review. *Journal of the American Medical Informatics Association*, 25(9), 1221–1227. https://doi.org/10.1093/jamia/ocy082.

[191] Ometov, A., Shubina, V., Klus, L., *et al.* (2021). A survey on wearable technology: history, state-of-the-art and current challenges. *Computer Networks*, 193, 108074. https://doi.org/10.1016/j.comnet.2021.108074.

[192] Phillips W. (2015). Ethical controversies about proper health informatics practices. *Missouri Medicine*, 112(1), 53–57.

[193] Raju, M. H., Ahmed, M. U., and Ahad, M. A. (2020). Health informatics: challenges and opportunities. *Intelligent Systems Reference Library*, 192, 231–246. https://doi.org/10.1007/978-3-030-54932-910.

[194] Rhodes, C. J. (2017). US withdrawal from the COP21 Paris climate change agreement, and its possible implications. *Science Progress*, 100(4), 411–419. https://doi.org/10.3184/003685017x15063357842600

[195] U.S. News Staffs. (2014). Online Health Informatics Master's Degree. *U.S. News*. Retrieved February 11, 2022, from https://www.usnews.com/education/online-education/health-informatics-masters-degree.

[196] Agarwal, R. and Dhar, V. (2014). Editorial—Big data, data science, and analytics: the opportunity and challenge for IS research. *Information Systems Research*, 25(3), 443–448. https://doi.org/10.1287/isre.2014.0546.

[197] Verma, S. (2021, January 11). How CMS has made progress on healthcare interoperability. *Healthcare IT News*. https://www.healthcareitnews.com/blog/how-cms-has-made-progress-healthcare-interoperability.

[198] Gupta, A., Gupta, V., Mittal, M., and Mittal, V. (2022). An efficient AR modelling-based electrocardiogram signal analysis for health informatics. *International Journal of Medical Engineering and Informatics*, 14(1), 74. https://doi.org/10.1504/ijmei.2022.10042702.

[199] Bath, P. A. (2008). Health informatics: current issues and challenges. *Journal of Information Science*, 34(4), 501–518. https://doi.org/10.1177/0165551508092267.

[200] Hung, M., Conrad, J., Hon, S. D., Cheng, C., Franklin, J. D., and Tang, P. (2013). Uncovering patterns of technology use in consumer health informatics. *Wiley Interdisciplinary Reviews: Computational Statistics*, 5(6), 432–447. https://doi.org/10.1002/wics.1276.

[201] Veinot, T. C., Mitchell, H., and Ancker, J. S. (2018). Good intentions are not enough: how informatics interventions can worsen inequality. *Journal of the American Medical Informatics Association*, 25(8), 1080–1088. https://doi.org/10.1093/jamia/ocy052.

[202] Riley, W. J. (2012) Health disparities: gaps in access, quality and affordability of medical care. *Transactions of the American Clinical and Climatological Association*, 123, 167–174.

Chapter 4

Role of Internet of Things and artificial intelligence for healthcare informatics: an overview

Ankit Garg[1], *Anuj Kumar Singh*[2] *and Mohit Garg*[3]

Abstract

The Internet of Things (IoT) and artificial intelligence (AI) must be integrated into the healthcare sector's rapidly expanding infrastructure in order to improve healthcare services. The successful integration of IoT-enabled devices into hospital networks increases patient safety, lowers costs, and raises the standard and accessibility of healthcare services. The development of AI has led to the observation that a variety of AI-based solutions help people do their tasks quickly and effectively. Healthcare personnel is always supported by AI, which also automates many administrative and healthcare chores. The chapter provides an overview of modern technological developments in healthcare systems, including the IoT and AI and contains information about efficient healthcare services that make use of IoT and AI technology. The chapter goes on to discuss further important uses of IoT and AI to address various problems in healthcare systems and describes several modern IoT and AI tools and technologies that are utilized to uphold safe data security and communication in IoT-based healthcare informatics. Various AI-based solutions with a higher impact on the healthcare industry are being employed to automate the repetitive tasks of healthcare informatics. IoT and AI-based healthcare architecture is presented in this chapter. The healthcare-related activities that are required for real-time patient monitoring and meeting the needs of healthcare professionals are all included in the architecture. The chapter also elaborates on a number of IoT and AI technology-related difficulties, particularly those that relate to the production of medical devices and the creation of efficient healthcare systems. The difficulties that healthcare system designers are now dealing with provide the groundwork for potential future improvements. The chapter also offers future prospects for researchers to create sophisticated IoT and AI-based plans to enhance the state of

[1]Apex Institute of Technology-Computer Science and Engineering, University Centre for Research and Development, Chandigarh University, Punjab, India
[2]Department of Computer Science and Engineering, Adani University, Ahmedabad, India
[3]Microsoft, Atlanta, GA, USA

healthcare informatics. The chapter offers a thorough analysis to researchers who are eager to conduct their studies and contribute their important efforts to create IoT- and AI-based healthcare systems. This chapter proposes a framework that makes use of IoT and AI as critical technologies to deliver safe medical services to patients in order to give a thorough analysis of contemporary healthcare systems. According to the framework, combining IoT and AI offers a number of features, including data protection, data management, effective diagnosis, and real-time health monitoring, which solve the shortcomings of the existing hospital information system. The hospital environment is improved by the inclusion of such features in the present system.

Keywords: Internet of Things (IoT); Artificial intelligence (AI); Healthcare; Security; Machine learning

4.1 Introduction

Modern healthcare industries use artificial intelligence (AI) technologies to enhance their daily operations. The expansion of the healthcare industry has included a variety of machine learning (ML) approaches, which gives academics new areas to investigate cutting-edge methods that can evolve over time. The performance of the healthcare sectors is significantly impacted by the accessibility of a vast quantity of digital data, cutting-edge technology, and the exponential rise of ML algorithms [1]. By applying various ML methods, the enormous volume of unstructured data produced by the low-computing healthcare equipment may be transformed into useful structured data in the healthcare industry. To make patient treatment easier, the researchers have proposed a number of cutting-edge ML-based techniques [2]. Furthermore, several studies have demonstrated that modern AI-based healthcare systems are more suited than conventional healthcare systems. Several crucial processes, including analysis, report production, encryption, and decision-making, have been integrated into the healthcare system using AI-based solutions. Various picture shrinking approaches are being developed in the field of image processing to minimize the size of medical images for effective transmission. When the aspect ratio of medical photographs is decreased in a content-aware manner, the amount of storage needed can be kept to a minimum [3,4]. The healthcare-related manufacturing businesses are making considerable contributions to market promotion of their medical products. Leading international information technology (IT) companies like Samsung, Google, and Apple are recently entering the industry alongside major medical equipment manufacturers like GE, Siemens, and Phillips. Microsoft and Amazon have made significant research advancements in the use of AI in the healthcare industry [5]. The steps that are done by manufacturing businesses and the medical community in the production of medical devices aid regulatory bodies in approving AI-based medical devices. The most recent AI-based healthcare systems can benefit from the expertise of people like doctors, nurses, radiologists, and pharmacists. As a result, this study examines the current level of domestic and global AI technology in healthcare as well as the issues that need to be resolved before AI can be used in healthcare [6].

In the past, patients would personally go to their assigned doctors for regular checkups to learn the state of their health. There is no means for continuous real-time monitoring of their health status in conventional healthcare systems. Following the development of contemporary technology, namely the IoT, IoT-enabled devices are now being utilized to provide their patients a variety of ideas and potential remedies. Even if they live in remote areas, patients have started to use these IoT-based tools to monitor their health problems. Now when they are satisfied, the patients get annoyed while talking to their doctors. Additionally, the availability of remote monitoring in healthcare lowers the cost of care and assists patients in avoiding hospital readmissions [7]. The use of IoT technology has altered the healthcare industry's current situation. The healthcare industry is using a variety of IoT-based medical devices to deliver emergency services in real time. IoT applications in healthcare benefit all people, including patients, families, physicians, hospitals, and insurance companies. According to their demands, patients can personalize wearable gadgets like fitness bands, blood pressure monitors, and heart rate monitors [8]. These wearable gadgets are simple to program and are able to deliver alerts about a patient's numerous health issues at an early stage of the disease [9]. The alarm system inbuilt in the IoT devices can send the message to the family member of the patient and health professionals if a person's health condition changes or is disturbed rapidly [10]. Doctors can also keep track of the transformation in the health status by adopting various wearable and other IoT-enabled home monitoring equipment. After analyzing the data received from these devices health professionals can get information on whether the patient is sticking to their treatment regimens and the prescribed medicine. They can also suggest the patient take emergency services that are being provided by the healthcare industries. IoT enables medical practitioners to collaborate with patients in a more attentive and proactive manner. In order to get the desired results for their patients, doctors may find it useful to use data collected by IoT devices to determine the best course of therapy. In addition, there are other ways that hospitals might benefit from IoT devices. IoT sensors make it simple to track various medical equipment that is often used in the healthcare sector in real time. These tools allow for real-time evaluation of the performance of healthcare professionals stationed at various points in healthcare informatics. Infection transmission among hospital patients is a particularly difficult problem. Various IoT-enabled hygiene monitoring tools can be used to overcome these obstacles and stop the spread of illness among patients. Asset management is another use for IoT-enabled medical equipment. The administration of healthcare inventories and environment monitoring, such as refrigerator temperature, humidity, etc., may be done on a regular basis [11]. By using IoT devices, insurance firms may grow their company. The insurance companies can utilize the information gathered from these devices for underwriting and claims processing. The insurance firms are also capable of identifying potential underwriters and identifying patient fraud claims. The insurer benefits from IoT-enabled devices in a number of ways, including underwriting, pricing, processing claims, and controlling major risk during claims. By sharing the health data generated by IoT devices, customers may also benefit from insurance firms. Concepts of cloud computing are now widely used to provide customers with scalability and security benefits. Users can make use of a variety of computer resources, including networks, storage, and on-demand services [12]. In recent years, cloud computing has assumed a greater significance in IoT-based healthcare informatics. To lower risk and improve healthcare system services, cloud computing models can exchange sensitive healthcare data with concerned healthcare professionals. These approaches aid in the systematic management of data and guarantee that no patient health-related information is ever lost. Fog computing may be used to relocate data processing tasks from the cloud to sensors, improving the healthcare sector's efficiency and lowering latency. Because of the reduction in latency, doctors and other healthcare providers may be able to provide patients with real-time healthcare. Due of their role in data processing, fog nodes are crucial in fog computing [13]. These nodes could also provide other services like storage, security, and monitoring. The fog node can continuously monitor the services and resources offered by the healthcare network [14]. A variety of healthcare gadgets are now being produced by manufacturing businesses using AI and IoT technologies. These gadgets are being utilized to gather beneficial real-time patient health data. The collected data is then kept in the cloud and examined by medical specialists to determine the best course of action for the patient's health. Patients visit their chosen doctors using modern healthcare equipment to get advice in real time about routine nutrition, exercise schedules, etc. AI-based wearables can communicate real-time patient health data to clinicians in addition to collecting information about patients' health. The goal of the project is to combine IoT and AI technologies to create cutting-edge algorithms that will enable several simultaneous tasks to be completed by a single wearable device. Due to the development of AI, ML, and IoT technologies, collaboration between healthcare organizations, hospitals, and healthcare practitioners is now feasible. Personal healthcare (PH) uses AI tools to collect pertinent patient health data. In order to make informed judgments, the acquired data is further examined to identify distinct patterns connected to the patients' behaviors [15].

The objectives of the chapter are:

1. To showcase several IoT and AI-based technologies that are applied for the creation of a successful healthcare system;
2. To explore various IoT and AI application areas in healthcare informatics;
3. To discuss several difficulties and problems that arise in the creation of an IoT- and AI-based healthcare system;
4. To showcase several healthcare designs based on AI and IoT technology;
5. To highlight the advantages of integrating AI and IoT technology in the healthcare industry;
6. And, to demonstrate a number of benefits of IoT and AI technology integration in healthcare systems.

4.1.1 Organization of chapter

The use of various healthcare technologies to automate the process of healthcare informatics is discussed in Section 4.2 of this chapter. The numerous noteworthy applications of healthcare IoT are categorized in Section 4.3. The seamless operation of different healthcare informatics operations is enabled by a variety of IoT technologies, including IoT-based sensors, cloud computing, fog computing, and Wireless Body Area Networks (WBANs). These technologies are discussed in Section 4.4 explores various IoT technologies such as IoT-based sensors, cloud computing, fog computing, and Wireless Body Area Networks (WBANs) that are responsible for the smooth functioning of various activities of healthcare informatics. Section 4.5 presents the IoT-based healthcare architecture and a comparison of different architectures based on parameters. Section 4.6 showcases several IoT challenges affecting various functionalities of healthcare industries. Section 4.7 explains the future scope of IoT in healthcare informatics. Section 4.8 elaborates on some significant applications of artificial intelligence in the healthcare sector. Section 4.9 discusses some important artificial intelligence technologies that are being widely used in healthcare. Section 4.10 focuses on various AI-based healthcare architectures and also discusses how AI technology facilitates various entities of the healthcare system to coordinate their routine activities. Section 4.11 focuses on various challenges and the future scopes of AI technology in healthcare informatics. Section 4.12 presents a case study based on AI-based healthcare. Section 4.13 presents various examples of AI and IoT-based healthcare systems. Section 4.14 elaborates future scope of AI in the healthcare sector. Section 4.15 presents the integration of AI and IoT technologies in the healthcare sector. Finally, section 4.16 concludes the chapter and provides future directions to the researchers to explore new ideas for developing advanced healthcare architectures.

4.2 Advancements in healthcare technologies

To revolutionize healthcare industries, healthcare informatics integrates data, information, communication, and medical services. The exponential advancement in recent healthcare technologies involves innovations in all the disciplines of healthcare. The inclusion of AI, blockchain, telemedicine, robotics, and automation transform the significant areas of healthcare to facilitate healthcare professionals and patients. The technologies that spark automation in healthcare industries are explained in detail in the subsequent sections.

4.2.1 mHealth

mHealth technology introduced various health applications that can monitor the health of the patient and provide real-time health data. These applications are capable to streamline the entire process of recent healthcare systems. The patient can utilize these applications in their daily life to schedule appointments and communicate with the doctors. These applications are being used as a tool through which the health of the patient can monitor even at remote locations. The information collected by these applications can be analyzed by the healthcare professionals to provide emergency medical services whenever desired [16].

4.2.2 Telemedicine

Telemedicine technologies enable virtual communication among patients and their designated doctors. These technologies empower healthcare professionals to treat their patients virtually through video conferencing without any physical visit to the hospitals and clinics. It is a booming technology that helps millions of patients and their caregivers living in remote areas. The emergency services such as transport and ambulance services can be provided to the patient at remote locations. In the COVID-19 pandemic in 2020, the telemedicine service has been widely used by doctors and patients. These services enable doctors and other healthcare workers to protect the human being from any kind of infection [17].

4.2.3 Electronic health records

Electronic health records play a significant role and support healthcare communities to diagnose patients based on their present and past health records. The manual entry of such data can mislead the doctors and affect the strong coordination among healthcare workers. The holistic picture of patient health can be analyzed by electronic health records (EHRs) that can be helpful to minimize the health discrepancy and streamline the entire healthcare process. The healthcare processes such as e-prescribing and telehealth are good examples of EHR. With the advancement of EHR, patients can make their own decisions regarding their health issues. These electronics can be accessed by the patient to know about their health status through the healthcare website developed under the guidelines of healthcare professionals.

4.2.4 The cloud and data analytics

Healthcare devices are generating a massive amount of healthcare data and other sensitive information that need to be stored in a database. The stored EHRs and other health information can be processed in the clouds. In recent years, cloud and data analytics provide trusted mechanisms to manage and sharing of data among healthcare professionals and patients. The healthcare communities are trying to centralize the healthcare data using cloud computing and data analytics mechanisms to enhance the accuracy of the data obtained from various medical devices. In healthcare informatics, data analytics programs and cloud computing provide innovative and faster optimal healthcare solutions. Healthcare professionals can get various opinions from the data analytics programs quickly to find out the optimal outcomes. The high speed in accessing the relevant information and enhanced decision-making capabilities can improve the entire process of the healthcare industry.

4.2.5 Wearables

In recent years, IoT-based wearable devices are being used in healthcare to collect the health record of the patient for future use. These devices empower the healthcare process and prevent the human being from chronic disease. Wearables such as smartwatches, and fitness tracking devices are some of the well-known wearable healthcare devices among human beings. In these devices, various significant features are introduced to enhance the capabilities of the healthcare professional to provide accurate diagnoses to the patient. The alarming system incorporated in these devices sends the information related to the health issues to the doctors so that adequate emergency services can be provided to the patients. Moreover, the information produced by these wearables can be in analysis and decision making [18]. The doctors may become active in advance based on the information emitted by these devices to provide immediate medical attention. Figure 4.1 shows different kinds of wearables that are widely used in healthcare.

4.2.6 Artificial intelligence

AI provides new capabilities to the healthcare providers to find out new ways of diagnosis so that potential risk factors in various healthcare activities can be

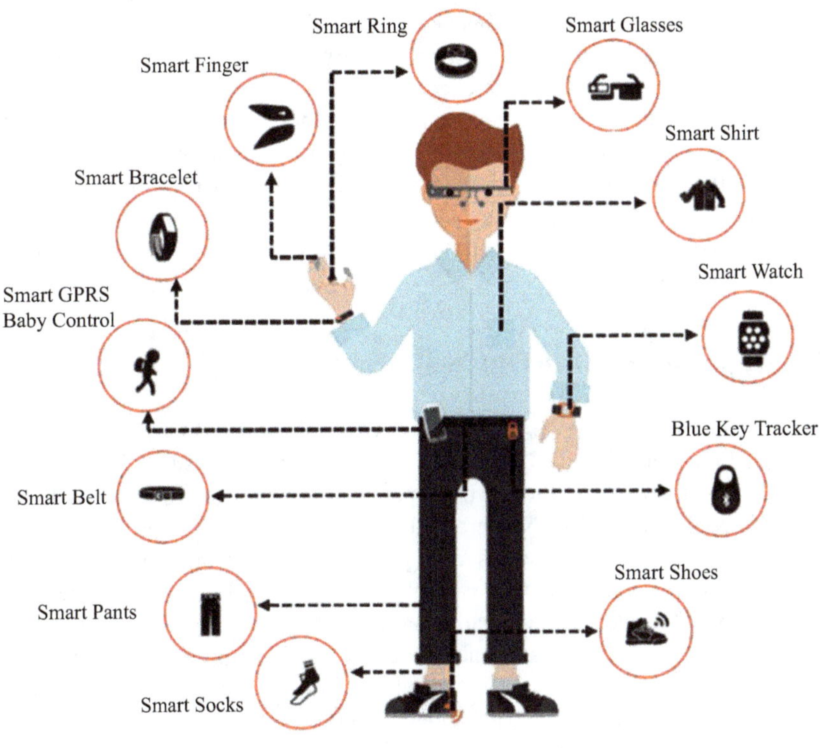

Figure 4.1 Various types of wearable devices

minimized. AI-based applications can be used to forecast the health of the patient based on past health data. The success of surgeries and other treatments can be improved by analyzing the complex patterns of the health data and symptoms of the disease. The human brain is not so capable in many areas of healthcare for accurate decisions based on historic healthcare data. The AI algorithms provide much deeper insight to the doctors on the patient disease to take significant steps to cure them [19].

4.2.7 Robotics

Nowadays AI, ML, and robotics are contributing their significant roles in governing various administrative activities related to the healthcare industries. In healthcare robotics equipped with AI technologies can act as surgical assistants to perform various activities during surgeries. In Sioux Falls, S.D., "Xena" which is a robot is being used as an employee of the hospital to perform various healthcare operations such as disinfect operating centers and extinguish superbugs. This robot played a significant role in the COVID-19 pandemic [20].

4.2.8 Blockchain

Blockchain technology is also being used in healthcare informatics across the world. In a business network, this technology facilitates keeping track of the transactions and assets for any kind of security breach. The tangible and non-tangible assets can be virtually tracked on a blockchain network [21]. The cost-effective use of blockchain technology can cut down the cost involved in the establishment of security mechanisms. Blockchain technology can be utilized to enhance EHR management and the insurance claim process [22]. Sensitive health information can be more secure by incorporating blockchain technology into the healthcare network. The patient can have greater control over the sharing of their health information [23].

4.3 Applications of IoT in healthcare sector

IoT applications have grown quickly in recent years to help healthcare professionals and medical experts. The idea of the Healthcare Internet of Things (HIoT) in the healthcare industry offers tremendous assistance for creating IoT-based healthcare applications [24]. Numerous theories put forth by researchers help programmers design healthcare apps and provide potential answers to problems. Healthcare workers may conduct a variety of tasks more efficiently with the help of user-friendly wearable medical gadgets that are offered by the manufacturing businesses. Figure 4.2 illustrates the procedure for storing data in the cloud and retrieving it to examine the patient's health. The IoT-enabled medical devices collect diverse patient health data and transfer it to the cloud for storage. The commercial healthcare devices are described in the subsections below. These gadgets can be used by the healthcare sectors to offer patients improved medical

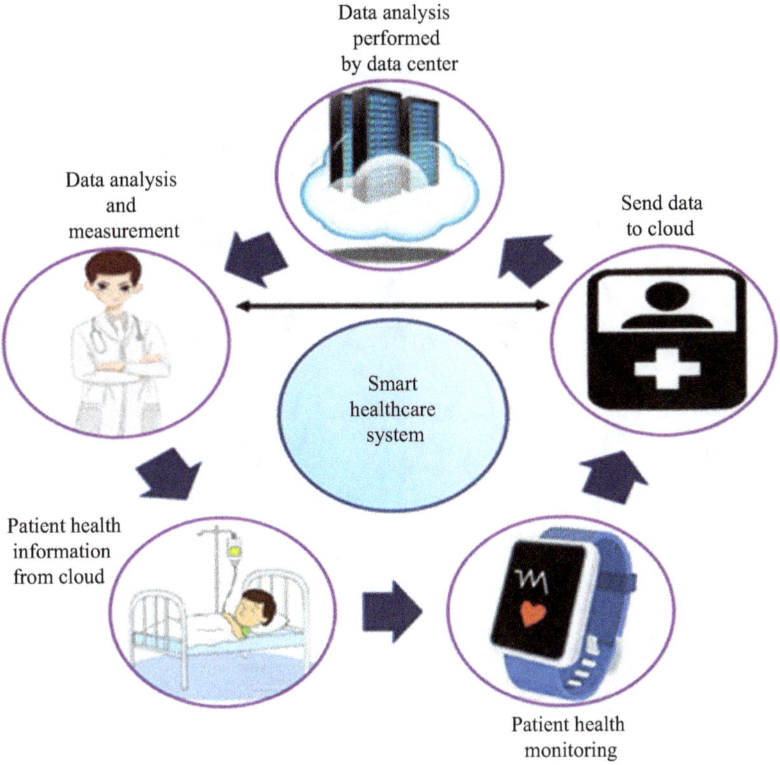

Figure 4.2 Process of information storage and retrieval in healthcare

treatment. The classification of various HIoT applications based on single and multiple circumstances is shown in Figure 4.3 [25,26].

4.3.1 Electrocardiogram monitoring

The electrocardiogram (ECG) monitoring system in the healthcare system detects changes in the patient's heart rate rhythm. These wearable Internet of Things-based gadgets function as an early-detection warning system for various heart problems. Drugs, poisons, and venoms can influence these irregularities in the patient's health. Researchers have discovered important uses for IoT-based ECG monitoring systems in the early diagnosis of cardiac problems in the literature [27]. A small-scale ECG monitoring system based on a neural network was proposed by Meng *et al.* [28].

4.3.2 Temperature monitoring

To identify other dangerous diseases, healthcare experts use a variety of temperature monitoring equipment. The body temperature fluctuating widely is a sign of sickness,

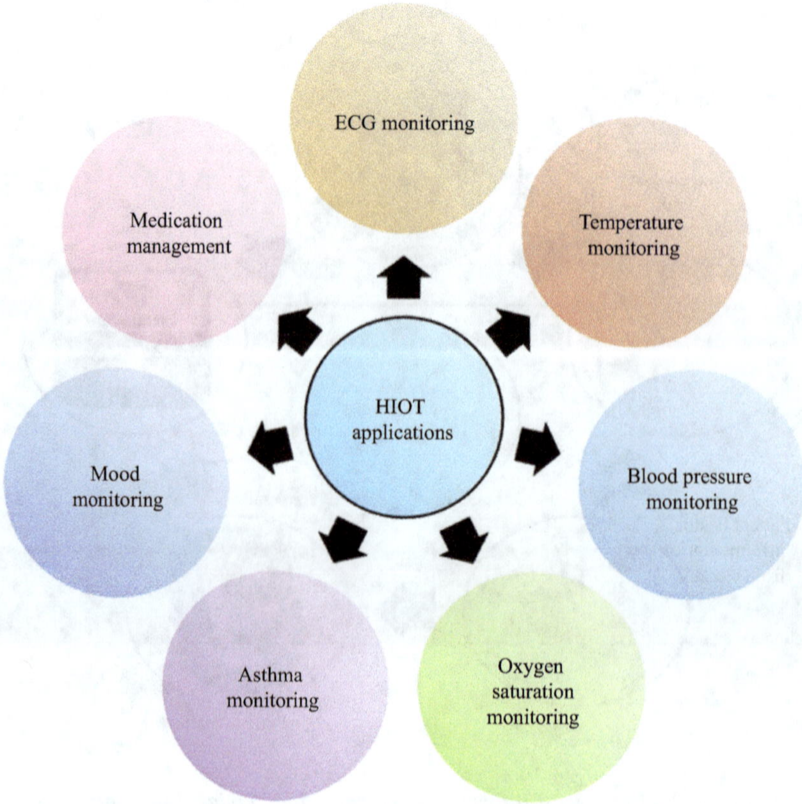

Figure 4.3 Applications of HIoT in healthcare

including trauma, sepsis, and other conditions. The doctors can draw conclusions about the patient's health status, thanks to routine body temperature monitoring. Early diagnosis of temperature changes helps save the patient from contracting another illness. Typically, a thermometer is connected to the patient's mouth, ear, and rectum in a typical temperature monitoring system. The patient may experience pain with the traditional temperature monitoring device. Multiple patients use the same thermometer, which increases the risk of infection, which is a constant concern. In order to address this problem, the researchers have suggested a variety of temperature monitoring tools. Tattoo temperature sensors are one type of sticky thermometer that has recently been widely accessible. These wearable thermometers are simple to use and can take temperature readings repeatedly. In the event of repeated peel-off, the adhesive substance employed in these devices may irritate the patient's skin.

4.3.3 Blood pressure monitoring

To identify any ailment, the blood pressure monitoring process is necessary. The blood pressure measurements must be recorded by at least one medical practitioner in the most common BP monitoring protocol. Blood pressure monitoring has undergone a significant transformation as a result of the incorporation of IoT technology. The wearable cuffless device, which can be used to measure both systolic and diastolic pressure, may be utilized by the patient. The data obtained with these technological devices can be kept in the clouds. Health specialists have examined the authenticity of the data collected from these devices and confirmed that it is correct in every way [29].

4.3.4 Oxygen saturation monitoring

Pulse oximeters are a common piece of medical equipment used nowadays to check a patient's oxygen saturation. Researchers have suggested a tissue oximeter that may be used to track a patient's pulse, fluctuations in heartbeat, and blood oxygen level in the literature [30]. The gathered data is then transmitted to the medical servers using cutting-edge technologies like Wi-Fi and Zigbee. Based on the information gained through these communication technologies, healthcare practitioners make a variety of judgments. Saha *et al.* [31] suggested an alarm system that notifies patients in real time when their oxygen saturation levels exceed certain levels.

4.3.5 Asthma monitoring

The primary cause of asthma illness today is the exponential rise in environmental pollutants. The patient experiences breathing difficulties as a result of this chronic condition because it narrows the airways. Shortness of breath and chest discomfort are common symptoms of asthma in patients. An inhaler or nebulizer can be used to treat an asthma attack, which can happen at any moment. These reasonably priced medical equipment options are readily available on the market. This condition needs to be treated at an early stage, which calls for effective real-time monitoring. Several IoT-based asthma monitoring systems have been developed in recent years by researchers and industrial enterprises. A smart HIoT solution has been made available to asthma sufferers [32] and the suggested gadget can track respiratory rate utilizing IoT-based smart sensors. These sensors gather data, which is then saved in the cloud for further study by medical experts and other caretakers. Effective decisions regarding the patient's health may be made using the examined data. A real-time respiratory rate monitoring alarm system that integrates a temperature sensor had been presented in the literature [33,34]. By measuring the difference in air temperature between inhaled and expelled breaths, the temperature sensors calculate the change in respiratory rate. The information gathered by these sensors is processed before being transferred to a healthcare facility or being made available for viewing on a hospital's web server.

4.3.6 Mood monitoring

Various IoT devices are being employed in the healthcare sector to track the patient's mental health. Significant real-time information on the patient's mood is provided by these gadgets. Due to the fast shift in the workplace, people nowadays are more likely to experience mental illnesses including stress and despair. An effective method of mood monitoring based on the CNN network was suggested by Fei *et al.* [35]. The suggested work keeps track of the patient's mental health using many categories, including happy, enthusiastic, sad, calm, disturbed, and furious. Similar to this, deep learning and fuzzy clustering may be used in literature [36] to measure a person's mood in real-time. The suggested application lists a number of crucial characteristics that medical personnel might utilize to gauge a patient's pleasure. The wearable technology that can assess a person's mental health based on emotions like anger, tension, panic, and melancholy can be used to assess the drivers' subconscious states.

4.3.7 Medication management

Healthcare workers should carry out their daily tasks in accordance with the established rules and guidelines. The negative impact on the patient's health might increase if the stated policies and timetable are not followed. Because elderly folks sometimes do not understand the prescriptions given by their caring doctors, medication disobedience is frequently seen in this population. Any age group of patients must adhere to the timetable and directions of medical specialists. Numerous IoT-based apps are being employed in the healthcare system to monitor patient medication compliance and remind them to take their medications according to schedule [37]. The medical applications that are used to monitor various health factors can record the conversations between the doctor and the patient. The server-stored health information may be utilized to examine changes in blood glucose, body temperature, oxygen saturation, and other parameters. Women undergoing IVF therapy can utilize a medical app called "Saathi" that was created by researchers. This software offers the ability to connect with caretakers and can track the ingestion of medications. The other IoT-based apps employ fuzzy logic to automatically alter the timing and amount of medication while monitoring the patient's body temperature using a temperature sensor.

4.4 IoT technologies in healthcare informatics

The standards are being used to mimic the disputes in a variety of IoT applications. The IoT network's embedded sensors are utilized to gather data, which is then processed further using cloud computing. Fog computing is used to handle short-term data, and a wireless body area network facilitates connection between integrated sensors positioned close to patients and other network components like routers and gateways. The following list includes the most recent IoT technologies being employed in healthcare informatics.

4.4.1 Sensors

The Internet of Medical Things (IoMT) is a phrase that has gained popularity in recent years due to its ability to connect different medical devices that are utilized in both clinical and non-clinical healthcare systems [38]. IoMT makes it easier to monitor a variety of patient health-related parameters in a clinical healthcare setting, such as body temperature, oxygen saturation, blood pressure, and heart rate. In order for doctors to make successful judgments based on the displayed health data, the sensors, which are essential IoMT components, are employed to present them with diverse health information. Remote monitoring of various patient healthcare services is made possible by the integrated sensors on IoT-based healthcare systems. By analyzing the blood chemistry, implanted sensors may be utilized to keep an eye on the patient's health. The micro particles that may be injected into the skin's dermis layer enable these sensors to work. The change in color of the micro particles makes it simple to see how the blood chemistry has changed. Only with the aid of a specific light source and the naked eye can the transition be clearly seen. The IoMT, on the other hand, may be utilized in a non-clinical setting to track the whereabouts of medical professionals such as physicians and other staff members, ambulance facilities in case of medical emergencies, and real-time data on logistics in the healthcare industry. Any device's appeal hinges on two factors: its cost-effective design and the accurate health information it generates. In Figure 4.4, the sensors utilized in the healthcare industry are divided into two groups i.e. clinical and non-clinical [39]. Wireless sensors and RFID tags, as well as other automatic identification and data collecting tools, are widely employed in healthcare systems. These devices have several security flaws because of their limited processing power [40,41]. To protect these devices from threats and assaults, certain low-cost security solutions have been put out in [42,43].

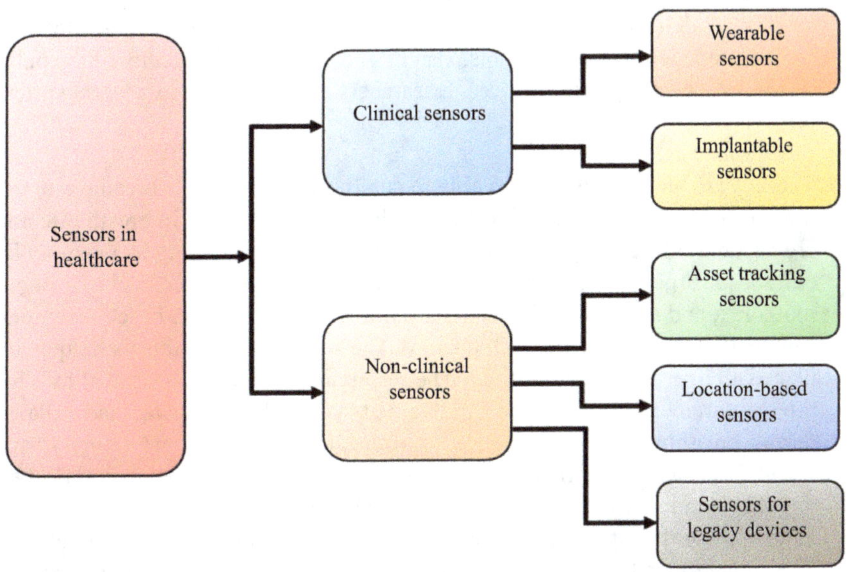

Figure 4.4 Different types of sensors in healthcare

4.4.1.1 Wearable sensors

Smart technologies that are being utilized to monitor various health vitals include wearable technology. These gadgets have clever sensors that can gather information on a variety of medical conditions. The patient's body or clothing can be attached to these devices to gather real-time health information. Numerous brands with various wearable device features are offered on the market, including Fitbit bands, MI bands, Samsung Galaxy Fit E, Fossil Gen 5, Apple smart Watch 5, etc. [44]. These gadgets have sensors that detect the patient's private health information and transmit it to the appropriate doctors or medical professionals so they may make informed judgments. These sensors may also be used for functions such as (a) sensing blood pressure and oxygen levels, (b) sensing breathing rate, (c) measuring temperature, and (d) sensing temperature and pulse rate [45].

- *Pulse sensors*
  Heart attack, cancer, HIV, diarrhoea, and other health-related eventualities are all monitored by the sensors that are employed to detect the patient's pulse. Humans may track their fitness with wrist sensors like the Fitbit PurePulse and TomTom Spark Cardio [45]. The businesses that make these gadgets stated that the wrist sensors are not ideal for identifying human health problems. Only the real-time health status can be measured by these devices, and the data they provide can only be utilized to help improve the health condition. The patient's blood pressure inside the body is another sensor that, like these ones, may be used to detect pulse rate. The photoplethysmographic (PPG), ultrasonic, and radio frequency (RF) sensors are the most well-known types of sensors. It is advised that PPG sensors can deliver accurate information about the patient's pulse in the literature [46]. To reduce the impact of noise that has been seen on the quality of the pulse signals produced by these devices, researchers are attempting to adopt a variety of already used strategies.
- *Respiratory rate sensors*
  These wearable sensors may be able to determine the patient's breathe rate per minute. These sensors show information based on variations in breath rate and are capable of identifying a variety of lung diseases, including as pneumonia, chronic obstructive pulmonary disease (COPD), lung cancer, asthma, pneumothorax, and atelectasis. In healthcare informatics, a contact-based technique can be used to find breathing frequency. The sensors used in this technique are placed all over the subject's body. The respiratory airflow technique allows for the measurement of the volume and velocity of air that is inhaled and exhaled during breathing using a variety of sensors. For measuring air velocity, different flow meters (DFs), turbine flow meters, hot wire anemometers (HWAs), and fiber-optic-based flow meters have all been widely utilized [47]. The patient's health status can also be determined by listening to their breathing. The sound made by the patient's throat while breathing is captured in order to better understand the inspiration and expiration phases of breathing. Acoustic sensors, such microphones, can be used to collect data on the fluctuation in air

pressure brought on by different sound waves in order to do this [48]. The nasal sensor, microphone, stretch sensor, ECG-derived respiration (EDR), and other sensors can also be used to measure the respiratory frequencies [45].

- *Body temperature sensors*
  To measure the body temperature of the patient various body temperature sensors can be used in healthcare systems. In the healthcare sector, professionals and health workers are adopting these sensors to check fevers, skin temperature, heatstroke, hypothermia, and more [39]. The thermistor-based sensors are used to sense the temperature of the body up to a certain range and the level of error is very low. These body temperature sensors can be used to improve the process of traditional healthcare monitoring systems. The reliability and accuracy of the sensed data obtained from these sensors are based on their position and closeness to the human body. A team of researchers from China and Singapore developed a patch that acts as a thermometer and can be glued to the human skin. The patch continuously measures the temperature of the human body at skin level and provides accurate data over a continuous period [49]. The other type of body temperature sensor is assembled with the cloth of the patient [50,51]. In comparison to the tattoo-like sensors, the sensors that are embedded in the cloth are more desirable due to the inconvenience of adhesive that is being coated at the backside of the patch.
- *Pulse oximetry sensors*
  These sensors provide the data related to the oxygen level in the blood. The data obtained from these sensors can be used to diagnose hypoxia. In the measurement, the PPG signals are analyzed that are made up of two LEDs. The light emitted from these LEDs penetrates the skin and is captivated by the hemoglobin. The photodiodes are used to measure the light that is not being captivated by the hemoglobin of the blood [45]. In a recent development, researchers have created a pulse oximeter that can be placed in the ear to measure blood oxygen [52]. Furthermore, the sensors that can be worn on the wrist are mostly preferable for the human being to measure the oxygen in the blood and the temperature of the body. In the next section, various technologies that pertain to cloud computing are discussed.

4.4.2 Cloud computing

In recent years, cloud computing contributed a significant role in IoT-based healthcare informatics [53]. To minimize the risk and to improve the services of the healthcare systems cloud computing models are capable of sharing sensitive healthcare information with the concerned healthcare professionals.

These models support the management of data in a structured manner so that the information related to the health of the patient cannot be lost [54]. Cloud computing provides various configurable resources that can be utilized by the user positioned at a remote location. The organizations that are not financially sound can invest their capital to purchase resources over cloud-like storage. These resources enable the organizations to perform their routine activities optimally. Various modes are available to deploy the cloud. The other significant functionalities that are being provided by cloud computing are self-manageable, robustness, and ubiquitous access. Figure 4.5 shows a five-layered IoT architecture. In the subsequent sections, the usefulness of fog computing in healthcare applications has been discussed.

4.4.3 Fog computing

The data processing processes may be moved from the cloud to sensors utilizing fog computing to increase the healthcare sectors' efficiency and reduce latency. The decrease in latency makes it possible for physicians and other healthcare professionals to offer patients real-time medical treatment. The fog nodes are crucial to fog computing since they handle much of the data processing. These nodes can offer a variety of different services, including storage, monitoring, and security. The services and resources that the healthcare network has to offer may

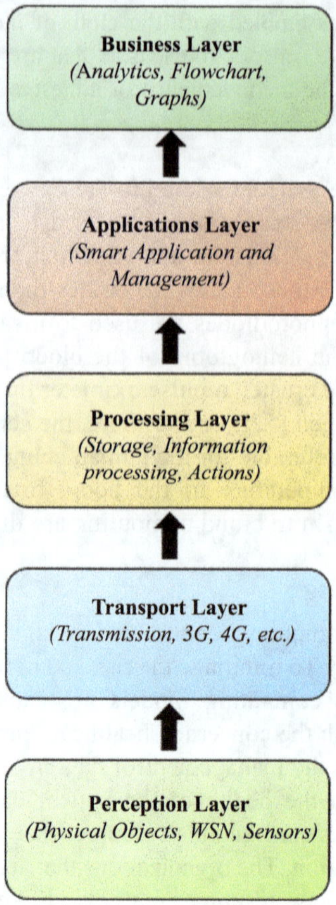

Figure 4.5 Five-layered IoT architecture

be continually monitored by the fog node [55]. The fog node can also carry out associated tasks including data analysis that can lead to an emergency solution. Facilities that offer the capability of data segregation and management are typically utilized to store sensitive healthcare data. The security facilities guarantee the privacy, consistency, and integrity of medical data. Fog computing may be implemented at the networking and perception layers in the IoT architecture. Fog computing also describes how edge computing operates and the numerous functionalities between endpoints and cloud data centers.

4.4.4 Wireless body area networks

Numerous issues are being faced by the healthcare industry as a result of the population's exponential increase. Recent information technologies have created wireless body area networks (WBANs) to address the new difficulties facing the healthcare sector. The patient may have died as a result of the lethal illness and a delayed diagnosis. Therefore, a quick and cost-effective health monitoring system must be used in the healthcare sector to deliver medical services to the patient at the earliest stages of the condition. The WBANs offer a number of apps that may be used to identify the most practical and effective treatment options. The patient does not need to stay in the hospital for an extended period of time after employing these technologies. WBAN applications are utilized in healthcare for a range of purposes, including blood glucose monitoring, cancer diagnosis, and the detection of cardiovascular disease [54,84].

4.5 IoT-based healthcare architecture

In some medical treatment, the patient needs immediate medical facilities so that he can easily fight any major disease like heart disorder. For the treatment of such diseases, the patient should get real-time medical facilities on a large scale in very little time. In a cloud computing environment, latency can be observed in three major steps such as the transmission of sensitive information to the cloud, data processing on the cloud to extract useful information, and generating the response to the raised queries. High latency in these steps is not acceptable while dealing with medical activities that require prompt action. In the field of healthcare informatics, a number of IoT-based healthcare architectures are suggested in literature [55–70]. In these healthcare architectures, different recent technologies such as IoT, fog computing, and cloud computing are being used to secure healthcare data and its management. Figure 4.6 shows a simple fog-based architecture to manage various activities related to healthcare informatics. The architecture explores the issues related to the data latency that is usually observed in cloud-based solutions. This architecture is best suited for healthcare applications that are time-critical and also support other applications that require real-time response. In the first layer i.e. sensing layer of the architecture, various sensors are used to monitor health-related activities and to measure the health vitals.

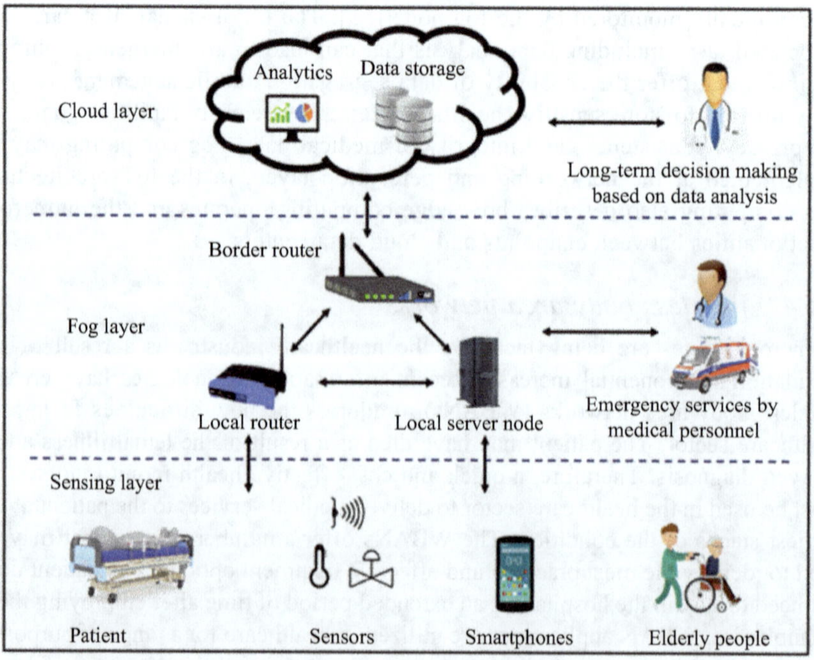

Figure 4.6 Fog-based architecture for IoT-based healthcare applications

The incorporated sensors in the architecture are used to transmit the sensitive health data of the patient to the device placed at nearby locations such as smartphones, Arduino [68], and Raspberry Pi [69]. The sensitive data and useful health information are further transmitted to the adjacent layer of the architecture i.e. fog layer. After that, the fog layer which is the second layer of the architecture comprises the local gateway and transitory servers in which the data can be analyzed and strained depending upon the necessities. To secure the entire architecture and to ensure the privacy of the system various policies are introduced at the local gateway. The data after the transformation is further transmitted to the subsequent gateway to be stored on the cloud layer. To manage the traffic in an efficient way, a firewall is incorporated at the local or border gateway that is usually available at the sensing and fog layer. The architecture guarantees secure communication through available communication channels among various network devices.

At the first layer i.e. cloud layer, the sensitive healthcare data is stored on the servers so that it can be further utilized by the healthcare professionals for effective decision making. The efficiency of the architecture can be observed based on different parameters. In Table 4.1, some parameters are selected to present a comparison of the existing architectures. The parameters are measured based on some values such as high, intermediate, and low. To measure the complexity of the architecture, the number of layers and available modules within them are

Table 4.1 Comparison of architectures based on parameters

References	Number of layers	Complexity	Data reliability	Real-time application support	Security
Cerina *et al.* [59]	4	Reasonable	Minimum	Reasonable	High
Verma and Sood [60]	5	High	Minimum	Reasonable	Reasonable
Azimi *et al.* [61]	3	Minimum	Minimum	Reasonable	High
Kumar *et al.* [62]	2	Minimum	Minimum	Minimum	Low
Nandyala and Kim [63]	4	Reasonable	Minimum	Reasonable	High
Plageras *et al.* [64]	3	Minimum	Minimum	Minimum	Reasonable
Villalba *et al.* [65]	2	Minimum	Minimum	Minimum	Reasonable
Mahmud *et al.* [66]	3	Minimum	Reasonable	Reasonable	Reasonable
Debauche *et al.* [67]	3	Reasonable	Minimum	Reasonable	Reasonable
Paul *et al.* [68]	3	Reasonable	Minimum	Reasonable	Reasonable
Awaisi *et al.* [69]	3	Reasonable	Moderate	Reasonable	Reasonable
Abdelmoneem *et al.* [70]	4	Reasonable	Moderate	Reasonable	Reasonable

computed. The reliability of the data at the fog layer can be measured based on the number of clusters. In many existing architectures and related healthcare applications, the reliability of data is not a major concern at the fog layer. To measure the real-time application support, the values can be obtained that are related to the presence of the fog layer and its performance. Similarly, security measures in the architecture can be obtained from the number of available layers within the architecture. From the above literature [59–70], it can be observed that the existing architecture does not pay much attention to security issues. In the existing healthcare systems, the incorporation of IoT is still in its immature stage. Hence, various challenges related to the healthcare systems and related architecture need to be focused more on by the research communities and healthcare industries. The major challenges and possible solutions related to the existing healthcare architectures are discussed in the subsequent section.

4.6 Challenges of IoT in healthcare

The researchers have carried out a lot of significant research in which the major concerns related to the healthcare IoT applications have been focused. Apart from this, many such challenges have not been addressed yet. This section showcases several challenges that are affecting various functionalities of healthcare industries. Moreover, possible solutions have also been addressed to resolve the issues.

4.6.1 Fault tolerance

The dependability of IoT-based healthcare systems is affected by the massive exchange of data across the cloud and processing layers by the sensors and communication nodes used in the system. The excellent and reliable functionality of these devices can make their unique contribution in emergencies. Gia *et al.* [73] proposed an IoT-based architecture that enables fault tolerance that can be achieved through backup routing among the nodes. The proposed architecture also supports some advanced services so that proper connectivity can be achieved while connections fail.

4.6.2 Latency

To improve the efficiency of the applications concerning healthcare, the network latency at two significant layers such as the transport layer and the communication layer should be reduced. The complexity and number of functionalities also effect of latency of the applications that are being used in healthcare sectors [71]. There are several real-time healthcare applications in which latency can have an adverse effect. Farahani *et al.* [75] discussed the architecture of the IoT healthcare ecosystem that elaborates the usefulness of the IoT in healthcare and medicine. The architecture focuses on patient-centric healthcare services in which all the components of the healthcare system can be integrated.

4.6.3 Energy efficiency

The batteries of the sensors are crucial to the real-time emergency services provided by healthcare systems. The type of medical applications determines how much electricity is used through the network. Between real-time services provided by the healthcare system and when the sensors' batteries run out, catastrophic circumstances may emerge [73]. A wearable sensor node was proposed by Wu *et al.* [77] to build the WBAN.

4.6.4 Interoperability

With the advancement of IoT applications, various standards have been established for network management. The inclusion of various IoT standards and their conflicts in a wide range of domains may lead to problems of interoperability. The interoperability hampers due to the strict regulations and standards that are required to be followed for the proper management of various functions related to the healthcare systems [72]. Touati *et al.* [78] proposed a real-time health monitoring system that highlights the problem of interoperability among various integrated nodes integrated within the network. The middleware can be used to fetch the complex details so that the problems associated with the interoperability can be minimized up to some extent [53]. Jabbar *et al.* [79] proposed an IoT-based semantic interoperability model (IoT-SIM). In healthcare systems, the proposed models can be utilized to ensure interoperability among a wide variety of IoT-enabled devices.

4.6.5 Availability

Doctors and other healthcare workers should have access to patient health data around-the-clock since this enables real-time monitoring, wise decision-making, and emergency care services [74,76]. The absence of the node and network might be the reason for the patient's passing at the perception layer and processing layer of the network. A European Union (EU) sponsored project strengthened the system so that the load over the connection can be properly handled in order to manage the communication data and to give a high degree of performance [78]. Cloud computing, big data, and other contemporary technologies are being employed to guarantee the security of the built infrastructure at the processing layer. The integrated cloud computing contains a variety of risks and susceptibilities to lessen the effectiveness and availability of the IoT-based healthcare system [21,79].

4.6.6 Servicing and maintenance cost

With the exponential growth in recent technologies, it is required to update the software, hardware, and other HIoT-based devices at regular intervals. The continuous up-gradation and service of the healthcare system require high maintenance costs. The IoT-based healthcare system is equipped with several medical devices that generate healthcare data in real-time. The high up-gradation and maintenance cost is a big challenge to the healthcare industries and end-users. The arrangement of the finance for the overall up-gradation of resources is mainly an issue for the small-scale healthcare industries. The cost of servicing and maintenance can be significantly reduced by adopting low-cost sensors and other IoT-based equipment.

4.6.7 Standardization

Manufacturing companies that specialize in healthcare informatics create a wide range of products that aim to enhance the patient experience. These products are designed to improve the overall quality of care and make healthcare services more accessible and efficient for patients. The government has defined some standards and protocols that are needed to be considered while designing these products. In the design process, various communication protocols, data validation, and interfaces need to be focused on by the selected group of healthcare professionals. The E-medical reports generated by HIoT devices are required to be validated according to the standards defined at the time of the design process. To define the standards for medical equipment, various organizations and bodies should work with the researchers and professionals to collaborate on various research related to healthcare. The other potential solutions, conflicts, and challenges can also be resolved by collaborating the research activities.

4.6.8 Privacy and security of healthcare data

The potential of cloud computing is used by healthcare systems to monitor patients in real-time and store health-related data for study. Cyber-attacks may be possible when cloud technologies are included into healthcare networks. Doctors and other healthcare professionals may be misled in their ability to make wise judgments if sensitive patient health data has been tempered. The data provided by IoT-based devices forms the basis of the patient's complete treatment procedure. The leakage of data on the IoT-based healthcare network can hamper the treatment process and delay medical services. In the IoT-based healthcare system, several users are connected and any kind of laxity can hamper the privacy of the patient. The more secure HIoT system can be implemented through advanced lightweight algorithms and cryptography techniques [80,81].

4.6.9 Environmental impact

The manufacturing companies concerned with healthcare are using various biomedical sensors with semiconductor-rich devices. In the manufacturing of such devices, various earth metals and toxic substances are being used. The heavy use of these substances can create an adverse effect on the environment. Therefore, proper standards and policies are required to be established to control the manufacturing process of medical devices. The manufacturing companies and researchers can make significant contributions toward the use of biodegradable materials to manufacture IoT-based sensors and other healthcare equipment.

4.7 Future scope of IoT in healthcare

The IoT has quickly acquired acceptance around the world. In addition, improvements in AI and ML have simplified IoT device automation. To enable proper automation, IoT devices are typically combined with AI and ML algorithms. The IoT has expanded its use across several healthcare businesses as a result. One of the most helpful technologies in the healthcare industry is the IoT. It offers a range of options and provisions for improving healthcare, particularly for patients, doctors, and researchers. Real-time services include smart diagnostics, patient management, and wearable technology for tracking health. Additionally, the widespread adoption of IoT devices relieves unneeded stress on healthcare workers' daily tasks. Healthcare devices may send doctors immediate access to patient health data through a secure network. This makes it possible for doctors to diagnose patients in far-off places. The key advantages of IoT in healthcare informatics are described in the subsections below.

4.7.1 Minimization of error

In healthcare informatics, the use of IoT-based devices can minimize human error in diagnosing patients. The patients can wear various medical IoT-based devices to send real-time health information to the doctors to get proper treatment. The IoT-based gadgets provide 24/7 diagnosis and provide a clearer picture of the patient's health to the physician. The real-time alarming system introduced in these devices can minimize the risk and error in emergency medical services.

4.7.2 Cost-effective treatments

Manual diagnosis takes time and involves the use of a variety of expensive medical equipment and additional hospital costs. As a result, the total cost of therapy significantly increases. The unnecessary hospital expenses can be reduced using IoT devices. Furthermore, medical services can be provided at the doorsteps of the patients to diagnose the patients which reduce the costs of the treatment and congestion in the hospitals.

4.7.3 Healthcare services at remote areas

The availability of physicians, particularly specialists, in distant areas is one of the main issues in the healthcare sector. IoT provides various ways through which the care of patients is possible in the absence of physicians. The wearable IoT-based devices can send data to the physical without any visit of the patient to the doctors. The appropriate health information is transmitted to the physician in real time to the concerned department of the healthcare industry for further analysis. In this manner, the IoT is assisting healthcare informatics in providing appropriate and reliable treatment to the patient.

4.8 Applications of AI in healthcare

With the advancements of AI, it has been observed that various AI-based tools facilitate human beings to accomplish their work on time in an efficient manner. This technology is not being used to replace the work of healthcare professionals and other healthcare workers. AI always supports healthcare staff and enables them to perform a variety of healthcare activities and other administrative tasks such as clinical documentation, patient diagnosis, and image analysis. Various AI-based medical devices are being used to automate the process of patient m-health monitoring in healthcare informatics. These devices are also being used to sense sensitive health information and increase the decision-making and thinking capabilities of healthcare professionals. In the subsequent sections, some significant applications of AI in healthcare have been discussed.

4.8.1 Support in clinical decisions

In healthcare sectors, healthcare professionals need to consider every piece of information at the time of diagnosis of their patients. The unstructured healthcare medical notes or records prepared by the doctors can mislead them during diagnosis. The complicated surgeries by doctors based on unstructured healthcare records can put the life of the patient at the risk. The AI-based system in healthcare improves the capabilities of decision support systems. The relevant healthcare data and knowledge base that is provided by an AI-based system can be used to make effective healthcare decisions after analyzing the healthcare data. The healthcare workers can be leveraged by the effective utilization of a decision support system to examine the individual patent. Moreover, AI-based healthcare systems can be used to manage unstructured healthcare data so that potential risks can be identified at the early stage of the disease. In healthcare informatics, IBM's Watson is a very good AI-based system that is being used to revolutionize the healthcare industry. This AI-based system is capable to assist both patients and healthcare professionals to predict heart attacks at an early stage. In healthcare industries, several practitioners utilize natural language processing (NLP) to transform the huge amount of unstructured data into a meaningful form to provide comprehensive care to patients [2].

4.8.2 Enhance primary care and triage through chatbots

Chatbots play a significant role in triage. Generally, it has been observed that people have a habit to book their appointments with the doctors for the treatment of the disease that can be cured by themselves. The patient that requires primary care can be easily identified through chatbots. AI-based healthcare systems can minimize the stress on doctors to automate the process of primary care and enable them to focus only on crucial and dire medical issues. The unnecessary visits to the doctors can be avoided and the money can be saved from these chatbots. These chatbots are leveraged with various crucial services supported by AI. The AI-based algorithm incorporated in these applications can provide instant answers to the queries of the patients. Moreover, the appropriate answers related to the raised queries can be provided to the patients which enabled them to take appropriate actions. These chatbots provide their services 24/7 and also have the ability to deal with multiple patients at the same time [85].

4.8.3 Robotic surgeries

In the healthcare sector, complicated surgeries can be performed by the collaboration of AI and robots. The process that is required in complex surgeries and dissections has now become fast. The utilization of robots can minimize the issue of fatigue during the crucial and time-consuming process of delicate surgeries. The new surgery procedures can be identified by AI-based machines from experience and available healthcare data. The accuracy of AI-based machines minimizes the chance of any unpredictable actions during the surgeries [83]. Moreover, Vicarious Surgical and Heartlander are two good examples of robots that integrate the concept of AI and virtual reality. These robots provide the facility for healthcare professionals to perform minimally critical surgeries. The Heartlander, the robot was developed by the Robotics Department at Carnegie Mellon University that facilitates therapy on the heart [86].

4.8.4 Virtual nursing assistants

AI-based virtual nursing assistants contribute a significant role in healthcare and are capable to perform several healthcare operations. The patients can directly interact with these systems to select effective healthcare units. These assistants are available 24/7 and can answer the queries generated by the patients. The optimal solutions related to their disease can be provided by these virtual nursing assistant applications. The first AI-based virtual nursing assistant namely Care Angle was devised to facilitate the routine checkup of the health of the patient [85].

4.8.5 Aiding in the accurate diagnosis

In healthcare sectors, AI-based applications and machines are enabling healthcare professionals more competent in diagnosing the disease of the patient. The doctors can detect, predict, and analyze the fatal disease more accurately and in a fast manner. The wearable healthcare devices are leveraged with AI algorithms that are efficient to provide accurate HER cost-effectively. For example, PathAI is a ML-based technology that enables the pathologist to provide precise predictions about the health condition of the patient. The healthcare industries can be adopting AI-based machines to detect cancer disease without any error in diagnosing. The buoy is an AI-based digital assistant that can be used by patients to check their health conditions. The data provided by these assistants can be used by the patient to take appropriate actions to cure their disease. This digital assistant listens to the symptoms of illness from the patient and suggests potential solutions based on its real-time diagnosis.

4.8.6 Minimizing the burden of EHR utilization

Although EHRs have played an important part to digitize the healthcare data management process. This drastic transition results in several problems such as cognitive overload, unending paperwork, and user fatigue. The adoption of AI-based systems to maintain the HER for the patient can reduce the significant amount of the user's time. The improved clinical documentation is now possible by incorporating the recent NLP technologies. The process of voice recognition and dictation may assist to improve the management of health records in digital form so that accurate data can be retrieved and inserted into the healthcare systems. AI-based assistants can provide regular alerts to patients regarding their health conditions. Healthcare professionals can set priorities for their routine work to facilitate their designated patients [86].

4.9 AI technologies in healthcare systems

AI is a collection of various emerging technologies such as ML, deep learning, and artificial neural network. These technologies have a greater influence in the healthcare sector and various operations of healthcare industries are being automated by adopting these technologies. In the following subsections, some important AI technologies are being discussed which are widely used in healthcare.

4.9.1 ML – neural networks and deep learning

ML is a statistical technique for creating training models that may be used to learn from data. This technology, which comes in a multitude of flavors, is at the core of many AI systems. ML plays a crucial part in precision medicine by predicting which therapeutic approaches and surroundings will be best given the patient's numerous medical conditions. A training dataset with a specified outcome variable is typically utilized in supervised learning, which is crucial for many ML and precision medicine applications [1]. An advanced kind of ML known as a neural network can be used in healthcare informatics to anticipate a patient's specific ailment. Based on a few key factors, the patient's sickness can be anticipated. The most complex sort of ML, which includes several layers of characteristics or variables that may be used to predict outcomes in healthcare, can be categorized as deep learning or neural network models. These models may have a variety of hidden qualities that the researchers have now made clear. The idea of deep learning is being utilized to detect certain significant traits in imaging data that were hard to see with the naked eye.

Different ML techniques are now used in the healthcare industry through the H-IoT to improve real-time services. To give patients with quality of service (QoS), these algorithms optimize the parameters. These algorithms can also help with defect identification in medical services, tracking fitness, and categorizing the input health vitals. By heavily utilizing ML techniques, the enhanced warning generating

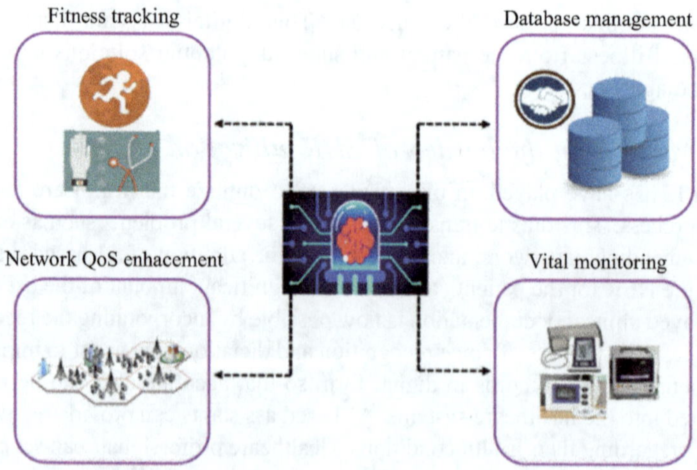

Figure 4.7 Use of ML in various healthcare activities

system may be deployed in the healthcare system [87]. Many other healthcare-related activities, including data management and preservation, data analysis, and network security, are well-known fields where ML techniques may be effectively used. Various healthcare system functions where modern technologies like AI, ML, and deep learning can be applied are shown in Figure 4.7.

4.9.2 Natural language processing

AI researchers have been working to grasp human language since the 1950s. NLP has been proven to have several important linguistic goals as well as voice recognition, text analysis, human voice translation, and other linguistic goals. There are two methods to tackle it: statistically and via semantic NLP. Different ML, deep learning, and neural network principles are included into the statistical NLP-based systems. These systems can carry out the process of recognition properly thanks to the inclusion of these modern technologies. The generation of health records, EHR interpretation, and document classification related to clinical activities are only a few of the tasks that the NLP technology may do in the context of healthcare informatics. The unstructured patient HER may be sorted in a well-structured manner for further analysis by the healthcare experts once NLP is adopted in healthcare systems. The creation of reports and the production of text-based transcripts of patient and designated doctor talks are two additional helpful healthcare tasks being carried out by NLP [88,90].

4.9.3 Rule-based expert systems

In the year 1980, expert systems were being developed based on collections of "if-then" rules. Nowadays these older expert systems are recognized as the leading AI technology. These systems are widely used to provide solutions to commercial applications. These rule-based expert systems are extensively used throughout the past several decades in healthcare informatics to support various clinical decisions and are still frequently used today. The healthcare industries are establishing various regulations, guidelines, and policies based on these software systems. The EHR of the patient is being generated to support the healthcare services based on the guidelines provided by these systems. To perform various significant operations in the healthcare sectors these systems require some set of rules in the defined area of knowledge that are created by the human and professional knowing specific domain. The efficiency of these systems can be affected by incorporating a big number of rules that can start clashing after some time. These systems also require a lot of effort and time from the professionals if there is a need to change the rules according to the transformation of the knowledge domain. To resolve this issue, various methods based on data and ML algorithms are being used to replace them gradually in healthcare informatics [89].

4.9.4 Physical robots

With the advancements in AI technologies, several physical robots are being deployed every year in industries to automate their systems around the globe. These physical robots accomplish various predefined activities such as lifting goods, repositioning heavy items, and assembling objects at various places of industries such as warehouses, healthcare units, and departments. Recently, the robots can be trained easily using various ML technologies so that they can collaborate on their designated task with humans intelligently. The inelegancy level of these robots can be improved by embedding various AI capabilities inside their brain manly in the operating system. The surgical robots that have been approved by the USA government in the year 2020 act as a superpower to support various clinical activities. The capabilities of the surgeons can be improved after collaborating on various tasks involved in complex surgeries, stitch wounds, and so forth. Various common surgical procedures such as head and neck surgery can be accomplished by these robots and in every surgical procedure, these robots provide a significant contribution to improving the decision-making capabilities of the surgeons when critical surgical conditions arise.

4.9.5 Robotic process automation

Healthcare industries leveraged this technology to automate various administrative processes. This technology is cost-effective and found easy to implement as compared to other categories of AI. The actions performed by this technology are very transparent in all the significant areas of healthcare. Instead of using robots, robotic process automation (RPA) technology utilizes computer applications that are executing on the servers. To act as a semi-intelligent user, this technology uses a mix of workflow, some identified business rules, and interaction of the presentation layer with information systems. In healthcare informatics, various repetitive operations can be performed by this technology such as authorization, modification of EHR of the patients, and maintaining invoice of the payment. This technology is capable to extract the data from the pictures and store them in the transactional systems of the healthcare industries. To identify the image and for extracting data from them this technology can be integrated with other technology such as image recognition. These technologies can be recognized as separate entities and can be progressively integrated with other recent technologies [90,91].

4.10 AI-based healthcare architecture

In the healthcare system, AI is widely being used and also contributing to the impact on modern business and society. This technology influences the healthcare practitioners, patients, caregivers, clinical process, and administrative process in a variety of ways. The majority of AI and healthcare technologies are useful in the healthcare area, but the strategies they assist might be rather different. It has been suggested by healthcare practitioners that AI can perform better than humans in some operations, such as illness diagnosis. Figure 4.8 shows an AI-based healthcare system in which various entities perform their routine healthcare activities. The inclusion of AI in the healthcare industries facilitates these entities to coordinate their activities to provide real-time healthcare services to the patient. The

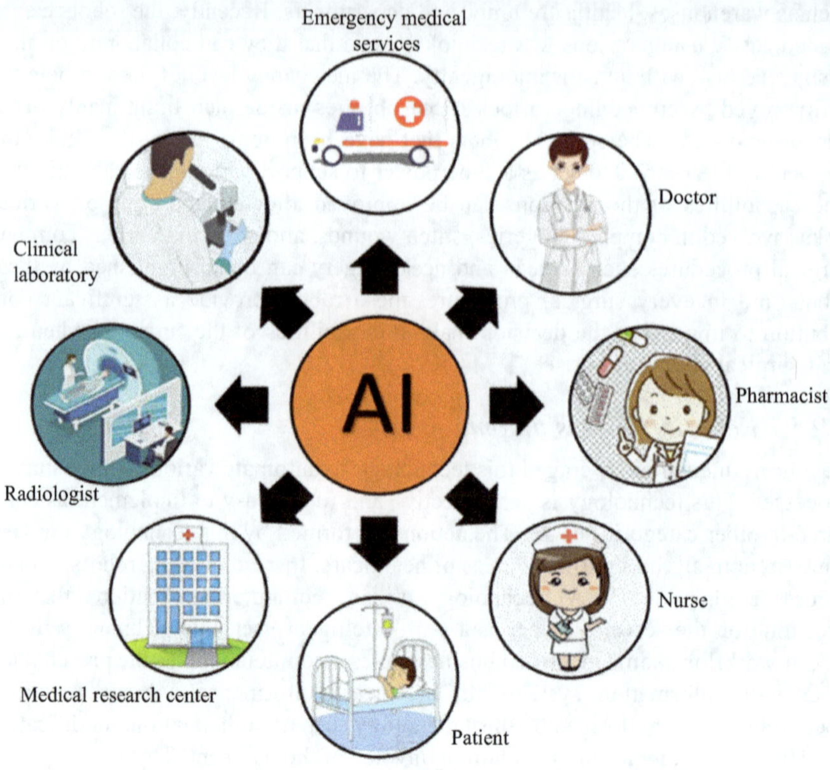

Figure 4.8 An AI-based healthcare system

subsequent sections discuss how these entities can be benefited from AI technologies in the healthcare system.

4.10.1 Patient

Hospitals in which the AI-based healthcare architecture is being used can leverage the sensor-based wireless devices to maintain the list of the patient. These devices can be positioned either at the home of the patient or in the hospitals. The data collected by these devices can be stored automatically in the centralized database systems. AI technologies sense the data and support the healthcare industries to make concrete decisions regarding patient health. These technologies also contribute their significant role to improve emergency medical services whenever required. Real-time monitoring through AI technologies can be helpful to diagnose the affected part of the patient's body in the early stage of the disease. Healthcare professionals and practitioners can leverage these technologies to provide proper treatment and prescriptions to their patients. The inclusion of AI makes entire healthcare systems smarter than the doctors and safeguards the patient from serious health disorders. The AI-based decision support system can make effective decisions and can provide future directions to the researchers to find out the possible solutions for the challenges that are being faced by the current healthcare industries.

4.10.2 Emergency medical service

The wearable sensor devices in the hand of the patient can be beneficial in emergency medical services. The data collected by the AI-enabled sensors make their significant contributions to the emergency medical service (EMS) and establish new policies and regulations to govern various healthcare functionalities. The analyzed data can be used by decision support systems of the healthcare industry. In automated healthcare informatics, the AI-based applications incorporate various algorithms that can receive various health parameters from the user such as body temperature, inputs related to the patient mood, blood pressure, stress level, etc. to provide real-time treatment. The data based on the parameters are analyzed by the medical applications and further transformed into a meaningful form. The transformation of the data in this way provides the right direction to the nurses in healthcare. The AI-based applications also provide knowledge to the patient about the designated hospital to get proper treatment.

4.10.3 Nurses

Nurses can effortlessly manage large amounts of patient data by adopting AI-based electronic gadgets. These AI gadgets reduce their efforts to maintain patient data through manual entries. The information collected by these AI-based gadgets always keeps nurses and other medical staff up to date regarding the health of the patient. AI also aids nurses in assessing illness severity and anticipating future interventions.

4.10.4 Doctors

In healthcare informatics, AI facilitates fast monitoring of patient health, keeps a record of sensitive information related to health and diagnosis, and also enhances the decision-making capabilities of healthcare professionals. The unstructured health data can be transformed into structured data with the use of AI technologies. The accurate and reliable transformation of data provides better results for diagnosing doctors and patients. To generate accurate clinical laboratory reports and radiologist reports, physicians can utilize AI-based tools. The AI-based decision support tools can be used to record the communication between nurses and patients to analyze the health condition of the patient in an effective manner. These tools facilitate the doctors to detect the high affected body part from the disease that degrade the patient's quality of life. The affected body parts can be targeted for proper treatment at an early stage so that patient can be saved from the development of any other chronic health implications.

4.10.5 Radiologists

In the healthcare industry, radiologists are also leveraged AI technologies to monitor the health of the patient. The disease can be identified from the medical images by introducing AI-based algorithms in various medical applications. The complicated radiological graphs can be analyzed to recognize the variation in the health of the patient with the use of AI. In recent years, AI applications are capable to detect various diseases that cannot be easily detected by the naked eye of healthcare professionals. To recognize the pulmonary nodules in the care of lung cancer AI-based healthcare applications can be used. The screening of the patient using these applications can provide rapid treatment to the patient in the early stage. The identified nodules can be easily categorized as malignant or benign. The long process of mammography screening is also a big hurdle in the treatment of the patient. The obtained results can be easily interpreted into meaningful information to gain knowledge about the calcium deposits in the affected body part of the patient. The detection of disease in the early stage using AI applications can minimize the development of any other chronic disease.

4.10.6 Clinical laboratories

The medical activities that are being performed in routine surgeries and clinical laboratories can be automated through AI. The recent technology namely digital pathology is now leveraged with the use of AI that identifies patterns using ML to provide detailed information to the pathologist. The full automation of clinical microbiology can be possible using various applications of AI. The AI-based clinical laboratories can perform the various test on collected blood samples quickly and provide concrete answers to all the queries of the healthcare professionals. The computer-aided diagnosis (CAD) is the most widely known AI application that can be used to detect the infection in the lungs and breasts of the patient. AI can understand and address a variety of clinical problems.

4.11 Challenges and future scope of AI in healthcare

The healthcare industry is facing many obstacles while utilizing various AI-based systems. In healthcare sectors, various aspects of AI are influencing the functionalities and, putting patients' lives in danger. Various challenges that are being faced by the healthcare industries in AI-based systems are given below.

4.11.1 Challenges in utilizing health care data

In healthcare data, various significant information are being stored to identify the right patient. Electronic health data comprises information like personal code, number, text, voice, and image. The AI-based medical equipment is collecting such sensitive data that can lead to legal concerns related to the privacy of the patient. In the healthcare sector, wireless AI devices are required to incorporate recent technologies such as IoT and cloud computing to improve their performance. Many security issues related to sensitive healthcare data are being observed by researchers in AI gadgets. To overcome this issue, researchers and manufacturing companies are contributing their efforts to revise the policies and regulations. To secure personal identity information, researchers have suggested various encryption schemes in their technical research. In addition, various nations are contributing their efforts to create some legal frameworks to protect personal data from any kind of security breach.

4.11.2 Injuries and error

The AI-based healthcare systems may generate incorrect results that can be harmful to the proper diagnosis of the patient that results in various other healthcare issues. The improper prescription of medicine and medication suggested by the AI-based healthcare system can fall the life of the patient into danger. Due to the representation of incorrect data by the AI-based systems can allot the bed to the patient who does have its necessity. The occurrence of an accident can be high due to inefficient AI-based healthcare equipment.

4.11.3 Data availability

In healthcare sectors, AI algorithms can be trained using various data sources such as EHRs, pharmaceutical records, insurance claims records, buying behavior of patients, and other relevant data produced by the consumers. The buying behavior of the patient can be changed over time or even they can change their insurance company to file their medical claims. The change of data and their formats may create some trouble in data management by the AI-based systems. The health statistics that are defined for a specific data format may not be applicable for the other format of data. The variety in the massive amount of healthcare data on different platforms and numerous systems may lead to the problem of generating inaccurate health data of the patient. The fragmentation of data at different platforms may affect the integrity and consistency of data. Managing different data in a variety of formats can increase the cost of data management and new difficulties can emerge in the development of a successful healthcare system.

4.11.4 Privacy concerns

In the healthcare sector, the patient can face various challenges concerned with the privacy of data related to their health. To develop an efficient AI-based healthcare system, developers are required to collect a huge amount of healthcare data from the healthcare centers. The collection of data without informing the concerned patient may violate their privacy. The patient can file lawsuits against the healthcare industry due to sharing of their sensitive health information among numerous AI-based systems. The AI-based system in healthcare can implicate the privacy of patients by predicting a new disease that was never found in any earlier diagnosis. For example, an AI-based system may identify that a patient is suffering from cancer disease based on the past data available in the healthcare systems while this information is never revealed to the patient in any diagnosis from healthcare professionals. In this care, the violation of a patient's privacy can be considered by the patient if such information would be shared among third parties such as banks and insurance companies.

4.12 Professional realignment

The smooth functioning of the healthcare sector can have a huge impact on the number of professionals changing their jobs. Research scholars have revealed in their research that excessive use of AI-based systems over some time can affect the ability of healthcare professionals to make decisions. Due to the widespread use of AI-based systems in the healthcare industry, healthcare professionals are not expanding their medical knowledge and are completely dependent on AI-based equipment.

4.12.1 Case study based on AI-based healthcare

Nowadays the process of smart healthcare systems is being automated using AI and IoT technologies. The case study mainly focuses on the recent healthcare system that is specifically located in remote areas and is getting benefited from the utilization of the fuzzy logic system. The implementation of this technique is very simple and also effective for concrete decision-making using sensor data regarding the health of the patient. The data collected using IoT-based sensors can be transmitted to the server and the result of the analysis can be generated on web-based applications. The patient and healthcare professionals can utilize the presented data to take certain actions and decision-making. The data obtained from the IoT-based sensors can be further utilized by the system in which fuzzy logic methods are already implemented. The classification of the data gathered from the sensors is

Table 4.2 Temperature monitoring

Temperature (°F)	Category
<99	No temperature
>99 and <=101	Temperature
>101 and <=103	High temperature
>103	Very high temperature

presented in Table 4.2. The data is classified into four different classes of temperatures ranging from 100°F to 105°F. In addition to this, Table 4.3 shows the three different classes of the pulse rate for patients. The pulse rates are also classified as low, normal, and high. To classify the pulse rate, two thresholds are defined i.e. 60 and 100. The low and high pulse rates can be considered while they are below 60 and greater than 100. The pulse rate that lies between the thresholds can be considered a normal pulse rate. Table 4.4 shows the range of blood pressure in the human body. The blood pressure can be considered normal and high it is 120/80. High blood pressure can be considered within the range of 129–140/81–89. The blood pressure above the range of high blood pressure can be considered in the category of very high blood pressure. In Table 4.5, the data obtained at regular intervals from IoT-based sensors is shown. Figure 4.9 shows the difference in gathered data using IoT-based sensors such as temperature sensor, pulse rate sensor, and blood pressure sensor. After obtaining data from Tables 4.2 to 4.5, the next step is considered to apply the fuzzy logic to take certain decisions based on the health condition of the patient. Table 4.6 shows the calibration of sensor data with

Table 4.3 Categories of pulse rate

Pulse rate	Category
<60	Low pulse rate
61–100	Normal pulse rate
<100	High pulse rate

Table 4.4 Categories of blood pressure

Blood pressure	Category
<100/<70	Very low BP
120–110/80–70	Low BP
120/80	Normal BP
130–139/80–89	High BP
>140/>90	Very high BP

Table 4.5 Data obtained from the sensor for experimentation

S. no.	Temperature (°F)	PR (%)	BP (low)	BP (high)
1	100	85	100	180
2	103	60	89	139
3	100	110	91	141
4	102	107	80	133
5	98	106	80	122

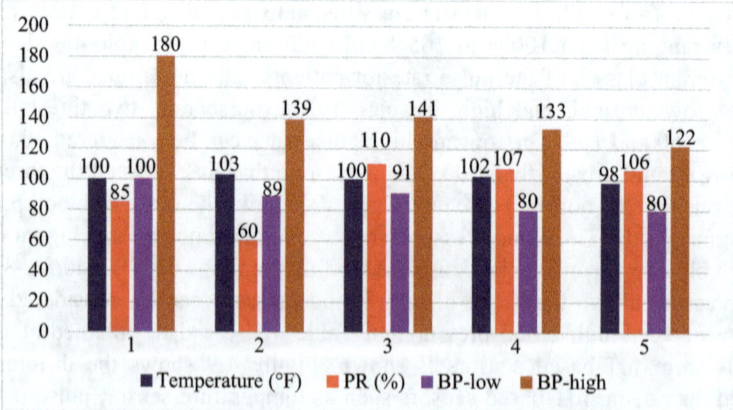

Figure 4.9 Variation of received data through IoT-based sensors

Table 4.6 Quantification of sensor data with FLS

S. no.	Temp	PL	BP	FLS	Precision (%)	Error rate (%)
1	H	L	VH	H	96.8	3.2
2	H	L	H	H	93.6	6.4
3	N	H	M	L	86.8	13.2
4	L	H	M	L	88.1	11.9
5	N	N	L	H	92.6	7.4
6	H	H	M	H	96.6	3.4
7	H	M	M	M	93.5	6.5
8	M	L	M	M	92.5	7.5
9	VH	H	H	H	96.6	3.4
10	M	H	M	L	86.9	13.1

the fuzzy logic system. The precision of the fuzzy logic-based system (FLS) is measured and presented in Figure 4.10. To measure the accuracy of fuzzy logic-based healthcare system, (4.1) is used [92]:

$$\text{Accuracy} = \sum \frac{\alpha(ki)}{m} \tag{4.1}$$

In (4.1), the accuracy of the system in percentage is represented by $\alpha(ki)$, m shows the number of experiments carried out to obtain the accuracy of the system. After observing the results obtained from the calibrated fuzzy logic system, it is found that the smart decisions can be feasible based on the sensor-based IoT system. To obtain the percent error from the obtained results, (4.2) is used:

$$PER_{Er} = \frac{Av - EV}{TV} \times 100 \tag{4.2}$$

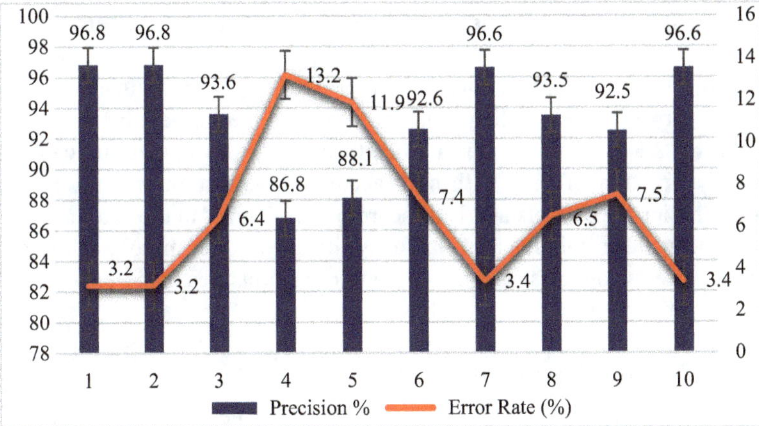

Figure 4.10 Outcome and accuracy of decision making based on fuzzy logic system

In (4.2), *Av* shows the accepted value, *Ev* shows the experimental value, and *Tv* shows the total value.

Temp, temperature; PL, pulse rate; BP, blood pressure; FLS, fuzzy logic system; H, high; N, normal; M, medium; L, low; VH, very high.

The data represented in Figure 4.10 shows the accuracy and reliability of the obtained results. The results presented in Tables 4.5 and 4.6 justify the performance of the fuzzy logic-based healthcare system by utilizing the sensors in decision making. The incorporation of fuzzy logic and sensor data makes the entire system intelligent, robust, and accurate. The obtained results show that the system saves time and is easy to handle by healthcare professionals.

The conventional healthcare systems are implemented based on simple decision-making to monitor the health of the patient at remote locations. The process of decision-making related to the health of the patient can be enhanced by adopting the fuzzy logic-based system in healthcare informatics. The percentage of error in diagnosis of the patients can be significantly deteriorated using such type of system. The IoT-based sensor data is more effective for better analysis as the data is obtained at regular intervals. The conventional system usually relies on video data for monitoring and making communication among patients and healthcare professionals. The conventional healthcare system provides a manual treatment system in which the error rate in the diagnosis can be very high. The fuzzy logic-based healthcare systems are smart enough to take intelligent decisions based on sensor-based data. In addition to this, the data obtained from the IoT-based sensors can be used to create a knowledge-based for further study and decision-making. In the future, more effective and robust fuzzy logic-based systems can be implemented to improve their performance based on the collected data available in the repository.

4.13 Examples of AI and IoT-based healthcare system

The use of AI and IoT in healthcare is improving medical treatment and patient experiences, from quicker diagnosis to surgeries performed by robots. After incorporating AI and IoT in various healthcare, various operations can be performed in a fraction of time with high accuracy and minimum cost. Really, AI makes life easier for patients and healthcare professionals. In the healthcare sector, a number of healthcare industries play a significant role to improve the traditional process of patient diagnosis. The list of healthcare industries that are equipped with AI and IoT technologies is given below.

4.13.1 AI-based healthcare system

AI has now become a key component for the global health organization to fight against pandemics. In the field of heath care, a number of leading organizations are now transforming their treatment process by leveraging AI/ML technologies. The list of organizations that are being leveraged with AI technologies is given below.

- *Viz.ai*
  It is a leading organization in the healthcare sector that applies AI in its healthcare process. The objective of this organization is to reduce the time of diagnosis using AI software so that better treatment can be provided to the patients in real-time. The healthcare product i.e. Viz LVO utilizes deep learning techniques to deliver sensitive information to healthcare professionals in a structured way. The delay in the treatment can put the life of the patient in danger. Therefore, by utilizing this software, cause of the disease can be diagnosed in a faster manner so that effective decisions can be taken by the doctors to save the life of the patient.
- *PathAI*
  This company incorporates ML and deep learning technologies in the healthcare software to facilitate the pathologists. This software improves the decision-making process of pathologists for more precise diagnoses. The main objective of this company is to minimize the percentage of error rate in cancer diagnosis. This company collaborates with other leading healthcare industries for the expansion of its AI technology.
- *Buoy Health*
  Buoy Health utilizes AI technology to accurately observe symptoms and causes of illness. Through AI-enabled Chabot patients can discuss their health issues and can get precise guidance in real-time. Buoy is powered by advanced ML algorithms that leverage the patient to communicate with their favorite doctors.
- *Enlitic*
  To modernize the process of radiology diagnoses, Enlitic develops a medical tool powered by deep learning algorithms. To provide better healthcare facilities to the patient this tool provides structured data related to medical images,

blood tests, and history of the patient's health. The unstructured healthcare data can be organized in a well-structured manner so that it can be analyzed by healthcare professionals quickly. Enlitic is transforming the process of healthcare informatics by leveraging AI and deep learning techniques and minimizing the challenges in the clinical workflow.

4.13.2 IoT-based healthcare system

There are many nations that do not offer the most cutting-edge methods for treating patients in the realm of health care. The patient's health is still being monitored by traditional healthcare institutions in these developing nations. The use of IoT in healthcare companies has the potential to significantly upgrade the current patient-centered system from a conventional treatment method. The doctor-to-patient ratio in the Indian healthcare system is 1:16,675. The Indian government has agreed to invest a significant sum, or about $15 billion, in the healthcare industry in order to correct this unfavorable ratio. Additionally, the government has opted to offer the healthcare sectors IoT-based healthcare solutions.

- *Spectral insights*
  It is a healthcare company that provides solutions to the patient and healthcare professionals through a multi-faceted platform that incorporates various recent technologies such as AI, ML, data management and visualization, and cloud computing. Healthcare professionals can be leverages this platform as it provides an advanced imaging system, advanced data analytics, and digital microscopy. The AI/ML-based analytical tools that are being utilized in this organization reduce time and increase the accuracy of diagnosis. The complexity of surgeries and biopsies can be reduced after utilizing cloud and AI/ML-based medical equipment.
- *Forus Health*
  It is a group of engineers that is being created in the year 2010. The main objective of this group is to minimize blindness. The group provides various potential technological solutions to the patient for the betterment of eye care. The group has developed a portable camera that is capable to capture the digital images of the cornea and retina. After analyzing the digital images common eye diseases such as glaucoma, diabetic retinopathy, and cataract can be easily identified. To provide remote treatment to the patient, the camera is integrated with the cloud-based telemedicine platform. The other objectives of this group are to reduce the time and cost of diagnosis and increase the awareness of eye care among the public.
- *Cardiac Design Labs (CDL)*
  CDL is established in the year 2011. The objective of this lab is to provide an automated cardiac monitoring system that is integrated with AI technologies. The artificially intelligent system MIRCaM (Mobile Intelligent Remote Cardiac Monitor) is capable to provide the report to the patient who is living in the remote locations. In India, the ratio of cardiologists and cardiac patients is not in balanced. The AI-based MIRCaM system can bridge this gap.

- *Bagmo*
  In the healthcare section, it is a private company which is established in the year 2017. The main objective of this company is to handle the lack of blood availability. The company has developed blood bag monitoring equipment that can sense the temperature of the blood at the time of its storage and transportation. The unique ID to each bag is provided through a radio frequency identification (RFID) card. The unique ID of the refrigerator is stored in the cloud. At the time of blood requirement, the blood monitoring equipment can trace the suitability of the bag and its associated refrigerator.

4.14 Future scope of AI in healthcare

In recent years, in the field of AI two approaches such as the human approach and the ideal approach, are generally being used. In the human approach, developers are mainly concerned with the development of AI-based systems that have the capabilities to think and act like human beings. While in other approaches AI-based systems are created that can think and act rationally. The researchers are trying to develop AI-based systems using both approaches and various significant milestones have been achieved by them. The AI-based systems are generally based on various concepts used in computer science and require a large database to store the information. The information stored within these systems is being used for analysis, problem-solving, and decision-making whenever required. Nowadays, various emerging technologies such as ML and deep learning are being used by developers to make the AI-based system effective. The emerging technologies are the sub-filed of AI to support various functionalities of AI-based systems. These technologies enable the developers to develop efficient AI-based algorithm that accepts the correct input from the machine. With the advancement of AI, many things are now being possible that were not possible in the past. For example, the concept of the self-driving car was only an idea but AI transformed this imaginary concept into reality. AI-based self-driving cars are easy in handling and safe in their use. The healthcare sectors are also being benefited from the use of AI. In the field of digital image processing many image retargeting algorithms are being used in medical science. The efficiency of these algorithms can be improved by incorporating the concepts of AI and ML [93–96]. With the incorporation of AI-based algorithms in healthcare systems, healthcare professionals can perform various complex tasks such as surgeries, clinical activities, and diagnosis. The wearable devices that are also developed using AI-based algorithms are helpful to the patient to get real-time healthcare services. The patient can select the right place to get appropriate treatment and can take better decisions regarding their health issues. In recent years, software-based robots have minimized the effort of doctors and required manpower in various critical surgeries. These AI-based robots are capable of taking decisions during the surgeries based on the past health records and other clinical information of the patient. In healthcare sectors, AI has a complete set through which revolution in healthcare industries is possible in the future.

4.15 Integration of AI and IoT in healthcare

The two main phrases now being used to forecast the direction of technology are AI and the IoT. In order to increase productivity, the companies making use of these technologies are redesigning their numerous procedures [97]. Various equipment used in the healthcare industry have IoT-capable sensors that people may wear. The patient can monitor a variety of health issues thanks to the sensors in wearable technology. These wearables are capable of carrying out several important duties including measuring steps, offering GPS capabilities, and calculating calories burnt. In addition to doing all of these activities, these gadgets make a substantial contribution to the patient's receiving health-related recommendations. This gadget is fully capable of providing all the essential information, such as what sort of food a patient needs to eat, the timetable of diet, and the quantity of food intake sometimes, after thoroughly analyzing all the health-related data. The simplest and most comfortable approach to collect data after assessing a patient's health is through wearables. The patient who is worried is also informed of the obtained data so they may find out their current health state. The functions and forms of IoT and AI-enabled equipment and gadgets are transforming the way that medicine is administered. IoT and AI platforms, which are essential for data collection, gathering, and administration, are used, for instance, in the production of armbands and ECG monitors [98]. A single wearable device that has all the features needed to deliver better healthcare outcomes when desired can be created by integrating many wearable devices to simplify life. The concept depends on wearable technology being accessible to everyone on a single platform. When these gadgets are equipped with AI and IoT characteristics, amazing things may happen. It is simple to create new-level wearable gadgets thanks to the combination of AI and IoT and their tendency to bring about transformations [99]. IoT will connect the gadgets, and AI has always played a role in data processing. Future wearable gadgets, in which a single device enables its users to accomplish everything they want to do at any location and at any time, are conceptualized as a result of the combination of AI and IoT. This gadget, which will improve not just healthcare or tracking but also humans through technology, will be created thanks to IoT and AI technologies. Wearable technology is pervasive and has taken over every aspect of our life. When healthcare professionals have access to data that has already been analysis by these sorts of devices, their decision-making efficiency will increase. Based on the information gathered by these gadgets, the important choice may be made. The use of sensors, trackers, fibers, AI, and IoT-based assistants is used by a variety of wearable IoT and AI-based products, such as virtual coaches, and fitness equipment. These devices' performance analysis and data analysis produce precise healthcare findings and a variety of helpful forecasts that may be delivered to patients at an early stage of their sickness.

In order to provide patients and medical professionals with improved healthcare services, Figure 4.11 illustrates the integration of AI and IoT in healthcare. The integration has significantly increased its presence in the healthcare industry, assisting patients in maintaining their health through individualized treatment,

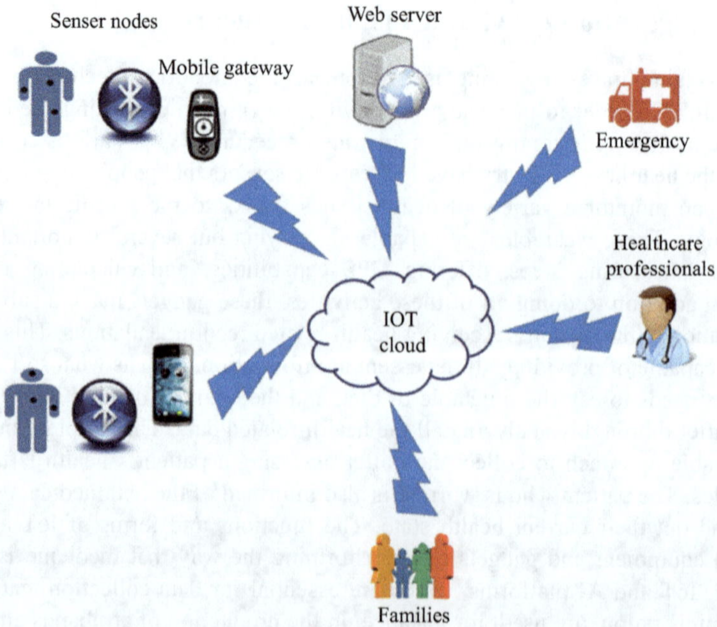

Figure 4.11 Wearable enabling services via AI and IoT

audio and video instructions, and other information about their surrounds and environment. The IoT and AI platforms and solutions provide the foundation for the algorithms and data sets employed by wearable technology. AI, ML, and IoT are being used by a number of healthcare-related manufacturing firms to evaluate the data they have gathered in order to facilitate collaboration across various healthcare organizations. To manage some of the patient's complicated medical issues, information exchange is essential. Healthcare firms may more effectively develop their AI-based decision-making system. The goal of the researchers' research is to investigate novel algorithms to enhance several important clinical procedures in the healthcare industry. AI is a recent technology that the PH may use to forecast disease, enhance clinical workflow, increase patient self-management, etc. ML techniques may be used to create healthcare analytic models. These models are being integrated into various healthcare applications and decision support systems to boost performance. In these models, information from multiple sensors is evaluated to identify the patient's behavioral patterns and various clinical problems. These models can identify the patient's routines, changes in eating, drinking, or sleeping patterns, among other things. The patient changes their lifestyle and treatment plans after recognizing trends in the healthcare apps and implementing a decision support system. The healthcare application advises include the specified medical professionals and cares in the recommended treatment strategy. When using ML techniques in this situation, though, there are a number of uncertainties that may be discovered. The interpretation of such data may be particularly difficult due to the possibility of numerous sorts of skewed data being produced by changes in clinical activity, patient behavior, and lifestyle. It is also conceivable that the training data model uses a dataset that is older. All of these issues might result in the system's decision support systems and healthcare apps producing inaccurate data without the user's knowledge.

4.16 Conclusion and future scope

The healthcare industry is changing its environment with the help of many cutting edge technologies. The application of AI, ML, IoT, and cloud computing improves healthcare services while lowering costs. These technologies allow the massive amount of healthcare data to be easily managed and assessed to aid in decision making. The IoT has in large part enabled the creation of IoT-connected devices that assist medical professionals in diagnosis (IoT). With the use of sensors and other wearable technologies, the patient is able to learn more about his current condition of health. To improve healthcare efficiency, IoT and AI technologies are being blended into healthcare design. The integrated architecture can handle all of the real-time demands of the doctors and patients while keeping an eye on the patient. Numerous study papers published by scientists have examined how well ML technology works for diagnosing and treating patients. Additionally, research has demonstrated that AI-based healthcare systems can perform a range of clinical activities more effectively. The design of medical gadgets to assist in decision making during treatments or for diagnosis, the analysis of medical imagery, and the encryption of medical data are just a few of the useful healthcare activities that these AI-based systems are capable of carrying out. The developer of AI-based solutions has conquered a number of challenges and issues by utilizing AI technology in healthcare. The knowledge of professionals like physicians, nurses, radiologists, and pharmacists may be used to improve the most current AI-based healthcare systems. As a result, this research looks at the state of domestic and international AI technology in the healthcare sector as well as the problems that need to be solved before AI can be used in the sector. Remote patient monitoring also shortens hospital stays and avoids readmissions. IoT significantly affects healthcare cost reduction and treatment effectiveness. Scalability and security advantages are now frequently offered to clients by use of cloud computing concepts. A wide range of computer resources, including as networks, storage, and on-demand services, are available to users. Several healthcare-related industrial companies are utilizing AI, ML, and IoT to assess the data they have been gathering in order to promote collaboration. Personal healthcare (PH) anticipates ailments by using AI algorithms on the obtained information to enable people to get ready for them in advance. In PH, a range of patterns linked to patient behavior are being researched so that healthcare professionals employing clinical decision support systems can make the right decisions. The focus of the researchers' substantial and significant research has been on the key problems with healthcare IoT applications. Additionally, there are a great deal more unsolved problems of a similar nature. IoT has a better possibility of having a positive impact on the healthcare industry in the future because to improved analytics, increased competition, and faster Internet connections. In recent years, a number of businesses have joined the IoT sector. A record number of businesses offering IoT healthcare solutions were on display at the 2020 Consumer Electronics Show (CES), including makers of care robots, advanced home security cameras, and, of course, smart speakers. By 2025, analysts estimate that the market for wearable technology will increase to $74.03 billion. This is a huge increase above the $27.91 billion value for 2019. Wearable technology is crucial to the healthcare sector since it will enable patients to monitor their activities, vital signs, and a variety of other health-related data. Along with the growing market for healthcare IoT, the COVID-19 pandemic has raised questions about the IoT's potential to safely connect patients and healthcare professionals in the future. In order to continue treating certain patients without bringing them into care facilities and risking their exposure to infection, hospitals and clinics were obligated to review telemedicine right away. Hospitals are also under continual pressure to find methods to save expenses. Wearable medical technology may lessen the amount of resources needed in hospitals by allowing some patients to get care and be monitored at home. With the development of 5G networks, data transfer is improved and more healthcare departments have faster access. Healthcare professionals, patients, and cares used the faster Internet connectivity that IoT devices had provided to securely exchange critical healthcare data. Using a variety of wearable devices, patients may communicate data about their health, such as sleep, heart rate, oxygen levels, and movements. The classic IoMT devices that were used for data collection have become smarter because to the application of AI in healthcare. These technologies allow for communication among medical professionals to discuss certain medical issues that can only be resolved via in-depth discussions. To lessen the administrative strain on healthcare workers, virtual agents are being used in the industry. To remind patients to take their meds, healthcare practitioners may use data collected by various IoT devices and voice-activated speakers. These virtual agents may be utilized to collect important patient data like pain levels, glucose levels, and oxygen levels and communicate it to the designated cares. The healthcare industry is utilizing IoT to manage its inventory. IoT devices incorporated into the healthcare sector may track the movement of several pieces of equipment used in clinical procedures. Equipment shortages are made feasible and their utilization is made more obvious with effective equipment tracking.

References

[1] Helm, J. M., Swiergosz, A. M., Haeberle, H. S., *et al.* (2020). Machine learning and artificial intelligence: definitions, applications, and future directions. *Current Reviews in Musculoskeletal Medicine*, 13(1), 69–76.

[2] Thaler, S. and Menkovski, V. (2019). The role of deep learning in improving healthcare. In *Data Science for Healthcare* (pp. 75–116). Springer, Cham.

[3] Garg, A. and Negi, A. (2020). Structure preservation in content-aware image retargeting using multi-operator. *IET Image Processing*, *14*(13), 2965–2975.

[4] Garg, A., Negi, A., and Jindal, P. (2021). Structure preservation of image using an efficient content-aware image retargeting technique. *Signal, Image and Video Processing*, *15*(1), 185–193.

[5] Maktoubian, J. and Ansari, K. (2019). An IoT architecture for preventive maintenance of medical devices in healthcare organizations. *Health and Technology*, *9*(3), 233–243.

[6] De Michele, R. and Furini, M. (2019, September). IoT healthcare: benefits, issues and challenges. In *Proceedings of the 5th EAI International Conference on Smart Objects and Technologies for Social Good* (pp. 160–164).

[7] Gupta, R., Shukla, A., and Tanwar, S. (2020). AaYusH: a smart contract-based telesurgery system for healthcare 4.0. In *2020 IEEE International Conference on Communications Workshops (ICC Workshops)*, Dublin, Ireland, pp. 1–6, doi:10.1109/ICCWorkshops49005.2020.9145044.

[8] Poongodi, T., Krishnamurthi, R., Indrakumari, R., Suresh, P., and Balusamy, B. (2020). Wearable devices and IoT. In *A Handbook of Internet of Things in Biomedical and Cyber Physical System* (pp. 245–273). Springer, Cham.

[9] Marques, G., Pitarma, R., M Garcia, N., and Pombo, N. (2019). Internet of things architectures, technologies, applications, challenges, and future directions for enhanced living environments and healthcare systems: a review. *Electronics*, *8*(10), 1081.

[10] Acharya, A. D. and Patil, S. N. (2020). IoT based health care monitoring kit. In *2020 Fourth International Conference on Computing Methodologies and Communication (ICCMC)* (pp. 363–368). IEEE.

[11] Islam, M. M., Rahaman, A., and Islam, M. R. (2020). Development of smart healthcare monitoring system in IoT environment. *SN Computer Science*, *1*, 1–11.

[12] Darwish, A., Hassanien, A. E., Elhoseny, M., Sangaiah, A. K., and Muhammad, K. (2019). The impact of the hybrid platform of Internet of Things and cloud computing on healthcare systems: opportunities, challenges, and open problems. *Journal of Ambient Intelligence and Humanized Computing*, *10*(10), 4151–4166.

[13] Tanwar, S., Vora, J., Kaneriya, S., *et al.* (2020). Human arthritis analysis in fog computing environment using Bayesian network classifier and thread protocol. *IEEE Consumer Electronics Magazine*, vol. 9, no. 1, pp. 88–94, https://doi.org/10.1109/MCE.2019.2941456.

[14] Negash, B., Gia, T. N., Anzanpour, A., *et al.* (2018). Leveraging fog computing for healthcare IoT. In *Fog Computing in the Internet of Things* (pp. 145–169). Springer, Cham.

[15] Ahamed, F. and Farid, F. (2018). Applying Internet of Things and machine-learning for personalized healthcare: issues and challenges. In *2018 International Conference on Machine Learning and Data Engineering (iCMLDE)* (pp. 19–21). IEEE.

[16] Paglialonga, A., Patel, A. A., Pinto, E., Mugambi, D., and Keshavjee, K. (2019). The healthcare system perspective in mHealth. In *m_Health Current and Future Applications* (pp. 127–142). Springer, Cham.

[17] Moazzami, B., Razavi-Khorasani, N., Moghadam, A. D., Farokhi, E., and Rezaei, N. (2020). COVID-19 and telemedicine: immediate action required for maintaining healthcare providers well-being. *Journal of Clinical Virology*, *126*, 104345.

[18] Banerjee, S., Hemphill, T., and Longstreet, P. (2018). Wearable devices and healthcare: data sharing and privacy. *The Information Society*, *34*(1), 49–57.

[19] Yu, K. H., Beam, A. L., and Kohane, I. S. (2018). Artificial intelligence in healthcare. *Nature Biomedical Engineering*, *2*(10), 719–731.

[20] Khan, Z. H., Siddique, A., and Lee, C. W. (2020). Robotics utilization for healthcare digitization in global COVID-19 management. *International Journal of Environmental Research and Public Health*, *17*(11), 3819.

[21] Hathaliya, J. J., Gupta, R., Tanwar, S., and Sharma, P. (2021). A smart contract-based secure data sharing scheme in healthcare 5.0. In *2021 IEEE Globecom Workshops* (GC Wkshps), Madrid, Spain, pp. 1–6, https://doi.org/10.1109/GCWkshps52748.2021.9681956.

[22] Kapadiya, K., Patel, U., Gupta, R., *et al.* (2022). Blockchain and AI-empowered healthcare insurance fraud detection: an analysis, architecture, and future prospects. *IEEE Access*, vol. 10, pp. 79606–79627, https://doi.org/10.1109/ACCESS.2022.3194569.

[23] McGhin, T., Choo, K. K. R., Liu, C. Z., and He, D. (2019). Blockchain in healthcare applications: research challenges and opportunities. *Journal of Network and Computer Applications*, *135*, 62–75.

[24] Pradhan, B., Bhattacharyya, S., and Pal, K. (2021). IoT-based applications in healthcare devices. *Journal of Healthcare Engineering*, 2021, 1–18.

[25] Rodrigues, J. J., Segundo, D. B. D. R., Junqueira, H. A., *et al.* (2018). Enabling technologies for the internet of health things. *IEEE Access*, *6*, 13129–13141.

[26] Anand, P., Singh, Y., Selwal, A., Alazab, M., Tanwar, S., and Kumar, N. (2020). IoT vulnerability assessment for sustainable computing: threats, current solutions, and open challenges. *IEEE Access*, vol. 8, pp. 168825–168853, https://doi.org/10.1109/ACCESS.2020.3022842.

[27] Xu, G. (2020). IoT-assisted ECG monitoring framework with secure data transmission for health care applications. *IEEE Access*, *8*, 74586–74594.

[28] Meng, L., Ge, K., Song, Y., *et al.* (2021). Wearable electrocardiogram signal monitoring and analysis based on convolutional neural network. *IEEE Transactions on Instrumentation and Measurement, 70*, 1–11.

[29] Swamy, T. J. and Murthy, T. N. (2019, January). Esmart: an IoT based intelligent health monitoring and management system for mankind. In *2019 International Conference on Computer Communication and Informatics (ICCCI)* (pp. 1–5). IEEE.

[30] Murali, D., Rao, D. R., Rao, S. R., and Ananda, M. (2018, September). Pulse oximetry and IOT based cardiac monitoring integrated alert system. In *2018*

International Conference on Advances in Computing, Communications and Informatics (ICACCI) (pp. 2237–2243). IEEE.

[31] Saha, J., Saha, A. K., Chatterjee, A., *et al.* (2018, January). Advanced IOT based combined remote health monitoring, home automation and alarm system. In *2018 IEEE 8th Annual Computing and Communication Workshop and Conference (CCWC)* (pp. 602–606). IEEE.

[32] Gholamhosseini, L., Sadoughi, F., Ahmadi, H., and Safaei, A. (2019, April). Health internet of things: strengths, weakness, opportunity, and threats. In *2019 5th International Conference on Web Research (ICWR)* (pp. 287–296). IEEE.

[33] Hasshim, M. H., Nordin, S., Hamzah, S. A., *et al.* (2020). IoT based health monitoring system for elderly patient. *Journal of Electronic Voltage and Application,* 1(2), 27–36.

[34] Fouad, H., Hassanein, A. S., Soliman, A. M., and Al-Feel, H. (2020). Analyzing patient health information based on IoT sensor with AI for improving patient assistance in the future direction. *Measurement, 159,* 107757.

[35] Fei, Z., Yang, E., Li, D. D. U., *et al.* (2020). Deep convolution network based emotion analysis towards mental health care. *Neurocomputing, 388,* 212–227.

[36] Lawanot, W., Inoue, M., Yokemura, T., Mongkolnam, P., and Nukoolkit, C. (2019). Daily stress and mood recognition system using deep learning and fuzzy clustering for promoting better well-being. In *2019 IEEE International Conference on Consumer Electronics (ICCE)* (pp. 1–6). IEEE.

[37] Pinto, S., Cabral, J., and Gomes, T. (2017). We-care: an IoT-based health care system for elderly people. In *2017 IEEE International Conference on Industrial Technology (ICIT)* (pp. 1378–1383). IEEE.

[38] Vishnu, S., Ramson, S. J., and Jegan, R. (2020). Internet of medical things (IoMT) – an overview. In *2020 5th International Conference on Devices, Circuits and Systems (ICDCS)* (pp. 101–104). IEEE.

[39] Naresh, V. S., Pericherla, S. S., Murty, P. S. R., and Sivaranjani, R. (2020). Internet of things in healthcare: architecture, applications, challenges, and solutions. *Computer Systems Science & Engineering*, *35*(6), 411–421.

[40] Singh, A. K. and Patro, B. D. K. (2019). Security requirements and attacks in wireless sensor networks. *International Journal of Applied Engineering Research 14* (*2*), 158–161

[41] Singh, A. K. and Patro, B. D. K. (2021). Security attacks on RFID and their countermeasures. In *Computer Communication*, *Networking and IoT* (pp. 509–518). Springer, Singapore.

[42] Singh, A. K. and Patro, D. B. (2019). A novel security protocol for wireless sensor networks based on elliptic curve signcryption. *International Journal of Computer Networks & Communications (IJCNC) 11*(5), 93–112.

[43] Singh, A. K. and Patro, B. D. K. (2020). Elliptic curve signcryption based security protocol for RFID. *KSII Transactions on Internet and Information Systems (TIIS)*, *14*(1), 344–365.

[44] Saganowski, S., Kazienko, P., Dziezyc, M., *et al.* (2020). Review of consumer wearables in emotion, stress, meditation, sleep, and activity detection and analysis. arXiv preprint arXiv:*2005*.00093.

[45] Baker, S. B., Xiang, W., and Atkinson, I. (2017). Internet of things for smart healthcare: technologies, challenges, and opportunities. *IEEE Access*, 5, 26521–26544.

[46] Tamura, T., Maeda, Y., Sekine, M., and Yoshida, M. (2014). Wearable photoplethysmographic sensors—past and present. *Electronics*, *3*(2), 282–302.

[47] Massaroni, C., Nicolò, A., Lo Presti, D., Sacchetti, M., Silvestri, S., and Schena, E. (2019). Contact-based methods for measuring respiratory rate. *Sensors*, *19*(4), 908.

[48] Sovijarvi, A. R. A., Dalmasso, F., Vanderschoot, J., Malmberg, L. P., Righini, G., and Stoneman, S. A. T. (2000). Definition of terms for applications of respiratory sounds. *European Respiratory Review*, *10*(77), 597–610.

[49] Piro, B., Mattana, G., and Noël, V. (2019). Recent advances in skin chemical sensors. *Sensors*, *19*(20), 4376.

[50] Muhammad Sayem, A. S., Hon Teay, S., Shahariar, H., Luise Fink, P., and Albarbar, A. (2020). Review on smart electro-clothing systems (SeCSs). *Sensors*, *20*(3), 587.

[51] Jakubas, A., Łada-Tondyra, E., and Nowak, M. (2017, June). Textile sensors used in smart clothing to monitor the vital functions of young children. In *2017 Progress in Applied Electrical Engineering (PAEE)* (pp. 1–4). IEEE.

[52] Venema, B., and Leonhardt, S. (2017, February). In-ear pulse oximetry in high altitude mountaineering. In *2017 13th IASTED International Conference on Biomedical Engineering (BioMed)* (pp. 254–259). IEEE.

[53] Sultan, N. (2014). Making use of cloud computing for healthcare provision: opportunities and challenges. *International Journal of Information Management*, *34*(2), 177–184.

[54] Dang, L. M., Piran, M., Han, D., Min, K., and Moon, H. (2019). A survey on internet of things and cloud computing for healthcare. *Electronics*, *8*(7), 768.

[55] Vora, J., Tanwar, S., Tyagi, S., Kumar, N., and Rodrigues, J. J. P. C. (2017). Home-based exercise system for patients using IoT enabled smart speaker. In *2017 IEEE 19th International Conference on e-Health Networking, Applications and Services (Healthcom)*, Dalian, China, pp. 1–6, https://doi.org/10.1109/HealthCom.2017.8210826.

[56] Odun-Ayo, I., Ajayi, O., Akanle, B., and Ahuja, R. (2017). An overview of data storage in cloud computing. In *2017 International Conference on Next Generation Computing and Information Systems (ICNGCIS)* (pp. 29–34). IEEE.

[57] Sethi, P. and Sarangi, S. R. (2017). Internet of things: architectures, protocols, and applications. *Journal of Electrical and Computer Engineering*, 2017.

[58] Anwar, M., Abdullah, A. H., Qureshi, K. N., and Majid, A. H. (2017). Wireless body area networks for healthcare applications: an overview. *Telkomnika*, *15*(3), 1088–1095.

[59] Cerina, L., Notargiacomo, S., Paccanit, M. G., and Santambrogio, M. D. (2017, September). A fog-computing architecture for preventive healthcare

and assisted living in smart ambients. In *2017 IEEE 3rd International Forum on Research and Technologies for Society and Industry (RTSI)* (pp. 1–6). IEEE.

[60] Verma, P. and Sood, S. K. (2018). Fog assisted-IoT enabled patient health monitoring in smart homes. *IEEE Internet of Things Journal*, *5*(3), 1789–1796.

[61] Azimi, I., Anzanpour, A., Rahmani, A. M., *et al.* (2017). HiCH: Hierarchical fog-assisted computing architecture for healthcare IoT. *ACM Transactions on Embedded Computing Systems (TECS)*, *16*(5s), 1–20.

[62] Kumar, N. (2017, August). IoT architecture and system design for healthcare systems. In *2017 International Conference on Smart Technologies for Smart Nation (SmartTechCon)* (pp. 1118–1123). IEEE.

[63] Nandyala, C. S. and Kim, H. K. (2016). From cloud to fog and IoT-based real-time U-healthcare monitoring for smart homes and hospitals. *International Journal of Smart Home*, *10*(2), 187–196.

[64] Plageras, A. P., Psannis, K. E., Ishibashi, Y., and Kim, B. G. (2016, October). IoT-based surveillance system for ubiquitous healthcare. In *IECON 2016 – 42nd Annual Conference of the IEEE Industrial Electronics Society* (pp. 6226–6230). IEEE.

[65] Villalba, M. T., de Buenaga, M., Gachet, D., and Aparicio, F. (2015, October). Security analysis of an IoT architecture for healthcare. In *International Internet of Things Summit* (pp. 454–460). Springer, Cham.

[66] Mahmud, R., Koch, F. L., and Buyya, R. (2018). Cloud-fog interoperability in IoT-enabled healthcare solutions. In *Proceedings of the 19th International Conference on Distributed Computing and Networking* (pp. 1–10).

[67] Debauche, O., Mahmoudi, S., Manneback, P., and Assila, A. (2019). Fog IoT for health: a new architecture for patients and elderly monitoring. *Procedia Computer Science*, *160*, 289–297.

[68] Paul, A., Pinjari, H., Hong, W. H., Seo, H. C., and Rho, S. (2018). Fog computing-based IoT for health monitoring system. *Journal of Sensors*, 2018.

[69] Awaisi, K. S., Hussain, S., Ahmed, M., Khan, A. A., and Ahmed, G. (2020). Leveraging IoT and fog computing in healthcare systems. *IEEE Internet of Things Magazine*, *3*(2), 52–56.

[70] Abdelmoneem, R. M., Benslimane, A., Shaaban, E., Abdelhamid, S., and Ghoneim, S. (2019, May). A cloud-fog based architecture for IoT applications dedicated to healthcare. In *ICC 2019—2019 IEEE International Conference on Communications (ICC)* (pp. 1–6). IEEE.

[71] Jayapradha, S. and Vincent, P. D. R. (2017, July). An IOT based human healthcare system using arduino uno board. In *2017 International Conference on Intelligent Computing, Instrumentation and Control Technologies (ICICICT)* (pp. 880–885). IEEE.

[72] Sunehra, D. and Ramakrishna, P. (2016, December). Web based patient health monitoring system using Raspberry Pi. In *2016 2nd International Conference on Contemporary Computing and Informatics (IC3I)* (pp. 568–574). IEEE.

[73] Gia, T. N., Rahmani, A. M., Westerlund, T., Liljeberg, P., and Tenhunen, H. (2015, April). Fault tolerant and scalable IoT-based architecture for health monitoring. In *2015 IEEE Sensors Applications Symposium (SAS)* (pp. 1–6). IEEE.

[74] Kishor, A., Chakraborty, C., and Jeberson, W. (2020). A novel fog computing approach for minimization of latency in healthcare using machine learning. *International Journal of Interactive Multimedia & Artificial Intelligence*, *1*(1).

[75] Farahani, B., Firouzi, F., Chang, V., Badaroglu, M., Constant, N., and Mankodiya, K. (2018). Towards fog-driven IoT eHealth: promises and challenges of IoT in medicine and healthcare. *Future Generation Computer Systems*, *78*, 659–676.

[76] Muhammed, T., Mehmood, R., Albeshri, A., and Katib, I. (2018). UbeHealth: a personalized ubiquitous cloud and edge-enabled networked healthcare system for smart cities. *IEEE Access*, *6*, 32258–32285.

[77] Wu, T., Wu, F., Redoute, J. M., and Yuce, M. R. (2017). An autonomous wireless body area network implementation towards IoT connected healthcare applications. *IEEE access*, *5*, 11413–11422.

[78] Touati, F., Tabish, R., and Mnaouer, A. B. (2013, April). Towards u-health: an indoor 6LoWPAN based platform for real-time healthcare monitoring. In *6th Joint IFIP Wireless and Mobile Networking Conference (WMNC)* (pp. 1–4). IEEE.

[79] Jabbar, S., Ullah, F., Khalid, S., Khan, M., and Han, K. (2017). Semantic interoperability in heterogeneous IoT infrastructure for healthcare. *Wireless Communications and Mobile Computing*, 2017.

[80] Fotouhi, H., Causevic, A., Lundqvist, K., and Björkman, M. (2016, June). Communication and security in health monitoring systems – a review. In *2016 IEEE 40th Annual Computer Software and Applications Conference (COMPSAC)* (vol. 1, pp. 545–554). IEEE.

[81] Suryateja, P. S. (2018). Threats and vulnerabilities of cloud computing: a review. *International Journal of Computer Sciences and Engineering*, *6*(3), 297–302.

[82] Singh, N., Kumar, S., Kanaujia, B. K., Choi, H. C., and Kim, K. W. (2019). Energy-efficient system design for internet of things (IoT) devices. In *Energy Conservation for IoT Devices* (pp. 49–74). Springer, Singapore.

[83] Gupta, R., Shukla, A., and Tanwar, S. (2021). BATS: a blockchain and AI-empowered drone-assisted telesurgery system towards 6g. In *IEEE Transactions on Network Science and Engineering*, vol. 8, no. 4, pp. 2958–2967, https://doi.org/10.1109/TNSE.2020.3043262.

[84] Pandey, C., Sharma, S., and Matta, P. (2021, April). Privacy techniques for body sensor network in healthcare internet of things (HIoT) – a critical survey. In *2021 5th International Conference on Computing Methodologies and Communication (ICCMC)* (pp. 385–389). IEEE.

[85] Pereira, J. and Díaz, Ó. (2019). Using health chatbots for behavior change: a mapping study. *Journal of Medical Systems*, *43*(5), 1–13.

[86] Dolph, E., Krause, C., and Oleynikov, D. (2019). Future robotic systems: microrobotics and autonomous robots. In *Robotic-Assisted Minimally Invasive Surgery* (pp. 329–335). Springer, Cham.

[87] Schuelke, S., Aurit, S., Connot, N., and Denney, S. (2019). Virtual nursing: the new reality in quality care. *Nursing Administration Quarterly*, *43*(4), 322–328.

[88] Lauritsen, S. M., Kristensen, M., Olsen, M. V., *et al.* (2020). Explainable artificial intelligence model to predict acute critical illness from electronic health records. *Nature Communications*, 11(1), 1–11.

[89] Utekar, R. G. and Umale, J. S. (2018, August). Automated IoT based healthcare system for monitoring of remotely located patients. In *2018 Fourth International Conference on Computing Communication Control and Automation (ICCUBEA)* (pp. 1–5). IEEE.

[90] Rajput, A. (2020). Natural language processing, sentiment analysis, and clinical analytics. In *Innovation in Health Informatics* (pp. 79–97). Academic Press.

[91] Ratia, M., Myllärniemi, J., and Helander, N. (2018, October). Robotic process automation-creating value by digitalizing work in the private healthcare? In *Proceedings of the 22nd International Academic Mindtrek Conference* (pp. 222–227).

[92] Kashif, H., Imran, S. B., Shabana, R., *et al.* (2020) An intelligent IoT based healthcare system using fuzzy neural networks. *Scientific Programming*, *2020*, 15.

[93] Garg, A. and Negi, A. (2020). A survey on content aware image resizing methods. *KSII Transactions on Internet and Information Systems (TIIS)*, *14*(7), 2997–3017.

[94] Garg, A. and Singh, A. K. (2022). Analysis of seam carving technique: limitations, improvements and possible solutions. *The Visual Computer*, 1–27. https://doi.org/10.1007/s00371-022-02486-2.

[95] Garg, A., Nayyar, A., and Singh, A. K. (2022). Improved seam carving for structure preservation using efficient energy function. *Multimedia Tools and Applications*, *81*(9), 12883–12924.

[96] Garg, A. and Singh, A. K. (2022). Performance analysis of seam diversion based image retargeting technique based on edge detection operators. *Multimedia Tools and Applications*, 1–44.

[97] Dwyer III, S. J., Weaver, A. C., and Hughes, K. K. (2004). Health insurance portability and accountability act. *Security Issues in the Digital Medical Enterprise*, *72*(2), 9–18.

[98] Iyengar, A., Kundu, A., and Pallis, G. (2018). Healthcare informatics and privacy. *IEEE Internet Computing*, *22*(2), 29–31.

[99] Shah, R. and Chircu, A. (2018). IoT and AI in healthcare: a systematic literature review. *Issues in Information Systems*, *19*(3), 87.

Chapter 5

Blockchain for security and privacy in healthcare informatics

Anuj Kumar Singh[1], *Ankit Garg*[2] *and Anand Nayyar*[3,4]

Abstract

The attention and interest of governments and related industry areas in the digitization of healthcare systems have been increased significantly in the recent past and the same is being evidenced by numerous initiatives taken by many countries and sectors. Healthcare informatics is a decisive component of a healthcare system that has demonstrated a growing specialization linking healthcare systems, communication systems, and information technology for improving the quality, storage, and safety of patient related information. Healthcare informatics facilitates organizations and service providers to discriminate between huge amount of big-data and the real meaningful data so as to integrate the meaningful data flawlessly into a health system.

Healthcare systems must protect the privacy and security of patient data, at the same time carrying quality patient care and following the essential regulatory requirements. With the evolution of technologies like Internet of Things (IoT) and Internet of Everything (IoE) and besides their advantages in today's highly pervasive computing era, there has been a significant demand of secure storage of a huge amount of patient data. New kinds of vulnerabilities and security attacks are threatening the IoT and IoE-based healthcare systems due to many reasons including the usage of various kinds of medical devices, nature of wireless communication, ignorance of security by manufacturers of medical devices, and many others. Electronic healthcare record systems must ensure the CIA requirements of security, i.e., confidentiality, integrity, and availability of the information stored. Moreover, this data must be communicated securely among all the authorized entities like medical practitioners, along with the privileges for accessing particular patient's specific data to expedite diagnosis and treatment.

Blockchain technology has evolved rapidly in past few years, and is making most of the centralized systems obsolete owing to its unique properties of

[1]Adani University, Ahmedabad, India
[2]Apex Institute of Technology-Computer Science and Engineering, University Centre for Research and Development, Chandigarh University, Punjab, India
[3]Graduate School, Duy Tan University, Vietnam
[4]Faculty of Information Technology, Duy Tan University, Vietnam

immutability, decentralized, transparency, and reduction in cost. Blockchain is going to revolutionize the area of healthcare informatics since it can facilitate the majority of functionalities required for healthcare information system like privacy of stored data, secure sharing of data, data audit, integrity checks, and identity management. In addition to integrity, confidentiality, and availability, blockchain-based security solutions can provide auditability, accountability, authenticity, and anonymity.

This chapter aims at exploring the different aspects of developing blockchain-based security solutions for healthcare informatics. The chapter identifies the security requirements and challenges in healthcare system. It also describes the security aspects in centralized/decentralized healthcare systems and also elaborates adaptions and impact of Blockchain Technology in Healthcare Systems. Security attributes and architecture of blockchain and its applicability in developing secure frameworks for healthcare informatics have also been explained here. How the issues of identity and trust management can be addressed in healthcare using blockchain, has also been presented in this chapter. The security of big data in healthcare using blockchain, along with the technologies like IoT and IoE have also been addressed in this chapter. Finally, the chapter speaks about the challenges, opportunities, and future insights of blockchain-based healthcare systems.

Keywords: Security; Privacy; Trust; Healthcare; Challenges; Opportunities; Blockchain

5.1 Introduction

Health Informatics is a relatively young, interdisciplinary discipline in the healthcare sector that employs information technology to consolidate and investigate health records for enhancing healthcare outcomes. Healthcare informatics which is a decisive component of a healthcare system, focuses on the study and use of approaches for ameliorating the clinical knowledge, management of patient data, demographic information, and other data useful for patient care and community health. It also empowers doctors to give better care in a faster way, and devote more time with their patients. Important health information can be exchanged all across the chain of care, from the patient to the healthcare administration and even to the pharmacist. With the advent of new technologies like Internet of Things (IoT), healthcare informatics has been further empowered with the usage of smart medical equipment and devices, leading to improved healthcare quality and patient progress reports for persons who require continuous and real-time medical information [1]. Using IoT in healthcare help with early disease detection, diagnosis support, and treatment for things including chronic diseases, elderly care, and fitness programs [2,3]. However, healthcare systems must be able to protect the privacy and security of patient data, at the same time carrying quality patient care and following the essential regulatory requirements. New kinds of vulnerabilities and security attacks

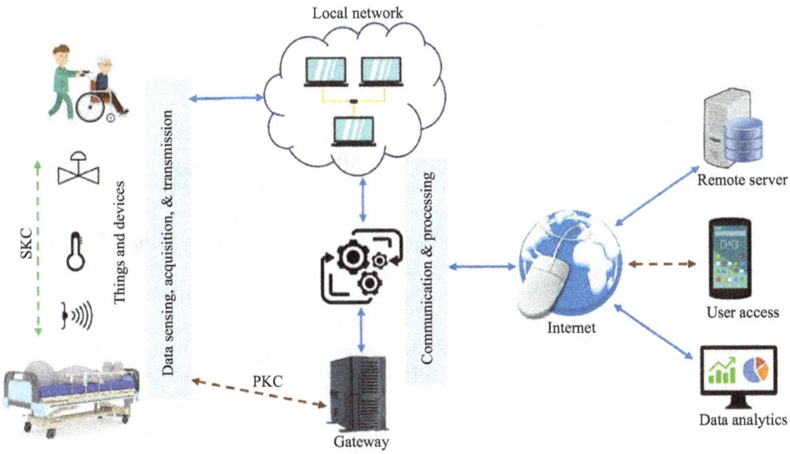

Figure 5.1 A secure IoT-based healthcare architecture

are threatening the IoT and IoE-based healthcare systems due to many reasons including usage of various kinds of medical devices, nature of wireless communication, ignorance of security by manufacturers of medical devices, and many others. A secure IoT-based healthcare architecture is shown in Figure 5.1 that includes three layers of processing – data sensing and acquisition, processing and communication, followed by data analytics [4]. Cryptographic techniques including both symmetric key cryptography (SKC) and public key cryptography (PKC) play a very significant role in ensuring the security of the information stored or transmitted in a healthcare system. It is to note that symmetric key cryptographic techniques are used to safeguard the data at data sensing and acquisition layer while public key cryptographic techniques are utilized to secure information at communication and processing layer [5]. Moreover, public key approaches are also used in back-end data sharing services through remote server. Further steganography using image processing can also be utilized in securing medical images through different image seam-carving techniques [6–8].

5.1.1 Security requirements in healthcare informatics

To ensure the security of any network and system the required security properties must be satisfied. The security requirements of an IoT-based healthcare system can be categorized as either CIA requirements or non-CIA requirements. CIA requirements consist of confidentiality, integrity, and availability while non-CIA requirements include authentication, identification, authorization, accountability, privacy, non-repudiation, freshness, and auditing [9]. The CIA and non-CIA security requirements of IoT-based healthcare system along with their elucidation have been listed in Table 5.1. These security requirements can be ensured by implementing appropriate cryptographic and information security techniques. In addition to these security requirements, an IoT-based healthcare system should also fulfil several requirements of cyber resilience which includes reliability, maintainability, safety, performability,

Table 5.1 Security requirements of an IoT-based healthcare system

Category	Security requirement	Elucidation
CIA requirements	Confidentiality	Ensures that confidential medical information is not disclosed to unauthorized parties
	Integrity	Ensures that medical information of patient is not modified/corrupted/removed by an attacker that lead to wrong diagnosis/prescription. It is concerned with accuracy and completeness of the data during the system lifecycle
	Availability	Ensures that patient's medical information and devices are available for the physicians all the time. It is concerned with the prevention from operational outage/failure of a medical equipment and the continuity of security-related operations
Non-CIA requirements	Authentication	Ensures that only the legitimate person/device is using the system resources
	Identification	Ensures the verification of identities of all the parties (doctors, patients, devices, etc.) in communication
	Authorization	Ensures that the access rights and privileges for all the users/parties are well defined in a way that users/parties can access the system resources according to their defined tasks
	Accountability	Ensures that all the parties are responsible and answerable to their activities and actions in case of unusual or damaging event
	Privacy	Ensures that the confidential medical information of the patient should not be revealed to anyone without the consent
	Non-repudiation	Ensures that the sender/receiver cannot deny the transmission/receipt of the message
	Data freshness	Ensures that only fresh messages are sent and received avoiding replayed and duplicated messages
	Auditing	Ensures the continuous tracking and monitoring of the system and the network

and survivability. Reliability refers to continuity of services of IoT network and system. Maintainability covers the aspects of adaptability, modifiability, configurability, reparability, and autonomy of the system. Safety requirement ensures the non-failure of the system, performability ensures continuous evaluation of system's performance, and survivability property guarantees the continuous operation of system components [9]. The requirements of cyber resilience for IoT-based healthcare system along with their explanation have been listed in Table 5.2.

5.1.2 Security challenges in healthcare informatics

Since IoT-based healthcare system integrates many devices and users, it poses many security challenges too. These includes mobile and medical device security,

Table 5.2 Requirements for cyber resilience in an IoT-based healthcare system

Cyber resiliency requirement	**Property**	**Elucidation**	
Reliability	Continuity	Ensures the continuity of services in the face of system disasters, dissimilar networks, and fluctuating environmental situations	
Maintainability	Adaptability	Ensures that in the course of design and implementation, the system is able to swiftly change and perform the correct function under a variety of operating circumstances	
	Modifiability	Ensures the capability of a system to incorporate new features or to modify existing features during the system design and implementation.	
	Configurability	Ensures the correct functioning of the system in varied operational situations even after adjusting system parameters for a set of processes	
	Reparability	Ensures that the system is able to discover and rectify its flaws, as well as goes back to its normal operating condition	
	Autonomy	• Self-healing	Ensures the ability of a system to detect and identify faults and recover them deprived of losing any information
		• Self-protecting	Ensures the capability of system to protect itself against different attacks and vulnerabilities
		• Self-optimizing	Ensures the ability of a system to optimally utilize the resources (memory, bandwidth, power, etc.) and improve its performance
		• Self-configuring	Ensures the ability of a system to install, configure, and integrate components automatically
Safety	Non-failure	Ensures that the system will not enter into failure if certain disastrous operations occur	
Performability	Performance measures	Ensures the performance evaluation of the system with respect to with respect to some critical parameters including accuracy, speed, memory, and others	
Survivability	Continuous operation	Ensures the completion of tasks and achieving objectives of the system timely even if some devices/nodes become corrupt or in case of some failures, disasters, and attacks	

data breaches, lack of security technologies, account hijacking, unsecured APIs, insider threat, viruses/malware, data violations, and wrong fixes [5].

(i) *Medical device security* – Connecting medical devices with the internal hospital network and the Internet facilitates the entire healthcare system but at the same time it gives rise to many cybersecurity vulnerabilities. Unauthorized access and control over medical device by an attacker and leaking of sensitive medical information that the device generates are two most common vulnerabilities in healthcare informatics. Food and Drug Administration (FDA) of USA has issued guidelines for securing medical devices and the details of these can be found in [10].

(ii) *Data breaches* – Healthcare information systems have been the primary targets for cyber attackers over the years. It has been found in a study that hacking is the foremost cause of exposing the sensitive and critical medical information [11]. Moreover, the most data breaches have been experienced in the healthcare informatics industry as compared to any other one. The main cause of the data breach in healthcare is the increasing value of Personal Health Information (PHI) in the black market and generally cyber criminals get higher incentive for stealing the medical databases.

(iii) *Lack of security technologies* – Many medical organizations fails to implement appropriate security techniques and protocols while their healthcare system is migrated over cloud. It is very critical to ensure the implementation of authentication and authorization procedures for cloud-based healthcare systems. But many organizations take it for granted and in thus the sensitive medical data is exposed to different kinds of threats.

(iv) *Account hijacking* – Through hijacking the accounts of medical individuals, cyber attackers get access to various information resources including sensitive healthcare records. Once an account is hijacked it can lead to serious vulnerabilities including tampering with the devices and health records that can be dangerous for the life of an individual patient.

(v) *Unsecured APIs* – A wide range of APIs and software are developed by the cloud service providers. The availability and security of the cloud services depend on these APIs and software and thus if poorly designed they can lead to infringement of critical medical information or even the misuse. Hacked, exposed, and broken APIs can have serious consequences of data breaches.

(vi) *Insider threats* – Insider threats can also exist in medical informatics where an inside individual unknowingly or intentionally perform some action that compromises the security of sensitive information. The action can be like clicking on some malicious link without knowing the seriousness of the activity or it can be dedicatedly leaking confidential PHI to get some money.

(vii) *Viruses and malware* – In a healthcare information system, there can be many computer terminals and devices holding critical information connected with the network and due to this if any terminal or the device is attacked by the virus or malware then it can lead to loss of data or stealing it. Cyber attackers can use virus and malware to attack the healthcare information system to steal valuable confidential information or to damage it.

(viii) *Phishing attacks* – Using phishing attacks the cyber attackers tries to obtain user password and personal information so that they can get unauthorized access to legitimate accounts. These attacks perform social engineering analysis and generally launched using fake e-mails that look legitimate to the users and once they respond by any action their secret information is revealed.

(ix) *Distributed denial of service (DDoS) attacks* – DDoS attacks attempts to target the network and server of the organization by generating a flood of fake requests and other traffic to make it unreachable and unusable for the legitimate requests. DDoS attacks are generally implemented using botnets and ransomwares that compromise the security of many hacked terminals and then using these as launch pads to send massive fake data. DDoS attacks are very harmful for healthcare information system in the case when healthcare providers want to retrieve PHI, communicating emails, filling prescriptions, and accessing records.

The objectives of this chapters are:

(i) To explore the impact of blockchain technology in healthcare with respect to security, privacy, and trust;
(ii) To identify the existing security solutions for healthcare informatics utilizing blockchain technology;
(iii) To highlight the challenges and opportunities in using blockchain for securing healthcare informatics;
(iv) And, to discuss the use of blockchain technology for big-data security in healthcare.

5.1.3 Organization of chapter

The remaining chapter is organized in eight further sections. Section 5.2 of the chapter discusses the security aspects in centralized and decentralized systems. Section 5.3 focuses on exploring adaptions and impact of blockchain technology in healthcare information systems while architecture and security attributes of blockchain have been presented in Section 5.4. Existing recent blockchain-based secure systems and authentication mechanism for healthcare informatics have been identified in Section 5.5 of the chapter. Section 5.6 discusses identity and trust management issues in healthcare using blockchain and Section 5.7 describes the big data security in healthcare through blockchain. Section 5.8 highlights the challenges and opportunities for the applicability of blockchain technology in healthcare sector. Section 5.9 presents "MedRec" a case study for securing healthcare records via blockchain, and finally Section 5.10 concludes the chapter with future scope.

5.2 Security aspects in centralized vs. decentralized systems

In the centralized systems, information is stored on one mainframe although it can be accessed from numerous locations. The information is acquired, saved, and controlled in one location, but the capability of a legitimate user to access the information from anywhere on the Internet is unaffected by centralization.

The advent of a central information system has a substantial impact on the healthcare sector. It enhances the high quality of therapy, improves the person's safety and security, protects patient information, and, last but not least, reduces operational and management costs. The communication tools provided by healthcare IT systems are simple to use and used by a large population all over the world, resulting in reduced in unwanted interactions and the transfer of viruses. As a consequence of the usage of centralized healthcare solutions, people now have easier access to information, and they also find it easier to obtain medical care in challenging time. The importance of addressing security and privacy breaches associated with centralized healthcare systems and information is growing. As per the study presented by Seh *et al.* [12], attackers steal and sell patient's data for up to 20 times extra the cost of banking data while, during last 20 years alone, security attacks on health records have faced a loss of approximately $30 billion. Researchers have recently been diverted from adopting centralized systems for healthcare due to these prevailing security issues. A decentralized system is needed to ensure stability and scalability, as well as to avoid many-to-one traffic patterns. We can also eliminate single points of failure and information disruptions by using such decentralized systems. Risk management and security compliance are easier to handle and enforce when security functions are implemented in decentralized systems. It provides every module greater control and makes security a shared responsibility. It integrates security into the system design so it can be interwoven into all aspects of the IT infrastructure, protecting the organization. Typically, information security has been handled by a single, centralized team of specialists. They established the rules, developed policy, and were in charge of approving and securing all new technology utilized by the organization. In today's cloud environment, decentralized security is more appropriate. Various teams have the authority to make technology and security decisions on their own. Rather than forcing a centralized staff to keep up everything being utilized across the business, decentralized security makes it their duty to protect the solutions they choose. Decentralized security is being used in a wide range of businesses to distribute security functionalities and keeping information safe everywhere while retaining a great degree of technical freedom [13].

5.3 Adaptions and impact of blockchain technology in healthcare systems

In today's computing era, blockchain is the fastest growing decentralized and distributed technology. Blockchain technology is based on distributed ledger technology for executing digital data transactions on peer-to-peer (P2P) network that can be disseminated in private or made public to all users of the entities, enabling the data in any form to be kept in a secure and verifiable manner.

Decentralization by attempting to make the ledger accessible to all entities, immutability by attempting to make the blockchain extremely difficult to tamper with censorship-resistant, availability by offering all entities with a copy of entire blockchain to obtain access to each transaction records that is time-stamped, and anonymity by allowing every user to communicate with the blockchain through a generated address that does not reveal the user's true identity are some of the key features of blockchain [14]. Better data interchange among healthcare providers implies more accurate diagnoses, better treatments, and a greater potential for healthcare organizations to provide cost-efficient care. By allowing them to track data provenance and any modifications made, blockchain technology can allow multiple players in the healthcare value chain to share access to their networks without jeopardizing data security and integrity [15]. Blockchain has the ability to drive preventative medicare and community-based healthcare model innovation. The capability of distributed ledger technology to maintain data integrity while communicating between parties helps enable coordination between rising trends in healthcare, which are critical to community health improvement around the world.

The most common worries about information transferred between different entities in the existing system are security and trust. Information can be entered at any point along the communication chain, causing trust concerns, particularly in the healthcare industry. There are also worries about numerous vendors holding invalidated versions of the same patient data, resulting in various inaccuracies, inconsistency, and incompleteness. It is no surprise that healthcare professionals are anxious, given stories of security breaches, data manipulation, and the ever-present hacking threat. Because blockchains are cryptographically secure and data authentication can be performed using digital signatures that are unique to each user, this technology could be the solution to many of these issues.

The emergence of blockchain technology as a transparent and responsible mechanism for storing and transmitting information is creating the room for new opportunities for data privacy, integrity, and security in healthcare informatics. In the last several years, blockchain technology has received a lot of attention from both industry and academia for applying it in the healthcare sector [16]. A blockchain possesses the power to remove the issue of counterfeit medications in the supply chain of medicare that can put patients at risk all over the world. In the Internet of Medical Things (IoMT), blockchain technology has now been examined in different of medical applications, including data storage, management, device connectivity, and security. Mostly the advantages paid by blockchain technology in the healthcare sector influenced the quality of experience of many of these stakeholders and end users, including patients, caretakers, academics, drug makers, and insurance firms, positively. One crucial step toward making the healthcare system smarter and enhancing the quality of medical services and users' experiences is the capability to communicate healthcare data without compromising users' privacy and security. Many significant developments in the healthcare field are built on blockchain technology. Sources of data, blockchain technology, applications for healthcare, and stakeholders are the four conceptual layers of emerging blockchain-based structured healthcare systems. Figure 5.2 shows a workflow for healthcare applications based on blockchain technology [16].

Organizational variables, technological factors, and environmental characteristics all have an impact on blockchain adoption in the healthcare sector. Organizational factors are those which contribute to a healthcare provider's internal environment. These factors will have a favorable or unfavorable impact on blockchain adoption in healthcare. Technological factors include technologies that

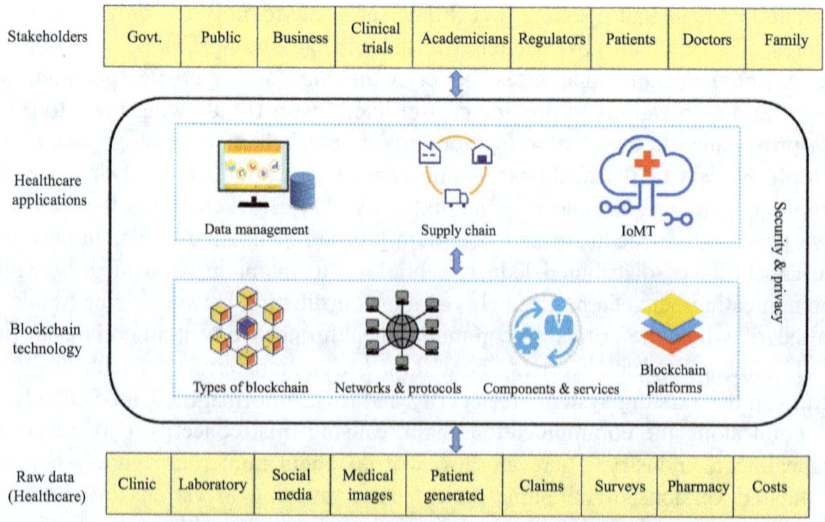

Figure 5.2 Blockchain-based healthcare system

businesses should be able to afford in order to use blockchain technology. These include reliability, transparency, durability, auditability, integrity, centralized registries, immutability, scalability, and speed. PEST (Political, Economical, Social, and Technological) subcategories are used to classify environmental factors. These variables, in contrast to organizational characteristics, cover the external factors that influence adoption [17].

5.4 Architecture and security attributes of blockchain

Blockchain can be best described by the phrase "immutable and distributed ledger". A ledger is significant since it keeps track of all transactions involving a set of business items. Organizations interact in a distributed ledger system to maintain a consistent record of a duplicated ledger through a process known as consensus. A bank account is the most common kind of ledger that can be encountered because it holds both a present value and a transaction history.

5.4.1 Architecture of blockchain

A blockchain network is composed of a number of participants, all of whom has a duplicate of the ledger. Because no single participant controls and maintains a blockchain ledger, it is decentralized. This ensures the distributed property of blockchain. Since every block of transaction is cryptographically hashed and referenced to the preceding block, it is computationally hard to manipulate the transactions in a blockchain ledger. Any transaction inside this chain cannot be altered without invalidating the hashes. As compared to a centrally controlled ledger, the decentralization and immutability properties of a blockchain offers

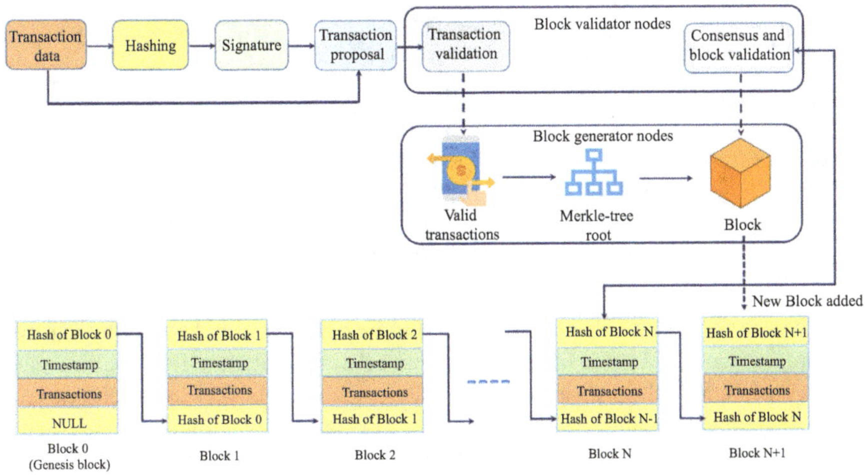

Figure 5.3 Typical architecture of blockchain

greatly improved trust, transparency, and resilience [18]. The key distinction between such a blockchain and a traditional database is that a blockchain enables encrypted data to be connected through blocks instead of being stored as raw data. Consensus and validation are two central tenets of blockchain technology. This is what identifies it as a cutting-edge technology. It is designed to function as a monetary system and has the same structure on all nodes. The main components in the architecture of blockchain (as shown in Figure 5.3) are block, transaction, Merkle tree root hash, previous block hash, timestamp, genesis block, and mining [19].

The transactions created by the users are first hashed then signed using the private key of the user. These hashed and signed transactions are proposed for validation by the validator nodes. If found valid, these transactions become valid transactions and are passed to the block generator nodes that takes the validated transactions and computes the Merkle-tree root which in-turn is used to create the new block to be added in the blockchain.

5.4.2 Security attributes of blockchain

The blockchain network is designed to ensure consistency of data and provenance, to be tamperproof and auditable, to be pseudonymized, to be immune to DDoS attacks, to be private, secure, and secret. The blockchain network is distinctive, transparent, and safe for processing online transactions because of these qualities [20]. This section addresses the security characteristics of blockchain-based systems. The primary security attributes of blockchain illustrated in Figure 5.4 have been explained below.

(i) *Consistency of data* – The notion of maintaining the same copies of data per each node of the network at all stages is known as data consistency in a blockchain network. The consistency of data in the context of Bitcoin is

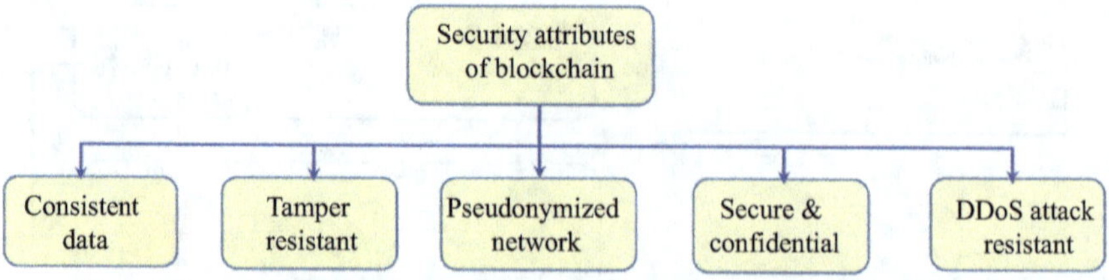

Figure 5.4 Security attributes of blockchain

controversial, as it has been claimed to be both weak and strong by various researchers [21]. For distributed networks, a paradigm for assuring data consistency (the subsequent consistency model) is presented that strikes a balance between availability of data and consistency, updates the data in a slow way, and provides the readers with the new values [22]. Consistency does not really imply that the data would be made consistent instantly; rather, it indicates that the information/data will be consistent ultimately, regardless of how long it takes. However, if the reader reads the node prior to the data update, the model could return stale data. A consistency model must ensure that no reading may be conducted until all nodes have been updated in order to cope with such scenarios. As a result, the blockchain saves the transactions in blocks, and when a new transaction request arrives, the miner accepts the request and initiates the proof of work method, whereby the block is distributed to other nodes for verification. Then just that block is added to the blockchain, and the other nodes accept it by putting its hash equal to the hash of the preceding block [23].

(ii) *Tamper-resistant data* – The capability of data to withstand any form of attack or tampering in the system, data, or any product, whether purposely or accidentally, is known as tamper-resistance. The blockchain is believed to be tamper-resistant, that means that the data recorded or generated on its blocks can never be manipulated or tampered with in any way, neither during nor after generation. The information can be tampered in two ways: first, whenever the miner attempts to alter the data; second, when the adversary strives to tamper with it. In a blockchain-based system, tampering can be prevented since transactions are protected using the hash function and digital signatures.

(iii) *Secure and confidential data* – Data security in the blockchain pertains to permitted access and secure storage, whereas data confidentiality speaks to a sensitive data's privacy. The blockchain was originally developed for the purpose of processing electronic currency, but its scope is far broader. It can be used for copyright management, contract management, and the digitization of a variety of other industries, like healthcare and sales. One of the most important aspects of the blockchain-based decentralized system is transaction secrecy. Pseudoidentities are utilized in the blockchain to make the nodes secure and secret. The purpose of blockchain security is to ensure that data is not subjected to malevolent or nonmalicious attacks.

(iv) *Pseudonymized network* – The users' pseudonymized credentials like identity or name are protected by the blockchain network. The user's public keys are kept as hashes in the blockchain, which are generally data footprints. These hashes (user's public keys) are used by nodes in the blockchain network to converse with the framework and safeguarding the users' actual credentials, making the users' addressed their pseudoidentity. Additionally, a person may own multiple pseudoidentities. In the applications such as healthcare, banking, insurance, news media, and others, the notion of a pseudonymized network in blockchain enabled systems makes the system more private and confidential for users [18].

(v) *Resistance from DDoS attacks* – Making resources unavailable for the legitimate users in a distributed blockchain is said to be a DDoS attack. The offender assaults the server with fake resource requests, producing a shortage of resources for valid users. DDoS assaults can be launched on the blockchain by the network's different nodes that are spread widely. By exploiting the system's shortcomings, the intruders seize control of some network nodes. It is tough to keep such attacks under control by just blocking nodes one at a moment. As a consequence, distributed peer-to-peer interconnections are the option for maintaining the blockchain, which guarantees that the blockchain processes transactions despite when several nodes fail.

5.5 Blockchain-based secure systems for healthcare informatics

The data is kept on a blockchain, which is really a decentralized node network. It is a great way of keeping confidential information safe inside the system. This technology aids in the secure and private communication of critical data. It is a perfect way to keep all the important records in one place safely. Using an individual health database of a patient, blockchain also accelerates search for candidates who meet certain trial criteria. In the healthcare sector, the necessity of innovation is growing at ever-increasing rates. Today's requirement is for high-quality healthcare facilities that are accompanied by innovative and cutting-edge technology. Block-chain is going to play a vital role in revolutionizing the health sector in. Furthermore, the health-care environment is transitioning toward a patient-centered strategy that emphasizes two primary aspects: always-accessible services and suitable healthcare resources [24,25]. The SWOT analysis of the applicability of blockchain performed in [26] has been shown in Figure 5.5. This study seeks to evaluate the usefulness of blockchain in healthcare sector in order to recognize any crucial aspects that future academics can investigate. Numerous schemes based on blockchain to secure the healthcare informatics have been proposed by different authors [27]. In addition to this many low cost authentication schemes have also been proposed by different authors in [28,29] but those have not utilized blockchain. The recent such approaches have been highlighted in this section of the chapter.

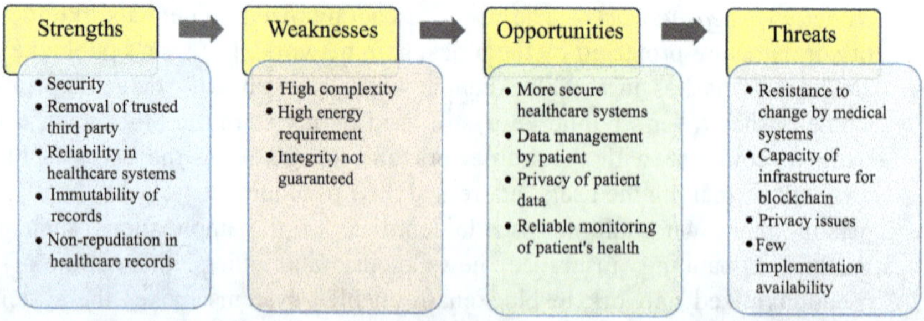

Figure 5.5 SWOT analysis of the applicability of blockchain in healthcare

Zyskind *et al.* [30] proposed the privacy platform based on Enigma that uses blockchain to control access and auditing logs, as well as security and privacy goals like a tampering proof transactions. Enigma uses a multi-party algorithmic approach and a provable secret exchange mechanism to ensure data privacy. Authors believed that Enigma can eliminate the requirement of a trusted third-party system, permitting personal data control to be done in private. Xia *et al.* [31] created a framework for data privacy and security using the blockchain. The authors developed a permissioned blockchain that enables access to data queries by supplying knowledge about the data kept in a cloud repositories. They used information grantors that allow for the accumulation and assessment of data, resulting in value synthesis. The system is lightweight, agile, and extensible, according to their experiments.

Yue *et al.* [32] developed the healthcare information gateways system, a decentralized vulnerability-control system using blockchain. Here, the patient can securely possess, administer, and redistribute his data without requiring complicated actions, providing an unique underlying approach to enhance healthcare systems while preserving patient private information. Due to the use of a decentralized network of individuals and a public ledger, the findings indicate that this system is reliable and auditable. In [33], authors have presented a solution for the Electronic Health Record termed as CP-ABE (Ciphertext-policy attribute-based encryption). For the patient, family, and the doctor, Ramani *et al.* [34] suggested a blockchain-based secure and efficient data access solution. This system also ensures the system's integrity.

Nguyen *et al.* [35] introduced a new EHR sharing mechanism on the mobile cloud environment combining the blockchain with a decentralized IPFS (interplanetary file system). Significantly, the authors developed a trustworthy access control technique using smart contract to confirm secure EHR dissemination amongst medical practitioners and patients. The study concluded that the presented work offered an efficient solution for trustworthy relevant records on the cloud, including the protection of critical medical data from potential threats. Chen *et al.* [36] devised a framework for storing blockchain-based private health data in the cloud. Authors utilized blockchain to establish a storage supply chain wherein all transactions were verifiable, immutable, and open. This protocol established the permissions for three types of activities, as well as the block generation and principal purpose of the healthcare blockchain. They also established an infrastructure for exchanging health records that safeguards health data management applications while without infringing on privacy policies.

Zhang and Lin. [37] established a blockchain-based secured and confidentiality preserving health records transfer mechanism for e-Health diagnosis advancements. Authors highlighted the blockchain consensus technique, which is a proof-of-conformance system for creating confirmed blocks. Furthermore, the authors used public-key cryptography with a blockchain-based search query. After obtaining trapdoors from the patient, a practitioner is able to discover and retrieve the expected history of health information in order to improve the diagnosis. Furthermore, authors stated that proposed eHealth system delivers privacy preservation, confidentiality, and secure medical data searches. Compared to the Bitcoin network, that partitions network contributors into clusters and preserves one copy of the blockchain for every cluster, Ismail *et al.* [38] suggested a new blockchain system for medical data management, which minimizes processing and communication overlaps. Their design requires the use of channels, which allow for safe and anonymous transactions among network participants. Additionally, the authors proposed a solution to the issue of counterfeiting in the Bitcoin network. By examining diverse threats, the authors confirmed the usefulness of their proposed framework in ensuring protection and confidentiality to the Bitcoin network. The authors concluded how the proposed design handles the threats in efficient manner. Radhakrishnan [39], presented a secure EHR authentication mechanism (shown in Figure 5.6) using blockchain that covers all the security aspects related to healthcare informatics.

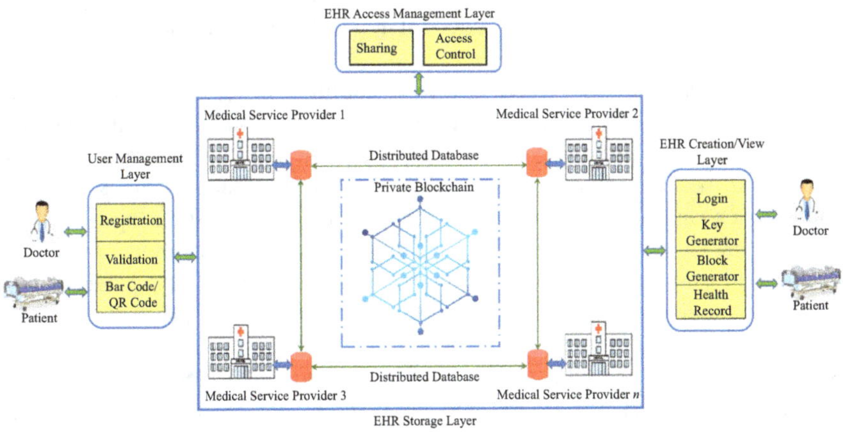

Figure 5.6 EHR authentication mechanism using blockchain [31,40]

5.6 Identity and trust management in healthcare using blockchain

IoT devices are generally used for the controlled, corporate environments these days. A homogeneous and permissionless ecosystem is expected to properly recognize the benefits of IoT applications. All IoT-based systems, even those that are acquaintances to one another, would have been free to transact activities and resources across domains. Uniquely resolvable identities are required to realize those applications. Due to the absence of a trusted authority, quantifiable trust in identities and their authentication is difficult to achieve in such an environment [41]. People, not the gadgets, are indeed the focus of identity and trust management frameworks. In contrary to IoT networks, they are critical for local area networks and are utilized in the conventional internet of computers, which has substantially more standardized software, hardware, and protocols. Existing frameworks, on the other hand, cannot be easily applied to IoT settings due to its distinctive IoT properties like as cross-compatibility, scalability, and portability. DLT has already been suggested as a solution to the challenge of identity management in trustless IoT systems. As the major issues in IoT are associated with security risks, DLTs can lead to a safe and secure way for storing and validating identities. Identity fraud and related offences such as IP spoofing or DDoS assaults can be decreased with DLT-base public key infrastructure solutions based on DLT, which do not have centralized failure points.

The growth of decentralized identity and trust management has been enabled mostly by recent emergence of blockchain technology. Customers are allowed to properly manage their identity on a blockchain that is not managed by a single centralized authority, owing to decentralized identity and trust management. Individuals can choose whatever they want to communicate, whenever they want to communicate it, and whom they want to communicate it. Various methods based on blockchain technology have recently been developed [42]. Decentralized trustworthy identity and self-sovereign identity and are two types of decentralized identity and trust management. Self-sovereign identification is a user-centric paradigm that does not necessitate a central authority, which could lead to identity exposure.

Trust management is an essential component for offering customers with a safe and reliable service. In a cloud virtualized environment, conventional trust management methods are centralized and lead to a single point of failure. Blockchain has evolved as a potential application case for an area that demands trust and reliability in a variety of ways. Numerous blockchain-based trust managing systems have been developed by different researchers. The standard trust model focuses on a centralized trust controller for third parties that causes congestion, delays, congestion and the risk of a single point of failure. Further, because proof of trust is not disclosed to every user in a trust management system, the outcomes of the trust assessment are fully verifiable by the users. Existing medical devices connected to IoT and cloud services can process and store health data on the cloud. In IoT and cloud-based healthcare, privacy, security, and trust challenges are emerging at an alarming rate [43].

Blockchain technology has gone mainstream and application in the areas of trust management, cryptocurrencies, immutability, and security. The resilience of the system is unaffected by the breakdown of a few nodes in the absence of a third party. Through digital signatures, as well as the data chains and consensus methods in the implementation of rules and data documents, provides transparency, non-repudiation, and privacy. The decentralized element of the blockchain is extremely crucial when it comes to the creation of a modern decentralized trust architecture. Blockchain is a cutting-edge way of establishing a trust-based cloud computing infrastructure. In this scenario, blockchain might be a viable option for providing trust, security, trust, authentication services [44]. Figure 5.7 depicts a high-level reference architecture for establishing trust in the healthcare context utilizing blockchain.

Organizational trust and technical reliability are inextricably linked when it comes to health care technology. While exploring Industry 4.0, one of the most significant takeaways is the need to discover more about how to establish trust in the framework of digitized and virtualized health. It has to do with humans having total control over information sharing and trusting systems and knowledge (human system). In both a conventional and virtualized medical systems, the patient must

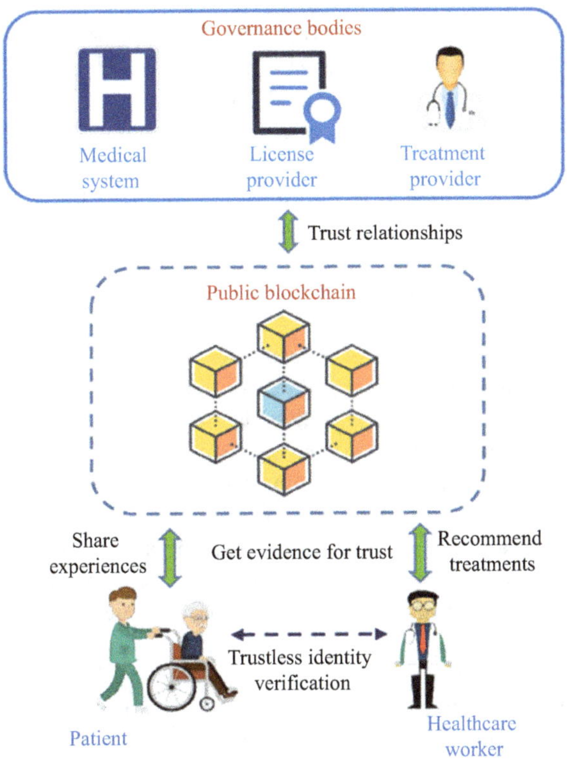

Figure 5.7 Establishment of trust in healthcare using blockchain [37,45]

trust that the healthcare provider has the necessary competence (and authority) effectively deal with his or her health issue. The healthcare provider or caregiver must convince to the patient that he or she is competent to manage the patient's specific health concern; alternatively, the patient may seek treatment elsewhere.

5.7 Big data security in healthcare through blockchain

Healthcare systems are now being revolutionized digitally in the form of e-health records as a consequence of technological developments. Healthcare big data allows health professionals to retrieve and analyze health information in a new dimension, in addition to the advancement of analytical tools in healthcare. Medical big data comprises patient health records, clinical notes, results of medical lab testing, and a variety of information recorded from IoT devices. Healthcare big data is exploding as a result of technological developments. To handle, extract, secure, and preserve healthcare big data, current technologies are necessary. Multiple electronic big data sources, such as patient reports, digital health records, clinical data, internet data, and e-health reports, are already available commercially. Many systems, applications, and communication methods have been developed during the last decade to establish a secure storage of health records. All of these solutions are essentially aimed at enhancing the privacy of medical data, detecting illnesses, and establishing disease prevention strategies around the globe. Each medical center preserves its patients' information in the form of an EHR in the traditional client-server health system [46]. Patients' health records, i.e. healthcare big data spanning many hospitals, are now stored digitally and made readily available to patients and healthcare providers. Patients' sensitive and private big data can indeed be encrypted before being transmitted to the cloud or a trustworthy cloud server, allowing physicians and healthcare institutions to safeguard their patients' data. The former demands a massive proportion of memory intensive computation, which really is undesirable in a healthcare context. Likewise, if patients are apprehensive about their information being exploited, the latter is incredibly difficult to work with. In virtualized systems, there still is the concern of a single point failure, as well as data security.

In [47], authors proposed a three-layered solution (depicted in Figure 5.8) to handle medical big data using blockchain. The application layer is the foremost layer, and it will serve as an administration gateway for access permissions provisioning, report creation, and other operations. The application layer will be the gateway to the blockchain's private execution. The blockchain is the second layer, consisting of a network of nodes (that can only communicate with the application layer), other nodes, a key generating entity, and the database. The resources that would create the public/private keys to be used in the blockchain are the key generating entity. The encrypted database layer is the third layer, and that's where all electronically confidential health Information will be kept. When a medical practitioner or a healthcare center generates a medical record, it is encrypted and then sent to the database for safekeeping. When data is stored to the database, a

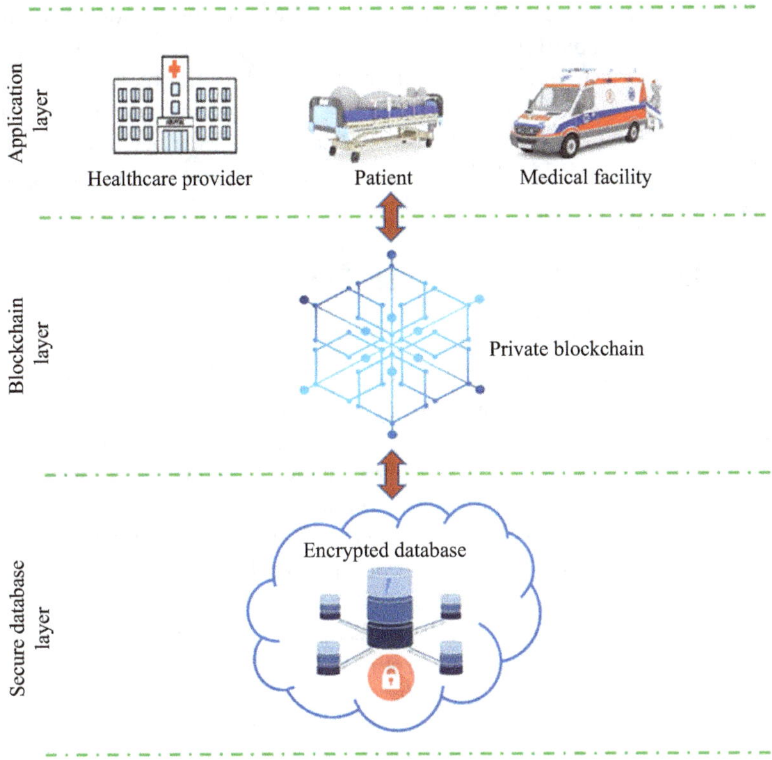

Figure 5.8 Three-tier blockchain-based solution to handle big-data in healthcare

reference pointer to that medical record is added to the blockchain, together with the distinct identity of the individual who generated it. The patient who owns this information will be informed that a new record has been added.

Blockchain enables us to structure data in a manner that transactions may be verified and logged when all parties involved reach a consensus. The system works on the principle of an authoritative ledger that captures all events. A data warehouse, EHR system, or a repository administered by a health information exchange are all instances of current ledger processing and storing data in a single, centralized location. Most of these systems are developed autonomously and may generate and store data in their own proprietary format, leading to data pyramids and interoperability concerns that frustrate patients, clinicians, and researchers. A blockchain, on the contrary, necessitates that each contributor, or node, retain a copy of the record. Every time a change happens, it must be validated against every node and confirmed; this improves the security and decreases the likelihood of someone committing an unauthorized modification [48].

Blockchain will enable appropriate verification that the medical industry demands that will be the thing that all stakeholders can trust. One entity is in responsible of data storage as part of the process, but everyone participating in the blockchain is accountable for ensuring the integrity. Apart from a public financial blockchain, the medical blockchain has to be restricted to the amount of permitted parties, that would include a healthcare professional, a patient, medical facilities, and, if essential, the patient's family. Since there are no intermediaries in possession of the data, this form of collaboration will alleviate many issues that plague the healthcare sector.

5.8 Challenges and opportunities in blockchain-based healthcare systems

When immutability and transparency are desired, blockchain is considered ideal for recording transactions with a minimal digital footprint. In other terms, it works best in circumstances where network participants lack security and data chunks are short. Validation of patients, clinicians, or suppliers; supply chain; and administrative structure of dynamic patient permission to information usage are all examples of situations in medical sector where blockchain could be extremely valuable. Due to the computational and capacity constraints of replicating the blockchain throughout every network node, blockchain is not very well suited for storing large huge amounts of data. Big records, such as extensive electronic medical records or genetic information records, would be expensive and inefficient to store on the blockchain.

5.8.1 Challenges in adopting blockchain for healthcare

Adopting blockchain technology is significant in the modern healthcare systems. Distributed ledger technology, for instance, might enable researchers to privately exchange information and contribute to solve complicated medical issues while also minimizing medical record errors. More significantly, blockchain technology can assist in the securing communication of sensitive healthcare data over a decentralized system. Because the sophisticated structure of blockchain, an attacker seeking to gain access to data would need to be able to regulate every machine on the network. Furthermore, since each patient owns the data in the system, it can only be accessible by those individuals or entities that have permission from the patient. This process adds yet another layer of protection. According to [49], despite the numerous benefits that blockchain provides, it nevertheless faces the following obstacles for healthcare organizations:

(i) Blockchain is a novel technology that has ties to the cryptocurrencies. Some medical research organizations are dubious of blockchain's application to their area since many people equate blockchain with Bitcoin.
(ii) Every entity involved in the care of patient and health research must be on board and ready to adapt to the multiple changes that this new technology will entail in order for blockchain to function.
(iii) How will patients who are unconscious provide the security key needed to access their records? Potential alternatives includes medical bands with biometric security.

(iv) Blockchain may struggle to support huge data sets (such as CAT scan and MRI images) in one transaction only due to its ledger history. In order to store and track huge amounts of data, organizations will need to use backend repository and encrypted modules.
(v) Since blockchain does not completely remove or replace modified records – rather, it merely adds more blocks to the blockchain, demand for more storage is increasing, posing both financial and technical issues.
(vi) Perhaps there is a way to incorporate blockchain into medical systems without introducing additional processes for already overloaded healthcare staff when entering the data.

Nonetheless, the positives of blockchain in health sector may exceed the downsides. Blockchain, when combined with artificial machine intelligence, has the potential to completely revolutionize the healthcare industry. Blockchain technology has already been used by some large healthcare organizations. Table 5.3 shows the existing difficulties and potential in blockchain-based healthcare [50,51].

5.8.2 Opportunities in adopting blockchain for healthcare

The real potential of interoperability might be unleashed via a blockchain-based healthcare information sharing. Current intermediaries' friction and expenses can be substantially reduced with blockchain-based systems. Blockchain's prospect has far-reaching ramifications for all participants in the healthcare system. Utilizing technologies to connect multiple systems and generate information and better measure the quality of healthcare has the promise to be revolutionary. In the long run, a countrywide blockchain network for digital medical data might boost efficiency and help patients achieve better health results [52]. Though blockchain technology is not really a silver bullet for data standardization or integration of systems, it would provide a new potential distributed framework for enhancing and promoting the integration of healthcare information across multiple applications and stakeholders. It overcomes a number of problems while also making the system more efficient, decentralized, and safe [53].

The major opportunities for before the blockchain in healthcare includes:

(i) *Disintermediation of trust* – Since all parties might well have access to the distributed ledger to establish a secure exchange without sophisticated managed trust, decentralization of trust would likely not demand the deployment of a health information exchange operator.
(ii) *Decreased transaction cost* – Because of decentralization and near-real-time execution, the system would be more efficient.
(iii) *Distributed architecture for electronic identities of patients* – A distributed architecture for patient digital identifiers that leverages cryptography to protect private and public identities generates a single, more secure means of securing patient identity.
(iv) *Data sharing* – Data sharing allows for relatively close real-time updates to all parties on the network.

Table 5.3 Challenges in blockchain-based healthcare system and opportunities [50]

Concern	Challenges	Opportunities
Accuracy of health data	• There is no such thing as a global medical information database • The data has been saved. The quality of a patient's diagnosis has an impact on whether or not they get treated	• Blockchain lists were created as a result of the blockchain's link with medical care • The diagnosis accuracy that is forwarded has to be improved
Interoperability of health data	• In medical data, interoperability is not investigated nearly as much • If information is passed from patient to caregiver, or from insurance presenter to presenter, the data becomes jumbled	• Blockchain models can easily be utilized by medical organizations to reverse digitalized health-related listings
Data security	• Medical data, on the other hand, is easily reformed with fake commitments • Furthermore, the data could have been lost due to a natural disaster	• The information lists that are stored in the blockchain are unaltered. This eliminates the chance of information leakage • Because there is no way of knowing when a natural disaster will strike, medical data stored on blockchain technology should be protected from destruction
Hiding of medical data	• It is possible that the maximum amount of resources and time will not be managed, as well as a non-organized medical organization dataset	• The data in the block chain network is saved and can be easily retrieved by connected groups at a later date
Drug treatment	• The number of fraudulent medical recommendations is high • The medicines are difficult to distinguish from the verified party • It's possible that it had a substantial impact on patients' health	• The data on the blockchain is immutable. It is not feasible to change the recommendation once it has been accepted • In terms of pressing revert on recommendations and medication circulation, positive influences are straightforward

(v) *Secure distributed access* – Accessibility to patient continuous health information through the distributed ledger in a decentralized and secure manner.

(vi) *Smart contracts* – Smart contracts establish a standardized, rule-based mechanism for obtaining access to patient information that may be granted to specified health agencies.

5.9 "MedRec" – a case study for securing EHR through blockchain

MedRec, a cutting-edge, decentralized record management system that uses blockchain technology to manage EHRs has been suggested in [54]. The MedRec solution provides patients with a comprehensive, unchangeable log and simple access to their health information across practitioners and treatment places. MedRec enables authentication, confidentiality, accountability, and information sharing factors for handling sensitive information by utilizing special blockchain capabilities. The four key challenges addressed by the MedRec blockchain implementation are – fragmented, delayed access to health data; system compliance; patient organization; and improved data quality and quantity for clinical research. In MedRec, the block information stands for the custody of the data and viewing privileges provided by participants in a personalized, peer-to-peer network. We can automate and trace certain state transitions using "smart contracts," which are enabled by blockchain technology. The ledger of patient–provider connections that link a health record with viewing rights and information retrieval procedures for execution on external data sources is kept up to date using smart contracts on an Ethereum blockchain. The blockchain includes a cryptographically generated hash code of the record to thwart tampering and preserve data integrity. Patients can authorize to the sharing of their records between providers, and providers can add additional records relevant to particular patients.

By offering a designated smart contract that compiles references to all of a user's patient-provider associations, MedRec prioritizes usability by establishing a single point of contact to inspect for any changes to the user's health history. MedRec WeM use a DNS-like solution to match an existing and frequently used form of ID to the user's Ethereum address. The system manages identity authentication via public key cryptography. Following a blockchain verification for permissions via our databases authentication mechanism, a synchronization algorithm handles data transfers "off-chain" between such a patient database and a provider database.

First, let us focus on security and robustness. The blockchain implementation in MedRec benefits from a lot of significant decentralization aspects [55]. MedRec has a robust backup model that relies on the system's different involved parties to eliminate a central point of failure. Local copies of authorization information are preserved on each network node, and health records are kept in independent provider and patient database. In an era of cyber-attacks and information leakage, the solution of MedRec avoids establishing a central target for information attack since both public authorization record and the underlying medical data remains scattered. Although a scaling limit on "block size" or storage capacity causes robustness issues for some blockchains, these parameters can be changed to optimize for other performance needs in a private blockchain. It should be underlined that MedRec makes no promises in ensuring the protection of specific provider database; this must still be adequately administered by the local IT systems engineer. As MedRec methodology assumes provider nodes that are restricted by external regulation restricting data copying in the healthcare use case, like HIPAA, it also does not seek to address the undesired data copying difficulties associated with Digital Rights Management.

The adoption of blockchain technology puts a lot of regulations on privacy. Transactions' pseudonymous aspect presently facilitates data forensics, or extrapolating treating patterns from frequency analysis. Through the study of network activity, it is reasonable to conclude that a certain entity has regularly communicated with another network element without revealing any names. It is still being investigated how to improve obfuscation on the blockchain while preserving auditability. Providing the blockchain a "permissioned" framework where only pre-approved, hardcoded nodes are allowed read entry to the ledger is one potential approach. By doing this, potential attackers would be unable to gather frequency-based insights from the blockchain data. In order to protect against unauthorized or harmful content access, encryption can also be added to the off-blockchain data synchronization processes. In order to provide the aggregated research data to medical research "miners," which is outside the purview of the initial prototype but unquestionably essential for future development, a comprehensive anonymity study of privacy-preserving query design is required.

5.10 Conclusion and future scope

Blockchain technology is now being utilized in almost all kinds of distributed computing environments and it becomes vital when used in sensitive applications like healthcare. This chapter has highlighted the major security aspects of the applicability of blockchain in healthcare informatics. First, the security requirements and challenges of healthcare informatics have been discussed here followed by the security aspects of centralized and decentralized systems. Then the adoption of blockchain in healthcare ecosystem and its impact has been revealed by highlighting the key points. Architecture of blockchain along with the security attributes of the blockchain have also been explained here. Then the recent works mentioning the blockchain based systems for healthcare informatics have been discussed. Owing to the significant importance, the issues of identity and trust management using blockchain in medical systems have also been explained in this chapter. Big healthcare data security using blockchain has also been discussed along with the three tier framework to provide security and privacy. Finally, the chapter elaborates the challenges and opportunities in adopting blockchain for healthcare informatics. Through this chapter it has been explored that although there are some existing challenges in applying blockchain for healthcare information system security but these challenges can be overcome in the near future and the blockchain technology is going to be vital in medical sector. The future scope of the work presented in this chapter is to devise the framework or mechanism for healthcare informatics while overcoming all the challenges highlighted here.

References

[1] Strielkina, A., Kharchenko, V., and Uzun, D. (2018). Availability models for healthcare IoT systems: classification and research considering attacks on vulnerabilities. In *2018 IEEE 9th International Conference on Dependable Systems, Services and Technologies (DESSERT)* (pp. 58–62). IEEE.

[2] Jaigirdar, F. T., Rudolph, C., and Bain, C. (2019). Can I trust the data I see? A physician's concern on medical data in IoT health architectures. In *Proceedings of the Australasian Computer Science Week Multiconference* (pp. 1–10).

[3] Garg, A. and Singh, A. K. (2021). Applications of Internet of Things (IoT) in green computing. In *Intelligence of Things: AI-IoT Based Critical-Applications and Innovations* (pp. 1–34). Springer, Cham.

[4] Vijayalakshmi, A. V. and Arockiam, L. (2018). A secured architecture for IoT healthcare system. In *International Conference on Computer Networks, Big Data and IoT* (pp. 904–911). Springer, Cham.

[5] Singh, A. K. and Patro, B. D. K. (2019). Security of low computing power devices: a survey of requirements, challenges and possible solutions. *Cybernetics and Information Technologies*, 19(1), 133–164.

[6] Garg, A. and Singh, A. K. (2022). Analysis of seam carving technique: limitations, improvements and possible solutions. *The Visual Computer*, 1–27.

[7] Garg, A., Nayyar, A., and Singh, A. K. (2022). Improved seam carving for structure preservation using efficient energy function. *Multimedia Tools and Applications*, 81(9), 12883–12924.

[8] Garg, A. and Negi, A. (2020). A survey on content aware image resizing methods. *KSII Transactions on Internet and Information Systems (TIIS)*, 14 (7), 2997–3017.

[9] Nasiri, S., Sadoughi, F., Tadayon, M. H., and Dehnad, A. (2019). Security requirements of internet of things-based healthcare system: a survey study. *Acta Informatica Medica*, 27(4), 253.

[10] Sivan, R. and Zukarnain, Z. A. (2021). Security and privacy in cloud-based e-health system. *Symmetry*, 13(5), 742.

[11] https://www.fda.gov/medical-devices/digital-health-center-excellence/cybersecurity

[12] Seh, A. H., Zarour, M., Alenezi, M., *et al.* (2020). Healthcare data breaches: insights and implications. In *Healthcare* (vol. 8, no. 2, p. 133). Multidisciplinary Digital Publishing Institute.

[13] Gupta, R., Reebadiya, D., Tanwar, S., Kumar, N., and Guizani, M. (2021). When blockchain meets edge intelligence: trusted and security solutions for consumers. In *IEEE Network*, vol. 35, no. 5, pp. 272–278, doi:10.1109/MNET.001.2000735.

[14] Shuaib, K., Abdella, J., Sallabi, F., and Serhani, M. A. (2021). Secure decentralized electronic health records sharing system based on blockchains. *Journal of King Saud University-Computer and Information Sciences*, 34(8), 5045–5058.

[15] https://www.thoughtworks.com/insights/decoder/d/decentralized-security#:~:text=Decentralized%20security%20enables%20risk%20management,infra-structure%20%E2%80%94%20helping%20safeguard%20your%20business (last accessed: 1 June 2022).

[16] Ben Fekih, R. and Lahami, M. (2020). Application of blockchain technology in healthcare: a comprehensive study. In *International Conference on Smart Homes and Health Telematics* (pp. 268–276). Springer, Cham.

[17] https://www.ibm.com/blogs/blockchain/2017/10/blockchain-in-healthcare-patient-benefits-and-more/ (last accessed: 1 June 2022).

[18] Khezr, S., Moniruzzaman, M., Yassine, A., and Benlamri, R. (2019). Blockchain technology in healthcare: a comprehensive review and directions for future research. *Applied Sciences*, 9(9), 1736.

[19] Mohammed, B. and Shadi, A. Z. (2020). Adoption of blockchain in healthcare sector analyzing the factors for successful deployment. (https://www.researchgate.net/publication/339850473_ADOPTION_OF_BLOCKCHAIN_IN_HEALTHCARE_SECTOR_ANALYZING_THE_FACTORS_FOR_SUCCESSFUL_DEPLOYMENT).

[20] https://www.ibm.com/cloud/architecture/architectures/blockchainArchitecture/ (last accessed: 7 June 2022).

[21] Ismail, L. and Materwala, H. (2019). A review of blockchain architecture and consensus protocols: use cases, challenges, and solutions. *Symmetry*, *11* (10), 1198.

[22] Idrees, S. M., Nowostawski, M., Jameel, R., and Mourya, A. K. (2021). Security aspects of blockchain technology intended for industrial applications. *Electronics*, 10(8), 951.

[23] Gupta, R., Nair, A., Tanwar, S., and Kumar, N. (2021). Blockchain-assisted secure UAV communication in 6G environment: architecture, opportunities, and challenges. *IET Communication,* 15, 1352–1367. https://doi.org/10.1049/cmu2.12113

[24] Xia, Q., Sifah, E. B., Smahi, A., Amofa, S., and Zhang, X. (2017). BBDS: blockchain-based data sharing for electronic medical records in cloud environments. *Information*, 8(2), 44.

[25] Reebadiya, D., Rathod, T., Gupta, R., *et al.* (2021) Blockchain-based secure and intelligent sensing scheme for autonomous vehicles activity tracking beyond 5g networks. *Peer-to-Peer Network Applications,* 14, 2757–2774. https://doi.org/10.1007/s12083-021-01073-x

[26] Haleem, A., Javaid, M., Singh, R. P., Suman, R., and Rab, S. Blockchain technology applications in healthcare: an overview. *International Journal of Intelligent Networks*, 2 (2021), 130–139.

[27] De Aguiar, E. J., Faiçal, B. S., Krishnamachari, B., and Ueyama, J. (2020). A survey of blockchain-based strategies for healthcare. *ACM Computing Surveys (CSUR)*, 53(2), 1–27.

[28] Singh, A. K. and Patro, B. D. K. (2020). Elliptic curve signcryption based security protocol for RFID. *KSII Transactions on Internet and Information Systems (TIIS)*, 14(1), 344–365.

[29] Singh, A. K. and Patro, B. D. K. (2020). Signcryption-based security framework for low computing power devices. *Recent Advances in Computer Science and Communications (Formerly: Recent Patents on Computer Science)*, 13(5), 845–857.

[30] Zyskind, G., Nathan, O., and Pentland, A. (2015). Enigma: decentralized computation platform with guaranteed privacy. arXiv preprint arXiv:1506.03471.

[31] Xia, Q., Sifah, E. B., Smahi, A., Amofa, S., and Zhang, X. (2017). BBDS: Blockchain-based data sharing for electronic medical records in cloud environments. *Information*, 8(2), 44.

[32] Yue, X., Wang, H., Jin, D., Li, M., and Jiang, W. (2016). Healthcare data gateways: found healthcare intelligence on blockchain with novel privacy risk control. *Journal of Medical Systems*, 40(10), 1–8.

[33] Ying, Z., Wei, L., Li, Q., Liu, X., and Cui, J. (2018). A lightweight policy preserving EHR sharing scheme in the cloud. *IEEE Access*, 6, 53698–53708.

[34] Ramani, V., Kumar, T., Bracken, A., Liyanage, M., and Ylianttila, M. (2018, December). Secure and efficient data accessibility in blockchain based healthcare systems. In *2018 IEEE Global Communications Conference (GLOBECOM)* (pp. 206–212). IEEE.

[35] Nguyen, D. C., Pathirana, P. N., Ding, M., and Seneviratne, A. (2019). Blockchain for secure EHRs sharing of mobile cloud based e-health systems. *IEEE Access*, 7, 66792–66806.

[36] Chen, Y., Ding, S., Xu, Z., Zheng, H., and Yang, S. (2019). Blockchain-based medical records secure storage and medical service framework. *Journal of Medical Systems*, 43(1), 1–9.

[37] Zhang, A. and Lin, X. (2018). Towards secure and privacy-preserving data sharing in e-health systems via consortium blockchain. *Journal of Medical Systems*, 42(8), 1–18.

[38] Ismail, L., Materwala, H., and Zeadally, S. (2019). Lightweight blockchain for healthcare. *IEEE Access*, 7, 149935–149951.

[39] Radhakrishnan, B. L., Joseph, A. S., and Sudhakar, S. (2019, March). Securing blockchain based electronic health record using multilevel authentication. In *2019 5th International Conference on Advanced Computing & Communication Systems (ICACCS)* (pp. 699–703). IEEE.

[40] Hathaliya, J. J., Gupta, R., Tanwar, S., and Sharma, P. (2021). A smart contract-based secure data sharing scheme in healthcare 5.0. In *2021 IEEE Globecom Workshops (GC Wkshps)*, Madrid, Spain, pp. 1–6, doi:10.1109/GCWkshps52748.2021.9681956.

[41] Luecking, M., Fries, C., Lamberti, R., and Stork, W. (2020, May). Decentralized identity and trust management framework for Internet of Things. In *2020 IEEE International Conference on Blockchain and Cryptocurrency (ICBC)* (pp. 1–9). IEEE.

[42] Gupta, R., Shukla, A., and Tanwar, S. (2020). AaYusH: a smart contract-based telesurgery system for healthcare 4.0. In *2020 IEEE International Conference on Communications Workshops (ICC Workshops)*, Dublin, Ireland, pp. 1–6, doi:10.1109/ICCWorkshops49005.2020.9145044.

[43] Javed, I. T., Alharbi, F., Bellaj, B., Margaria, T., Crespi, N., and Qureshi, K. N. (2021, June). Health-ID: a blockchain-based decentralized identity management for remote healthcare. In *Healthcare* (vol. 9, no. 6, p. 712). Multidisciplinary Digital Publishing Institute.

[44] Rahmani, M. K. I., Shuaib, M., Alam, S., *et al.* (2022). Blockchain-based trust management framework for cloud computing-based internet of medical things (IoMT): a systematic review. *Computational Intelligence and Neuroscience*, 2022, Article ID 9766844.

[45] Bhavin, M., Tanwar, S., Sharma, N., Tyagi, S., and Kumar, N. (2021). Blockchain and quantum blind signature-based hybrid scheme for healthcare 5.0 applications. *Journal of Information Security and Applications*, 56, 102673.

[46] Siddiqui, S. T., Shuaib, M., Gupta, A. K., and Alam, S. (2020, September). Implementing blockchain technology: way to avoid evasive threats to information security on cloud. In *2020 International Conference on Computing and Information Technology (ICCIT-1441)* (pp. 1–5). IEEE.

[47] Hasselgren, A., Rensaa, J. A. H., Kralevska, K., Gligoroski, D., and Faxvaag, A. (2021). Blockchain for increased trust in virtual health care: proof-of-concept study. *Journal of Medical Internet Research*, 23(7), e28496.

[48] Sharma, P., Borah, M. D., and Namasudra, S. (2021). Improving security of medical big data by using blockchain technology. *Computers & Electrical Engineering*, 96, 107529.

[49] Bhuiyan, M. Z. A., Zaman, A., Wang, T., Wang, G., Tao, H., and Hassan, M. M. (2018, May). Blockchain and big data to transform the healthcare. In *Proceedings of the International Conference on Data Processing and Applications* (pp. 62–68).

[50] https://www.devprojournal.com/technology-trends/blockchain/what-stands-in-the-way-of-healthcare-blockchain-adoption/ (last accessed: 15 June 2022).

[51] Gupta, R., Shukla, A., and Tanwar, S. (2021). BATS: a blockchain and AI-empowered drone-assisted telesurgery system towards 6G. In *IEEE Transactions on Network Science and Engineering*, vol. 8, no. 4, pp. 2958–2967, doi:10.1109/TNSE.2020.3043262.

[52] Sharma, V., Gupta, A., Hasan, N. U., Shabaz, M., and Ofori, I. (2022). Blockchain in secure healthcare systems: state of the art, limitations, and future directions. *Security and Communication Networks*, 2022, 3–13.

[53] Tanwar, S., Tyagi, S., and Kumar, N. (Eds.) (2019). *Security and Privacy of Electronic Healthcare Records: Concepts, Paradigms and Solutions*. Institution of Engineering and Technology.

[54] Ekblaw, A., Azaria, A., Halamka, J. D., and Lippman, A. (2016, August). A case study for blockchain in healthcare: "MedRec" prototype for electronic health records and medical research data. In *Proceedings of IEEE Open & Big Data Conference* (vol. 13, p. 13).

[55] Gupta, R., Thakker, U., Tanwar, S., Obaidat, M. S., and Hsiao, K. -F. (2020). BITS: a blockchain-driven intelligent scheme for telesurgery system. In *2020 International Conference on Computer, Information and Telecommunication Systems (CITS)*, Hangzhou, China, pp. 1–5, doi:10.1109/CITS49457.2020.9232662.

Chapter 6

Unification of machine learning and blockchain technology in healthcare industry

Megha Gupta[1], Mandeep Singh[2], Anupam Sharma[3], Namrata Sukhija[2], Puneet Kumar Aggarwal[4] and Parita Jain[5]

Abstract

Smart healthcare systems are useful for the prediction of diseases to provide a proper diagnosis, prevention, and cure by analyzing them. Blockchain (BC) is one of the emerging and trending technologies for various smart and intelligent health care systems that are very much required in the current scenario to expand the medical boundaries. The mechanism used in it gives assurance of security of data of patients which is the most important aspect of a healthcare system. But with one issue resolved, cannot stop the other issues to rise like attacks and double spending. To pick up these issues, analytics of data is required on the secured data for better results in the diagnosis and analysis in the healthcare industry. Till now, machine learning algorithms for recognizing the patterns in healthcare, data analyzing, and making some decisions based on the data have also shown their value among different sectors. The basic characteristic of ML is to modify the accuracy of results by involving the rational amount of healthcare data for precise diagnosis or decisions. To make positive decisions, the proper amount of data is required. So, the combination of these two technologies – machine learning and BC technology – can improve the results to a better precision of improvement to the next level which may be very important and useful for medical or healthcare systems. Support vector machines (SVM), deep learning (DL), convolution neural network (CNN), and long–short-term memory (LSTM) are some of the traditional ML techniques which can be useful for the analysis of the issues like the attack on the BC network. In this study, we will discuss about a detailed comparison of smart applications in the

[1]Department of CSE, AKG Engineering College, India
[2]Department of CSE, Raj Kumar Goel Institute of Technology, India
[3]Department of IT, HMR Institute of Technology and Management, India
[4]Department of IT, ABES Engineering College, India
[5]Department of CSE, KIET Group of Institutions, India

healthcare sector that mainly includes applications like smart grid, smart cities, etc. Data security increases by combining both machine learning and block chain.

Keywords: Healthcare systems; Blockchain; Machine learning; Data security and privacy; Data analytics; Smart applications

6.1 Introduction

In healthcare sector, machine learning (ML) gives better results such as identification of body organs from various medical images, categorization of lung diseases, cardiac problem discernment, and many more. Medical healthcare repositories create a lot of heterogeneous data every day that raise issues for the "conventional techniques" to dissect or to handle it. ML methods help to viably investigate this information for significant experiences. Furthermore, there are heterogeneous sources of information that can expand medical services information, for example, genomics, clinical information, information from online media, natural information, and so forth a portrayal of these sources of information appear in Figure 6.1. The four significant applications of medical care that can profit from ML strategies are forecast, determination, treatment, and clinical work process, which are portrayed straightaway [1].

In this chapter, focus is on how healthcare systems used two technologies (ML and blockchain technology (BT)) together for prediction, analysis of many health issues through data and records. Training datasets are stored in a decentralized database of blockchain (BC) as it provides data security. Data security is worried with viable attacks that can be mainly done on datasets. In the healthcare sector, we use the following types of learning i.e., supervised, unsupervised, and reinforcement [2].

Supervised learning is the type of learning where the input values are mapped with output values through labelled training dataset. Classification is a type of

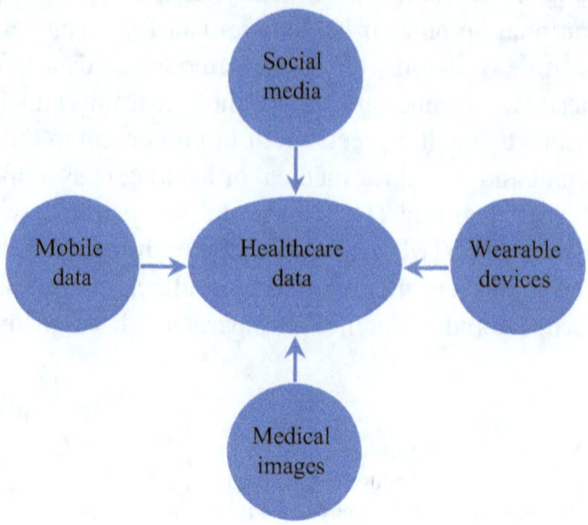

Figure 6.1 Heterogeneous sources contributing to healthcare data

supervised learning technique that is used in many fields of the healthcare sector and some of the examples are classification of lung diseases, medical image recognition, etc. [3]. Unsupervised learning is the learning technique where we use labelled data for mapping input and output data. Clustering techniques are used in many fields of the healthcare sector and some of them are prediction of heart diseases, hepatitis diseases prediction, etc. and reinforcement learning is the learning method which includes action taken by agents to maximize the rewards. Many healthcare applications have been transformed by RL with great potential and recently, it has been unused for disease diagnosis in the healthcare sector. In addition, an example of Go Game where computer defeated a human player by combining RL with supervised and unsupervised algorithms is considered as the future of using RL.

6.1.1 Applications of ML in healthcare

Healthcare applications like prognosis and medical image analysis (MIA) are done with ML algorithms. Prognosis is a method of clinical practice in which the prediction of the predicted development of a disease is done. Also, it identifies the symptoms for a particular disease and gives an idea about the signs related to the disease either problem increasing or decreasing for example classification of different types of cancers based on symptoms. MIA was also done by ML algorithms so that extraction of successful and structured information gathered by many imaging center's data which includes magnetic resonance imaging (MRI), computed tomography (CT), ultrasound, and positron emission tomography (PET), etc. and information is used for early detection and diagnose the abnormalities in different parts of our body.

To enhance the intelligence of blockchain-based applications, machine learning techniques can be employed. This can lead to improved security of the distributed ledger by utilizing machine learning. The available healthcare datasets could be stored on a BC decentralized database for avoiding unauthorized access. Data-related issues in ML models may be eliminated by BC. As in the technological age, remote monitoring machines are essential for health issues so BC improves the quality of machines by providing security to data and this data has been provided to patients for further action [4].

BT has also been employed to store the information of medical issuance of people and records of clinical trials. It can easily detect information accuracy because of its transparency. For the clinical trial, data BC is considered as security providers. BT gives a significant degree of security and trust. By incorporating BC with ML models has been in brilliant agreements to guarantee that information is well protected and not available publicly.

BT facilitates the effective implementation of incentive-based frameworks, thereby empowering users to contribute data. Decentralized networks across the globe are managed by blockchain tools such as Ethereum. This platform can also be used to provide patients with access to knowledge and information. In terms of use cases, blockchain technology simplifies secure data protection, approval control, medical tools, and product monitoring.

- One more application of healthcare is implemented by ML and BC is clinical speech and audio processing.
- It includes a solution of documentation work at the time of patient's discharge.

- By clinical speech and audio processing, doctors directly interact with patients, and transcription is automatically done by the ML algorithms.
- This approach will make the work easier and interactive.

The objective of the chapter is to make healthcare industry more secure by using the blockchain with the ML and the following are the specific contributions:

1. It is overviewed that ML techniques are used for four major pieces of work in healthcare industry i.e., prognosis, diagnosis, treatment, and clinical workflow.
2. A survey of security and robustness challenges explained and then it is also explained that how ML models with the help of blockchain used for the applications in healthcare industry.
3. Explanation of different solutions are defined which mainly solve the security issues.
4. At last, research open issues are highlighted.

6.1.2 Organization of chapter

The rest of the chapter is organized as: Section 6.2 includes related work; Section 6.3 discusses the use of ML techniques in healthcare industry; Section 6.4 elaborates the challenges faced in healthcare system and its applications; Section 6.5 introduces the working of ML algorithms in BC with the basic applications, Section 6.6 contains the future research and open issues and Section 6.7 concludes the chapter with future scope.

6.2 Related work

In 2019, Kim *et al.* [3] provided an introduction of different confrontational attacks and resistance with their applications in ML-based MIA.

In 2019, Yuan *et al.* [5] provided a survey of literature on confrontational attacks and resistance in general.

For searching of the existing research work, several databases like Google scholar, IEEE Xplore, Science direct, etc. are used. Keywords like ML in healthcare or ML with BC for different applications are used and many results are produced based on the same. Now in the next phase, ML model-based solutions are used in this model for applications of healthcare systems. By analyzing of many journals, magazines with keyword ML-based applications some important points are observed. By studying different parts of the searched article like abstract, introduction, conclusion, etc., then divide the article into different parts related to ML with BC in healthcare systems.

In [2], the paper which gave an explanation of examples in clinical applications and presented the concept of confrontational patients in the context of false models in clinical trials.

Rahouti *et al.* [6] provided a detailed study that explores the key security concerns. Authors started by presenting a global introduction of the Bitcoin protocol and components of bitcoin. Next, depth description of recent threats and weak points of the bitcoin system and its solution using ML techniques was presented. Last, authors discussed the current security studies and solutions.

Meng *et al.* [7] provided a detailed study about the intrusion detection which is used to detect cyber-attacks on data over network. This survey was included the combination of block chain for data security with many intrusion detection systems which mainly known as intrusion detection systems (IDSs). The main disadvantage faced in this survey was that it only focused on the trust management.

Conti *et al.* [8] provided a survey regarding the bitcoin security using BT. Bitcoin and its components also explained majorly but there was a lack of description of issues that occur in bitcoin security.

Ucci *et al.* in [9] explained the usage of ML algorithms in the field of malware detection. In this survey, the author mainly focused on the types of malware analysis and ML algorithms that are used in learning. Complexities of different algorithms were also explained but there was no proper explanation of BC given in this survey.

Salah *et al.* [10] provided a survey of usage of BT for artificial intelligence applications. In this paper, the explanation of AI-based applications with block chain advantage is given as AI makes machine smart for decision making and BC provides security on data so it also explains the healthcare applications based on the same but there is no proper explanation of privacy given in this survey.

Casino *et al.* [11] explained the different blockchain applications and frameworks. Here a great explanation was given that how blockchain-based applications come in future for use. It also included applications based on many fields like health, business, bitcoin, etc.

Kumar *et al.* [12] explained the BT for smart healthcare applications. Here authors took data from many healthcare electronic data centers and apply simulation optimization for data security, integrity, and develop a smart healthcare system.

Mahapatra *et al.* [13] provided a detailed study about the Internet of things (IoT)-based smart healthcare cities and included the concept of cloud computing and electronic health records (EHRs) which mainly include lab results, clinical notes, patients listing, etc.

Ren *et al.* [14] described the ML techniques with the advancement of baseline and deep learning technology. It included morality prediction, estimation of length of stay, etc. by many algorithms.

Ghassemi *et al.* [15] described the detection of vocal hyper function done which means detection of wrong voice created by someone i.e., voice misuse or, in medical terms, it is known as voice disorders. In this survey, it was corrected by using supervised ML algorithm to detect the two distinct voices using acceleration signals.

Juneja *et al.* [16] explained how to solve and diagnose the chest problems which are increasing now a days due to pollution. In this research, the chest issues were solved by using ML algorithms like CNN.

Singh *et al.* [17] showed the combination of BC, IoT, and artificial intelligence (Table 6.1). In this survey, comparison of many studies related to fog and edge was done on the basis of BC and AI [18].

Table 6.1 Comparison between the proposed work and existing work

Authors	Year	Objective	Advantage	Disadvantage	Architecture	Security	Challenges	Applications
Rahouti *et al.* [6]	2018	He provides ML-based solutions for bitcoin	Great explanation of the ML techniques and classify the threats	Blockchain other advantage are missing	No	Yes	Yes	No
Meng *et al.* [7]	2018	BT in intrusion detection	Scope of intrusion detection was classified and discussed	Only trust management issues are discussed	No	Yes	Yes	Yes
Conti *et al.* [8]	2018	Bitcoin security and privacy issues are discussed	A detailed survey of workable attacks to bitcoin and provided corrective measures	There are no proper issues described based on BC	Yes	Yes	Yes	No
Ucci *et al.* [9]	2018	ML techniques related to malware analysis	Complexities for different methodologies have been described in detail	Not explained these techniques in block chain	No	No	Yes	Yes
Salah *et al.* [10]	2019	Applications and protocols for blockchain in AI	Decentralized feature of blockchain is explained specially for AI	Discussion on privacy is not mentioned thoroughly	Yes	No	Yes	Yes
Casino *et al.* [11]	2019	Survey of many applications based on blockchain	Basics of BC applications are discussed in detail	Open issues are not discussed properly	No	No	Yes	Yes
Proposed survey	–	To survey how ML used with BC in healthcare-based applications	Discussion of technologies and architecture at basic level	–	Yes	Yes	Yes	Yes

6.3 Healthcare systems implemented by ML

A large amount of data and reports is generated by healthcare services daily and it has become difficult for "traditional methods" to analyze and process the data. ML methods are used to analyze the data. There are many applications of healthcare like prognosis, diagnosis, treatment, clinical workflow, etc. These all can be analyzed by the various ML techniques (Figure 6.2).

6.3.1 ML in prognosis

Prognosis is the process where the prediction of a development of a disease is done. Symptoms, signs, and activity related to specific disease are also predicted. ML techniques are used in different healthcare datasets. Datasets include patients' data from different sources like phenotype, genomic, proteomic, pathology test results, medical images, etc. [1]. ML models are also used to detect lung nodules [19] and different types of cancers like brain tumor. Thus, ML models used for predicting the symptoms of various diseases, repetition, risks, and substitutability have been used under research processes that can act as medicine [20]. So, the field of personalized medicine is derived that needs a broad research of bioinformatics, strong validations of ML so to accomplish conveying and big impact.

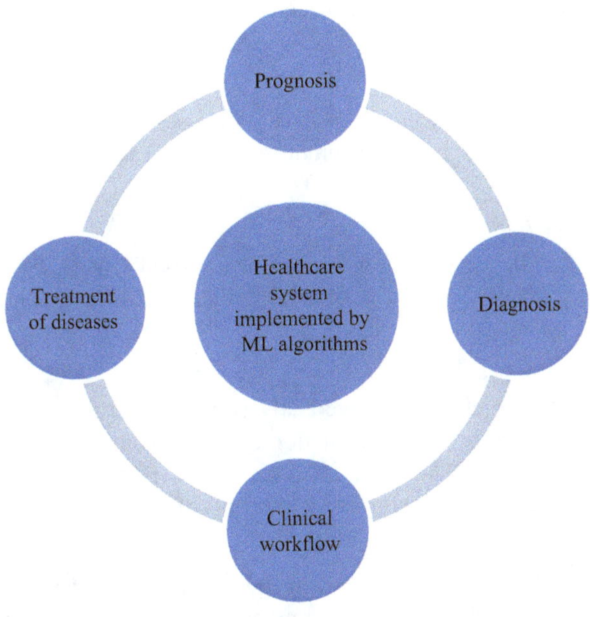

Figure 6.2 ML algorithms uses in healthcare system

6.3.2 *ML in diagnosis*

6.3.2.1 EHRs

A large amount of EHRs are generated by the hospitals and healthcare centers on a daily basis. This data includes a complete medical history of patients. For decreasing the diagnosis process, ML-based techniques have been exploited for getting the medical features [21]. In [22], semi-supervised machine learning techniques were utilized to extract diagnostic data from unstructured EHRs. ML models used for features of robustness using EHRs as models are trained for past data and tested on future data [23]. For the diagnosis of diabetes from EHRs, ML techniques were used in [24].

6.3.2.2 MIA

It includes the usage of different ML techniques for the extraction of data from many medical images that are collected by the different imaging like MRI, CT, ultrasound, X-ray, positron emission tomography (PET), etc. These datasets give important information about the body organs and help in the detection and diagnosis of aberration. It includes enhancement, detection, reconstruction, classification, and image registration.

6.3.2.2.1 Enhancement

It is a pre-processing step that directly affects the diagnosis process. Sometimes the digital images are disturbed by the many noises so that the quality of images decreases. Let us take an example of MRI, image generation is a lengthy process as good quality images are needed of patients. Patients have to sit straight and at the same place for a long time as false outputs can be produced with movement in the body [25]. The use of ML algorithms makes it easy to get output images with fewer errors.

6.3.2.2.2 Detection

It is processes where we detect or identify the diseases of patients. ML algorithms make it easy to detect diseases like tumor, cancer, etc. as traditional methods are very time consuming.

6.3.2.2.3 Classification

In this technique, ML classification algorithms are used to classify or analyze the different medical images so that classification of abnormalities is done on different body organs.

6.3.2.2.4 Segmentation

It is the process by which quantitative analysis of medical images is done. Brain cancer images analyzed by the segmentation in which measurement of volume and shape of cancer is done. These are all done by the partitioning of images into multiple parts using different parameters like color, texture, etc.

6.3.2.2.5 Registration of images

It is the process where linking of input images is done with the reference image. Unsupervised ML algorithms are used for mapping work. It is mainly used in spinal surgery and brain tumor removal process where localization is done.

6.3.2.2.6 Reconstruction

In this process, raw data obtained from an imaging sensor is illustrated and new images are reconstructed. For example, a lot of time is taken by many medical sensing for MRI and CT raw data. So, in this process, the main goal is to decrease the image reconstruction time and storage space. Many ML algorithms use this process like neural networks.

6.3.2.2.7 Retrieval

As a lot of data is generated by many healthcare services, it becomes difficult to manage a huge database so the usage of ML algorithms makes it easy by comparing the previous data with current data [26,27].

6.3.3 Applications of ML in treatment of patients

6.3.3.1 Real-time health monitoring

ML algorithms are used in real-time analysis of diseases. Continuous monitoring is done by many wearable devices and then data is stored on clouds for analysis.

6.3.3.2 Interpretation of image

For preparing high-quality medical report images, many radiologists and pathologists devote much time. So, to avoid all these, ML algorithms are used for the same. Recurrent neural network algorithm is an example of ML algorithm which is used in getting high-quality images [28].

6.3.4 Applications of ML in clinical workflow

6.3.4.1 Disease diagnosis and prediction

Disease prediction is one of the best applications of ML in healthcare sector. The prediction of cardiovascular risk is studied in [29], in which various ML techniques are used. Different surveys were conducted for the diagnosis and prediction of various diseases like liver, heart, dengue, diabetes, etc. [30]. In [31], for the prediction and diagnosis, various ML techniques were used.

6.3.4.2 Computer-aided diagnosis or prediction

Radiologists are assisted by many computers-aided diagnosis (CAD) systems. Traditionally, image processing and computer vision are used but now ML techniques are used for the same. Example of CAD is IBM's Watson which is used to diagnose diseases related to liver [32]. Many more applications are also detected by CADs such as the detection of fatty liver using ML techniques.

6.3.4.3 Clinical reinforcement learning

As reinforcement learning is the learning in which learning is done and tries to maximize the reward. Reinforcement learning is used to produce optimal treatment for patients. In [33], for the treatment of sepsis (it is an infection) in ICU, an evaluation is done with different techniques of reinforcement learning. It includes the patient's state of mind and provides the solution and result for the disease. It is concluded that for the treatment of sepsis, the simple Q learning method is more effective than the deep Q learning method under reinforcement learning.

6.3.4.4 For time-series data

Healthcare time-series data can be modeled and some of the applications are the prediction of brain-related diseases and blood pressure prediction. In [14], exploration of problems done which are related to unexpected respiratory trauma diseases. In [34], for the management of ICUs forecasting tasks, attention models are used by combining time series measurement data and clinical notes.

6.3.4.5 Natural language processing

For improving the recent clinical data of patients, natural language processing (NLP) is used as it stores previous data related to the patient. It also helps in doing research and clinical practice. The opposition and chance of clinical NLP for languages other than English is studied in [35]. NLP in clinical text analysis is presented in [36].

6.3.4.6 Clinical audio and speech processing

In the healthcare sector, more data is prepared related to patients like notes, discharge summary, reports, medicine prescription, etc. Due to administrative work, there is a lack of motivation among clinicians as suggested by Wallace [37], so clinical audio and speech processing can be used in which less time is spent on documentation as audio automatically creates notes and summary related to patients. It is also used to detect disorders in speech and audio [15] which results in benefits for the many stakeholders i.e., patients, doctors, healthcare industry as it decreases the cost and time and increase the efficiency and productivity.

6.4 Challenges faced in healthcare systems and its applications

In this section, many challenges are discussed which occur in healthcare systems.

6.4.1 Safety challenges

Phenomenal execution in a controlled lab environment (which is a typical ML people group practice) is not proof of security. Security of ML is the guarantee of how safe the ML framework is for patients. There ought to be a consistent belief of security all through the ML lifecycle. Dominant part of routine clinicians' work is ordinary, and patients they experience have the same health issues. It is their part of diagnosing uncommon, unpretentious, and covered up medical issues which happen once in millions. Allowing ML to perform well on secret layers, anomalies, edge, and inconspicuous cases is vital to guarantee the security of current AI systems [38].

6.4.2 Ethical challenges

In client-driven applications of ML like medical care, it is critical to guarantee the moral utilization of data. For gathering data for building ML models, direct measures ought to be taken to comprehend the user population and their sociological angles. Additionally, seeing how data assortment can hurt a patient's prosperity and respect is a significant consideration in this manner. Applications of ML give unfavorable results if moral concerns are not considered into account. It is crucial to have an entire understanding of the AI systems in complex and unknown situations so that ML guarantees fair and moral operations of automated systems [39].

6.4.3 Privacy challenges

Privacy is one of the significant difficulties in data-driven medical care which is worried about the utilization of clients' information by the ML frameworks for making predictions. The users (i.e., patients) expect that their medical care administration suppliers are following essential security measures to shield their intrinsic right to the protection of their private data, e.g., age, sex, date of birth, and clinical data. There are two types of potential privacy threats: one is unveiling confidential information and another is malicious use of data [40].

Protection relies on the qualities and nature of the information being gathered, the environment it has been made in, and patients' socioeconomics. Accordingly, relief of security penetrates utilizing the suitable technique(s) that is critical as contravention can straightforwardly hurt the patients. The private information ought to be unknown to stop contravention, for example (re-)recognizable proof of the people [41]. In addition, vital consideration ought to be paid to comprehend security worries at each phase of data processing and the exchange of information among various departments inside a clinic ought to be imparted in a secure environment. Protection challenges additionally emerge with reception of ML strategies for building biometric medical care frameworks either disconnected (e.g., face or unique finger impression acknowledgment-based framework to ensure clinical offices and hardware [42]) or online frameworks, e.g., continuous clinical frameworks [43] and utilization of biometrics for the validation of clinical IoT gadgets [44], and so on. The security and protection of such frameworks are of most extreme significance; subsequently, most pessimistic scenario heartiness test ought to be performed for biometrically secure medical services systems. Worst-case testing is an amazing asset that can give sufficient proof about a framework's power and can be recognized from a framework that never fizzles and a framework that bombs once in billion paths.

6.4.4 Accessibility of good quality data

The accessibility of delegates, and great information is one of the major challenges in medical services. For example, the measure of information accessible to the examination local area is little in size and restricted in scope when contrasted with the heterogeneous assortments of huge scope multi-modular patient information being produced on every day premise by various little and enormous size medical services organizations. In any case, the advancement of good quality information that looks like genuine clinical settings is on the other testing and requires assets for the executives and upkeep. The accessibility of great information can adequately fill the planned need of sickness expectation and dynamic for arranging therapy. The information gathered practically speaking experience the ill effects of various issues such as subjectivity, excess, and predisposition. As the ML models perform inductions by exclusively learning the dormant variables of the information on which they are prepared, along these lines, the impact of information created by the unfortunate past practices of emergency clinics will be reflected in the results of the calculation. For instance, most individuals with no medical coverage are denied medical services administrations. What's more, if AI gains from that information, it will do likewise. It has been shown that a model could portray racial inclination by creating differing results for various subpopulations [45] and the preparation information can likewise present its own modeling challenges [46,47].

6.4.5 Causality is a challenge

Understanding causality is significant in medical care on the grounds that a large portion of the vital medical care issues require causal thinking, i.e., "consider the possibility that?" [48]. For model, posing an inquiry about what will happen if a specialist recommended treatment A rather than treatment B. Such inquiries cannot be misused through old style learning calculations and, to answer them, we need to examine the information from the focal point of causality [49]. In medical care, learning is regularly exclusively dependent on observational information and posing causal inquiries by gaining from observational information is very difficult which requires building causal models. ML models are discoveries which need a central hidden hypothesis and these models basically work by abusing designs furthermore, connections without considering any causal connection [50]. All in all, this cannot be considered as a limit since expectation does not need any causal connection. In prescient medical services, the shortfall of causal connection can bring up issues about the ends that can be drawn from results of ML models. Moreover, decency in dynamic can more readily be upheld through the perspective of causal thinking [51,52]. The assessment of the causal impact of some variable(s) on an objective yield (e.g., target class in multi-class characterization issue) is imperative to guarantee reasonable forecasts.

6.4.6 Distribution shifts

The issue of information dispersion shifts is one more significant test and may be quite possibly the most provoking issue to address [53]. In clinical work on, preparing and testing information dispersions can separate because of numerous reasons, e.g., clinical information is created by various foundations utilizing various gadgets for patients having confounded cases. Because of this issue, ML models created utilizing accessible public datasets (by established researchers and academicians) do not give anticipated execution when sent in a real clinical climate. Dispersion shifts are successive in the clinical area, specifically, clinical envisioning where various conventions and boundary decisions can bring about pictures of altogether various disseminations. ML models are ordinarily prepared under the guideline of experimental danger minimization (ERM) which gives great learning limits and certifications if its presumptions are fulfilled. For example, one of the first and solid suspicions is that both the preparation and test datasets are obtained from a comparable area (i.e., information circulations). Be that as it may, this supposition that is not legitimate practically speaking and models prepared under such a presumption neglect to sum up to different areas conversely, the basic nature of life of clinical applications requests a smooth and safe activity of ML procedures.

6.5 ML adaptation in BC: basic application

As it is known that ML is the field where applications are developed through many learning techniques and BC is the field which is somehow similar to accounting ledger. Blockchain is the technique where linking of blocks is done. Current block is created by linking of hash of preceding block and so on. BC are immutable as older blocks cannot be modified. So, BC technique is applied to ML learning which makes applications advanced and secure. By using BC, security may be improved in ML-based applications. Decentralized database of BC has become an advantage for the ML systems. By making better data sharing routes, ML algorithms are used to increase the time taken to reach consensus. In Figure 6.3, an integration of ML with BC is represented.

Collection of data is done from different healthcare sources like sensors, smart devices, and IoT devices. BC works as an integral part of these applications. ML algorithms are used to predict and diagnose the smart application

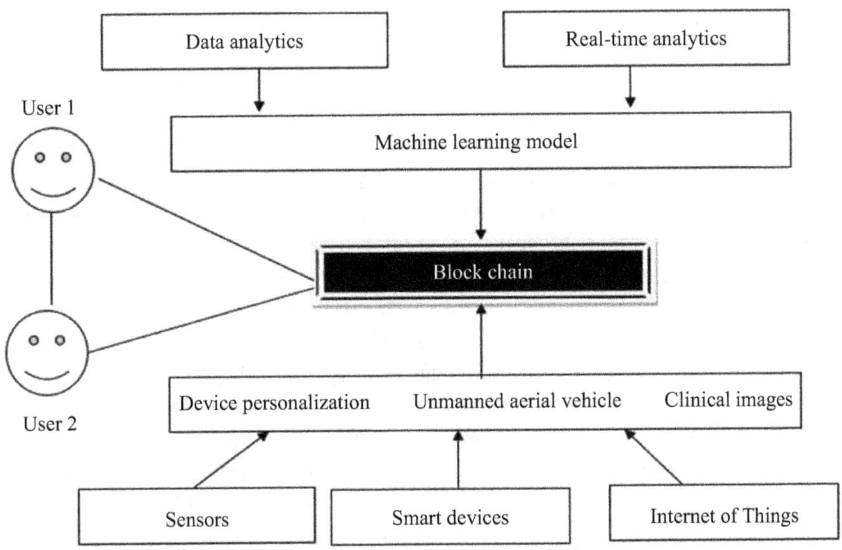

Figure 6.3 ML adoption in BC-based smart applications

data. BC networks are used to store datasets used by ML algorithms. This decreases the errors in the data like noise, missing values, and redundancy. Healthcare data-related errors in ML models may be removed by using BCs as they are focused on data. ML models can be applied on particular chains rather than the entire dataset. This could give models for mainly fraud detection in healthcare systems

There are some benefits of unification of ML with BT. First, BC provides security and trust at a high level. Second is user verification as a valid user for performing and requesting any transaction in the BC network. Third, ML models can be modernized without any price on the field of BC. Fourth, BC incorporates public ML models into contracts to guarantee that the conditions and terms which were earlier concurred are supported. Fifth, BC helps in the solid execution of a motive-based framework, i.e. it empowers users to give information. This tremendous information will assist in improving ML model execution. Sixth, great information commitments can occur from users/customers, this information reliably enumerates, and gifts can be given to the users. Seventh, tamper proof agreements can be assessed by various machines (having distinctive equipment configuration), ML models would not separate from their capacity and give results precisely as it should do. Eighth, BC instruments, for example, all over the world Ethereum, manage many decentralized machines everywhere. This ensures users that it is never totally inaccessible or unplugged. So, integration of ML with BC results in a secure smart application for healthcare systems.

Case study

In order to explain the proposed work, let us consider a case study which is basically an energy trading system based on smart grid system for energy transaction using cryptocurrency. It is a system which is a person-to-person energy system and uses BC and deep learning approach to fulfil an energy transaction. In order to get the maximum throughput, Byzantine algorithm is used which is a fault tolerance algorithm. This algorithm works in five phases (i) setup phase, (ii) agreement phase, (iii) consensus-making phase, (iv) block creation phase, and (v) a change view phase. By using hash functions and a short signature, generation of blocks should be done. It is an IDS works on recurrent neural network (RNN) which is used to detect the fraudulent transactions and network attacks. The performance of IDS is checked by datasets. It has four entities for communications: (i) energy buyer, (ii) energy seller, (iii) BC, and (iv) IDS.

The energy seller produces energy and use that energy for his own and after that the left energy will be published on BC network. The buyer looks for published price and unit of energy and then buyer checked has an account that consists of sufficient cryptocurrency for trading of energy. Then if requirement matched with buyer balance, a request is generated by the buyer to the network. After validating the buyer information, the seller sells the energy and get the respective cryptocurrency in his account. IDS can help person to person energy trading system to find the fraud transactions in the network.

6.6 Future research and open issues

The integration of ML and BC can be resisted by the many barriers [17]. In Figure 6.4, discussion of many issues and challenges of ML is integrated with BC. The challenges are the following: suitability, infrastructure, privacy, security, memory, implementation, and quantum resilience.

6.6.1 Framework

The performance of applications of BC may be improved by BC network and hardware. In [54], many things include administration of the network, mining hardware, decentralized storage, and communication protocols. Major tech companies are still involved in the investigation of BC with ML.

6.6.2 Privacy

Data generated by many ML models stored on BC is visible to other nodes. It helps to keep the data secure and confidential. Many related issues could be resolved by the use of different types of BC, cryptography, etc. To impose barriers for prediction and analytics, ML models are adopted [55–62].

6.6.3 Security

In [63], BC is copied and spread across a network. Due to security issues like different attacks done by hackers, the consensus protocols may be compromised. So when new blocks are added to the chain, then mining power will control it. It is present in public BC [64]. As each node is identified in a private version of BC, it is not affected by attacks [65–70].

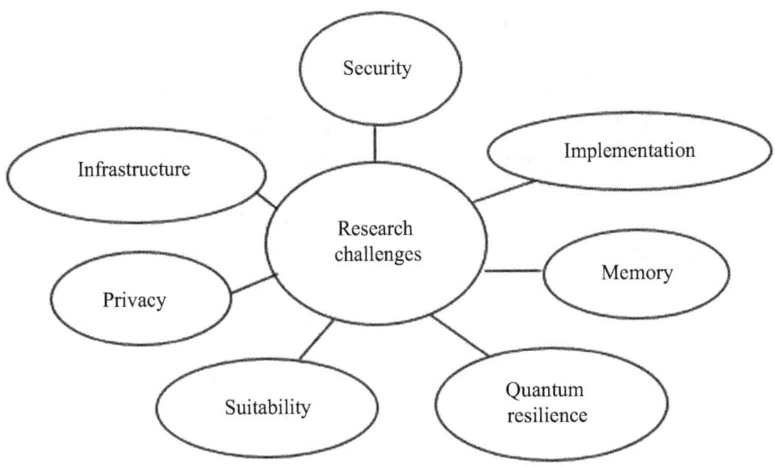

Figure 6.4 Future research issues

6.6.4 Implementation

A large number of transactions will be needed to a large-scale BC network. Many potential issues are occurring by the execution of this size of BC. Due to heavy transaction traffic, the network requires high bandwidth which cannot be easily reduced. Therefore, if more blocks and transactions are added in the chain, then it may decrease the unavoidable demand [71].

6.6.5 Memory

If the size of BC increases, then new blocks are needed to add to it. Hence, all nodes must be stored in the whole chain. An increase in chain size performance was also affected. Many computational resources are wasted by storage of irrelevant data. In most execution, storage is the major problem because of immutable BC [72–80].

6.6.6 Quantum resilience

Quantum computers will break the hashing algorithms used by BC. BC uses one-way functions for encryption, so they are at high risk with quantum computing. This may create hurdles and makes the chain outdated. Quantum computing offers to enhance the security of BCs by providing quantum cryptography whereas quantum communication is that where a user cannot mimic another user as it encrypt, person to person communication and covert the old signatures into BC network. In [81], there is research which includes quantum with BC discussed.

6.6.7 Suitability

If source of data is not trusted, then BC is the best technology for suitability. When requirement of performance is important, then the usage of simple databases is a better option. In [82], the study of the architecture of BC is understood.

6.7 Conclusion and future scope

In this survey, we have presented the enhanced concept of usage of ML algorithms with BC in the healthcare system. Comparisons of different surveys which already exist are also presented. So presentation of unification of machine learning with block chain in the healthcare sector is done which mainly includes treatment and detection of different diseases. Research challenges are also explained which are faced during the survey which requires solutions. Future research issues are also explained and concluded in the survey. Algorithms of ML with the combination of BC are used for the detection of many health problems and also for treatments of patients. In the near future, we will analyze the possibility of usage of BT in order to create a decentralized and safety mechanism of management privileges and consents with more accuracy. Probably only the decentralized environment protects users from losing control of their data or being dependent on a single provider. Finally, intend to create a decentralized platform in which we can create and distribute a validated and comparable ML algorithms.

References

[1] A. Collins and Y. Yao, "Machine learning approaches: data integration for disease prediction and prognosis," in *Applied Computational Genomics*. Berlin, Germany: Springer, 2018, pp. 137–141.

[2] K. Papangelou, K. Sechidis, J. Weatherall, and G. Brown, "Toward and understanding of adversarial examples in clinical trials," in *Proceedings of Joint European Conference on Machine Learning and Knowledge Discovery in Databases*, 2018, pp. 35–51.

[3] H. Kim, D. C. Jung, and B. W. Choi, "Exploiting the vulnerability of deep learning-based artificial intelligence models in medical imaging: adversarial attacks," *J. Korean Soc. Radiol.*, vol. 80, no. 2, pp. 259–273, 2019.

[4] R. Gupta, A. Shukla, and S. Tanwar, "AaYusH: a smart contract-based telesurgery system for healthcare 4.0," in *2020 IEEE International Conference on Communications Workshops (ICC Workshops)*, Dublin, Ireland, 2020, pp. 1–6, doi:10.1109/ICCWorkshops49005.2020.9145044.

[5] X. Yuan, P. He, Q. Zhu, and X. Li, "Adversarial examples: attacks and defenses for deep learning," *IEEE Trans. Neural Netw. Learn. Syst.*, vol. 30, no. 9, pp. 2805–2824, 2019.

[6] M. Rahouti, K. Xiong, and N. Ghani, "Bitcoin concepts, threats, and machine-learning security solutions," *IEEE Access*, vol. 6, pp. 6718967205, 2018.

[7] W. Meng, E. W. Tischhauser, Q. Wang, Y. Wang, and J. Han, "When intrusion detection meets blockchain technology: a review," *IEEE Access*, vol. 6, pp. 1017910188, 2018.

[8] M. Conti, E. S. Kumar, C. Lal, and S. Ruj, "A survey on security and privacy issues of bitcoin," *IEEE Commun. Surveys Tuts.,* vol. 20, no. 4, pp. 34163452, 4th Quart., 2018.

[9] D. Ucci, L. Aniello, and R. Baldoni, "Survey of machine learning techniques for malware analysis," *Comput. Secur.*, vol. 81, pp. 123147, 2019.

[10] K. Salah, M. Rehman, N. Nizamuddin, and A. Al-Fuqaha, "Blockchain for AI: review and open research challenges," *IEEE Access*, vol. 7, pp. 1012710149, 2019.

[11] F. Casino, T. Dasaklis, and C. Patsakis, "A systematic literature review of blockchain-based applications: current status, classification and open issues," *Telematics Inform.*, vol. 36, pp. 5581, 2018.

[12] A. Kumar, R. Krishnamurthi, A. Nayyar, K. Sharma, V. Grover, and E. Hossain, "A novel smart healthcare design, simulation, and implementation using healthcare 4.0 processes," *IEEE Access*, vol. 8, pp. 118433–118471, 2020.

[13] B. Mahapatra, R. Krishnamurthi, and A. Nayyar, "Healthcare models and algorithms for privacy and security in healthcare records," In *Security and Privacy of Electronic Healthcare Records: Concepts, Paradigms and Solutions* (p. 183). IET, 2019.

[14] O. Ren, A. E. W. Johnson, E. P. Lehman, *et al.*, "Predicting and understanding unexpected respiratory decompensation in critical care using sparse

and heterogeneous clinical data," in *Proceedings of IEEE International Conference on Healthcare Informatics*, pp. 144–151, 2018.

[15] M. Ghassemi, J. H. Van Stan, D. D. Mehta, *et al.*, "Learning to detect vocal hyper function from ambulatory neck-surface acceleration features: initial results for vocal fold nodules," *IEEE Trans. Biomed. Eng.*, vol. 61, no. 6, pp. 1668–1675, 2014.

[16] S. Juneja, A. Juneja, G. Dhiman, S. Behl, and S. Kautish, "An approach to thoracic syndrome classification with convolutional neural network," *Comput. Math. Methods Med.*, vol. 2021, Article ID 3900254, 2014, https://doi.org/10.1155/2021/3900254

[17] S. K. Singh, S. Rathore, and J. H. Park, "BlockIoTIntelligence: a blockchain-enabled intelligent IoT architecture with artificial intelligence," *Fut. Gener. Comput. Syst.*, vol. 110, pp. 721–743, 2019. to be published, doi:10.1016/j.future.2019.09.002

[18] S. Tanwar, J. Vora, S. Kaneriya, *et al.*, "Human arthritis analysis in fog computing environment using Bayesian network classifier and thread protocol," *IEEE Consum. Electron. Mag.*, vol. 9, no. 1, pp. 88–94, 1 January 2020, doi:10.1109/MCE.2019.2941456.

[19] W. Zhu, C. Liu, W. Fan, and X. Xie, "Deeplung: deep 3D dual path nets for automated pulmonary nodule detection and classification," in *Proceedings of IEEE Winter Conference on Applications of Computer Vision*, 2018, pp. 673–681.

[20] Afshar, A. Mohammadi, and K. N. Plataniotis, "Brain tumor type classification via capsule networks," in *Proceeding of the 25th IEEE International Conference on Image Processing*, 2018, pp. 3129–3133.

[21] P. B. Jensen, L. J. Jensen, and S. Brunak, "Mining electronic health records: towards better research applications and clinical care," *Nature Rev. Genetics*, vol. 13, no. 6, pp. 395–405, 2012.

[22] Z. Wang, A. D. Shah, A. R. Tate, S. Denaxas, J. Shawe-Taylor, and H. Hemingway, "Extracting diagnoses and investigation results from unstructured text in electronic health records by semi-supervised machine learning," *PLoS One*, vol. 7, no. 1, 2012, Art. no. e30412.

[23] B. Nestor, M. B. A. McDermott, W. Boag, *et al.*, "Feature robustness in non-stationary health records: caveats to deployable model performance in common clinical machine learning tasks," In *Machine Learning for Healthcare Conference* (pp. 381–405). PMLR. 2019, *arXiv:1908.00690*.

[24] T. Zheng, W. Xie, L. L. Xu, *et al.*, "A machine learning-based framework to identify type 2 diabetes through electronic health records," *Int. J. Med. Informat.*, vol. 97, pp. 120–127, 2017.

[25] M. Lustig, D. L. Donoho, J. M. Santos, and J. M. Pauly, "Compressed sensing MRI," *IEEE Signal Process. Mag.*, vol. 25, no. 2, pp. 72–82, 2008.

[26] M. Singh, N. Sukhija, A. Sharma, M. Gupta, and P. K. Aggarwal. Security and privacy requirements for IoMT-based smart healthcare system. *Big Data Analysis for Green Computing*, pp. 17–37, 2021. https://doi.org/10.1201/9781003032328-2

[27] A. Qayyum, S. M. Anwar, M. Awais, and M. Majid, "Medical image retrieval using deep convolutional neural network," *Neurocomputing*, vol. 266, pp. 8–20, 2017.

[28] X. Wang, Y. Peng, L. Lu, Z. Lu, and R. M. Summers, "Tienet: text-image embedding network for common thorax disease classification and reporting in chest x-rays," in *Proceedings of IEEE Conference on Computer Vision and Pattern Recognition*, 2018, pp. 9049–9058.

[29] S. F. Weng, J. Reps, J. Kai, J. M. Garibaldi, and N. Qureshi, "Can machine learning improve cardiovascular risk prediction using routine clinical data?" *PLoS ONE*, vol. 12, no. 4, 2017, Art. no. e0174944.

[30] M. Fatima and M. Pasha, "Survey of machine learning algorithms for disease diagnostic," *J. Intell. Learn. Syst. Appl.*, vol. 9, no. 1, pp. 1–16, 2017.

[31] J. A. Cruz and D. S. Wishart, "Applications of machine learning in cancer prediction and prognosis," *Cancer Informat.*, vol. 2, pp. 59–77, 2006.

[32] S. Sharma, M. Singh, R. Singh, and R. Prajapati. "ASD screening using machine learning," *Int. J. Sci. Res. Manag. Stud.*, vol. 5, no. 7, pp. 1–8, 2021.

[33] A. Raghu, "Reinforcement learning for sepsis treatment: baselines and analysis," in *Proceedings of the 36th International Conference on Machine Learning*, Long Beach, California, USA, 2019.

[34] H. Song, D. Rajan, J. J. Thiagarajan, and A. Spanias, "Attend and diagnose: clinical time series analysis using attention models," in *Proceedings of Thirty-Second AAAI Conference on Artificial Intelligence*, 2018.

[35] A. Névéol, H. Dalianis, S. Velupillai, G. Savova, and P. Zweigenbaum, "Clinical natural language processing in languages other than English: opportunities and challenges," *J. Biomed. Semantics*, vol. 9, no. 1, p. 12, 2018.

[36] E. Soysal, J. Wang, M. Jiang, *et al.*, "Clamp—a toolkit for efficiently building customized clinical natural language processing pipelines," *J. Amer. Med. Informat. Assoc.*, vol. 25, no. 3, pp. 331–336, 2017.

[37] D. S. Wallace, "The role of speech recognition in clinical documentation," *Nuance Commun.*, 2018, Accessed: Dec. 14, 2019. [Online]. Available https://www.hisa.org.au/slides/hic18/wed/SimonWallace.pdf

[38] M. Bhavin, S. Tanwar, N. Sharma, S. Tyagi, and N. Kumar, "Blockchain and quantum blind signature-based hybrid scheme for healthcare 5.0 applications," *J. Inf. Secur. Appl.*, vol. 56, p. 102673, 2021.

[39] J. Zhang and E. Bareinboim, "Fairness in decision-making—the causal explanation formula," in *Proceedings of the Thirty-Second AAAI Conference on Artificial Intelligence*, 2018.

[40] S. Tanwar, S. Tyagi, and N. Kumar (eds.). *Security and Privacy of Electronic Healthcare Records: Concepts, Paradigms and Solutions. Institution of Engineering and Technology*, 2019.

[41] M. Al-Rubaie and J. M. Chang, "Privacy-preserving machine learning: threats and solutions," *IEEE Secur. Privacy*, vol. 17, no. 2, pp. 49–58, 2019.

[42] J. Chaudhry, "Securing healthcare data using biometric authentication," in *Security and Privacy in Communication Networks. Berlin, Germany*: Springer, 2018, pp. 123–132.

[43] A. H. Mohsin, A. A. Zaidan, B. B. Zaidan, *et al.*, "Real-time medical systems based on human biometric steganography: a systematic review," *J. Med. Syst.*, vol. 42, no. 12, p. 245, 2018.

[44] Y. Sun, F. P.-W. Lo, and B. Lo, "Security and privacy for the internet of medical things enabled healthcare systems: a survey," *IEEE Access*, vol. 7, pp. 183 339–183 355, 2019.

[45] I. Chen, F. D. Johansson, and D. Sontag, "Why is my classifier discriminatory?," in *Proceedings of Advances in Neural Information Processing Systems*, 2018, pp. 3539–3550.

[46] M. Ghassemi, T. Naumann, P. Schulam, A. L. Beam, I. Y. Chen, and R. Ranganath, "Practical guidance on artificial intelligence for health-care data," *Lancet Digit. Health*, vol. 1, no. 4, pp. e157–e159, 2019.

[47] T. Panch, H. Mattie, and L. A. Celi, "The 'inconvenient truth' about AI in healthcare," *NPJ Digit. Med.*, vol. 2, no. 1, pp. 1–3, 2019.

[48] P. Schulam and S. Saria, "Reliable decision support using counterfactual models," in *Proceedings of the Advances in Neural Information Processing Systems*, 2017, pp. 1697–1708.

[49] M. Ghassemi, T. Naumann, P. Schulam, A. L. Beam, and R. Ranganath, "A review of challenges and opportunities in machine learning for health," in *Proceedings of the AMIA Joint Summits on Translational Science*, 2020, p. 191.

[50] E. Begoli, T. Bhattacharya, and D. Kusnezov, "The need for uncertainty quantification in machine-assisted medical decision making," *Nature Mach. Intell.*, vol. 1, no. 1, pp. 20–23, 2019.

[51] A. Khademi, S. Lee, D. Foley, and V. Honavar, "Fairness in algorithmic decision making: an excursion through the lens of causality," in *Proceedings of the World Wide Web Conference*, 2019, pp. 2907–2914.

[52] N. Kilbertus, M. R. Carulla, G. Parascandolo, M. Hardt, D. Janzing, and B. Schölkopf, "Avoiding discrimination through causal reasoning," in *Proceedings of the Advances in Neural Information Processing Systems*, 2017, pp. 656–666.

[53] C. S. Perone, P. Ballester, R. C. Barros, and J. Cohen-Adad, "Unsupervised domain adaptation for medical imaging segmentation with self-ensembling," *NeuroImage*, vol. 194, pp. 1–11, 2019.

[54] S. Velankar, S. Valecha, and S. Maji, "Bitcoin price prediction using machine learning," in *Proceedings of the 20th International Conference on Advanced Communication Technology (ICACT)*, Feb. 2018, pp. 144–147.

[55] B. Fisch, R. Pass, and A. Shelat, "Socially optimal mining pools," in *International Conference on Web and Internet Economics*, 2017, pp. 205–218. Springer.

[56] A. Gervais, G. Karame, S. Capkun, and V. Capkun, "Is bitcoin a decentralized currency?" *IEEE Security & Privacy*, vol. 12, no. 3, pp. 54–60, 2014.

[57] M. Glickman, J. Balthrop, and S. Forrest, "A machine learning evaluation of an artificial immune system," *Evol. Comput.*, vol. 13, no. 2, pp. 179–212, 2005.

[58] M. A. Harlev, H. Sun Yin, K. C. Langenheldt, R. Mukkamala, and R. Vatrapu, "Breaking bad: de-anonymising entity types on the bitcoin blockchain using supervised machine learning," in *Proceedings of the 51st Hawaii International Conference on System Sciences*, 2018.

[59] J. Hirshman, Y. Huang, and S. Macke, Unsupervised approaches to detecting anomalous behavior in the bitcoin transaction network. *Technical Report*, Stanford University, 2013.

[60] A. Holub and J. O'Connor, "Coin hoarder: tracking a Ukrainian bitcoin phishing ring dns style," in *APWG Symposium on Electronic Crime Research (eCrime)*, 2018, pp. 1–5. IEEE, 2018.

[61] G. O. Karame, E. Androulaki, M. Roeschlin, A. Gervais, and S. Capkun, "Misbehavior in bitcoin: a study of double-spending and accountability," *ACM Trans. Inf. Syst. Security (TISSEC)*, vol. 18, no. 1, p. 2, 2015.

[62] P. K. Kaushal, A. Bagga, and R. Sobti, "Evolution of bitcoin and security risk in bitcoin wallets," in *Proceedings of the International Conference on Computer, Communications and Electronics (Comptelix)*, pp. 172–177. IEEE, 2017.

[63] A. Kumari, S. Tanwar, S. Tyagi, and N. Kumar, "Verification and validation techniques for streaming big data analytics in Internet of Things environment," *IET Netw.*, vol. 8, no. 2, pp. 92–100, 2018.

[64] R. Gupta, U. Thakker, S. Tanwar, M. S. Obaidat, and K. -F. Hsiao, "BITS: a blockchain-driven intelligent scheme for telesurgery system," in *2020 International Conference on Computer, Information and Telecommunication Systems (CITS)*, Hangzhou, China, 2020, pp. 1–5, doi:10.1109/CITS49457.2020.9232662.

[65] A. Kiayias, E. Koutsoupias, M. Kyropoulou, and Y. Tselekounis, "Blockchain mining games," in *Proceedings of the Conference on Economics and Computation*, pp. 365–382. ACM, 2016.

[66] J. A. Kroll, I. C. Davey, and E. W. Felten, "The economics of bitcoin mining, or bitcoin in the presence of adversaries," in *Proceedings of the Workshop on the Economics of Information Security (WEIS)*, vol. 2013, p. 11, 2013.

[67] S. Mishra, A. Shukla, S. Arora, H. Kathuria, and M. Singh, "Controlling weather dependent tasks using random forest algorithm," in *2020 Third International Conference on Advances in Electronics, Computers and Communications (ICAECC)* (pp. 1–8). IEEE, 2020, December.

[68] A. R. S. Sharma, M. Singh, R. Singh, and R. Prajapati, "ASD screening using machine learning," *Int. J. Sci. Res. Eng. Manag. (IJSREM)*, vol. 5, no. 7, pp. 1–8 , 2021.

[69] M. Singh, A. Verma, A. Parasher, N. Chauhan, and G. Budhiraja, "Implementation of database using python flask framework," *Int. J. Eng. Comput. Sci.*, vol. 8, pp. 24894–24899, 2019.

[70] S. Juneja, M. Gahlan, G. Dhiman, and S. Kautish, "Futuristic cyber twin architecture for 6G technology to support internet of everything," *Sci. Program.*, vol. 2021, https://doi.org/10.1155/2021/9101782 (SCI, Impact Factor: 1.025).

[71] R. Gupta, A. Shukla, and S. Tanwar, "BATS: a blockchain and AI-empowered drone-assisted telesurgery system towards 6G," *in IEEE Transactions on Network Science and Engineering*, vol. 8, no. 4, pp. 2958–2967, 1 October–December 2021, doi:10.1109/TNSE.2020.3043262.

[72] A. Jain, and A. Nayyar. "Machine learning and its applicability in networking," in *New Age Analytics* (pp. 57–79). Apple Academic Press. 2020.

[73] A. Nayyar, L. Gadhavi, and N. Zaman, "Machine learning in healthcare: review, opportunities and challenges," in *Machine Learning and the Internet of Medical Things in Healthcare*, pp. 23–45, 2021.

[74] P. K. D. Pramanik, A. Solanki, A. Debnath, A. Nayyar, S. El-Sappagh, and K. S. Kwak, "Advancing modern healthcare with nanotechnology, nanobiosensors, and internet of nano things: taxonomies, applications, architecture, and challenges," *IEEE Access*, vol. 8, pp. 65230–65266, 2020.

[75] S. Ali, Y. Hafeez, N. Z. Jhanjhi, *et al.*, "Towards pattern-based change verification framework for cloud-enabled healthcare component-based," *IEEE Access*, vol. 8, pp. 148007–148020, 2020.

[76] J. Vora, A. Nayyar, S. Tanwar, *et al.*, "BHEEM: a blockchain-based framework for securing electronic health records," in *2018 IEEE Globecom Workshops (GC Wkshps)* (pp. 1–6), IEEE, 2018.

[77] R. Krishnamurthi, D. Gopinathan, and A. Nayyar, "A comprehensive overview of fog data processing and analytics for healthcare 4.0," in *Fog Computing for Healthcare 4.0 Environments*, pp. 103–129, 2021.

[78] S. Dey, Securing majority-attack in blockchain using machine learning and algorithmic game theory: a proof of work, 2018. arXiv preprint arXiv:1806.05477.

[79] I. Eyal and E. G. Sirer, "Majority is not enough: bitcoin mining is vulnerable," *Communications of the ACM*, vol. 61, no. 7, pp. 95–102, 2018.

[80] A. Feder, N. Gandal, J. Hamrick, and T. Moore, "The impact of DDoS and other security shocks on Bitcoin currency exchanges: evidence from Mt. Gox," *Journal of Cybersecurity*, vol. 3, no. 2, pp. 137–144, 2018.

[81] A. K. Fedorov, E. O. Kiktenko, and A. I. Lvovsky, "Quantum computers put blockchain security at risk," *Nature*, vol. 563, pp. 465–467, 2018.

[82] K. Salah, M. H. U. Rehman, N. Nizamuddin, and A. Al-Fuqaha, "Blockchain for AI: review and open research challenges," *IEEE Access*, vol. 7, pp. 1012710149, 2019.

Chapter 7

Robotics and process automation technologies for healthcare informatics

Vivek Kumar Prasad[1], Rachana Mehta[1], Madhuri Bhavsar[1], Sudeep Tanwar[1] and Mahima Bakshi[2]

Abstract

Healthcare is the backbone of the nation and is a patient-centered industry. The introduction of robots into the health sector will play a vital role. Robots were frequently seen as "advanced technology," and patients may find them indifferent. Nonetheless, robots are starting to appear in a variety of functions in hospitals, i.e., from health monitoring of the patients to the management of the hospitals. Even the management can generate lots of revenues if the robots work perfectly. Robotics' automation can be readily designed to do simple tasks across apps in the same way those humans do. A process with several stages (like monitoring of the patient's health-related parameters) and operations, such as receiving forms, delivering a reception message, validating the form for compliance, putting the form in a folder, and maintaining a spreadsheet with the identity of the document, the date submitted, and so forth, can indeed be explained through the program. The training of the robots to work efficiently in the healthcare systems faces many issues, as its dysfunction can result in a disaster and the life-threatening to the patients. The chapter discusses the various challenges of the use of robots in the healthcare industry. The challenges are like what training the robots should get to work perfectly with the given environment. How the robots assist in the scarcity of nursing, doctors, and other healthcare providers by supporting the providers with repetitive duties, allowing the practitioner to undertake higher-level functions. How the robots minimize the need for "outsourcing production" management operations in healthcare. These intense observations and literature conclude that the Health Robotics categorization outlines how to offer high-quality patient care, effective medical operations, and a comfortable workplace for both doctors and healthcare staff. The chapter

[1]Department of CSE, Institute of Technology, Nirma University, India
[2]Arizona State University, USA

concludes with the case study where the robot ensures to protect the privacy and confidentiality of the patient's data.

Keywords: Healthcare systems; Robotics; Monitoring; Automation; Privacy; Confidentiality

7.1 Introduction

The incorporation of the Artificial Intelligence (AI) based on robots idea into healthcare units through the use of clever intelligent gadgets and the use of high-speed data transmission networking techniques has raised the bar in healthcare ideology to a new level [1]. Patient care is the primary service a hospital provides. A lot of support services have to be provided in the hospital in order to deliver this service. The way patients visit the hospital has a significant impact for the robotics culture in the healthcare organizations [2].

People live a healthier life when new technologies are developed and advanced in healthcare institutions, and, as a result, people's quality of life improves. AI embedded machines are termed as smart wearable devices with fully interconnected efficient sensors that involve monitoring, collecting, and diagnosing patients based on symptoms collected from sensory data [3]. On the other hand, a robot physician could timely investigate and document a patient's heart problem when the medical professionals are not available and help people to identify the health condition regardless of geographic location. AI and automation are two emerging concepts in healthcare 5.0 that are ready to alter any conditions of service [4].

Figure 7.1 shows how the industrial robot market has increased during, before, and after the pandemic hits the world [5]. Table 7.1 shows the metrics which are

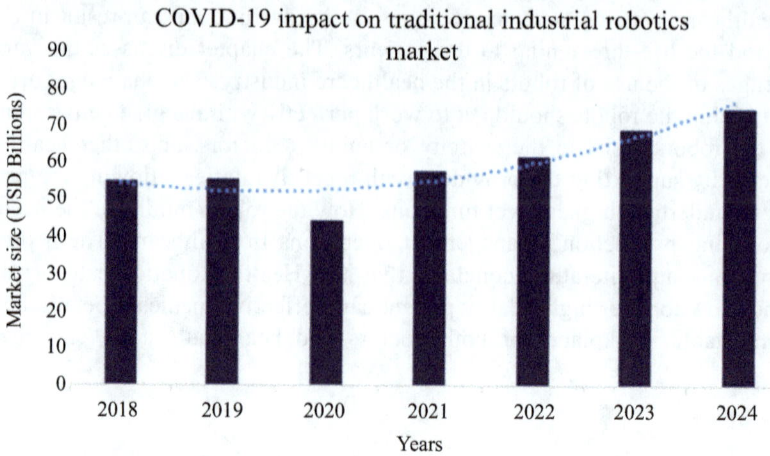

Figure 7.1 Observations of pandemic into the traditional industrial robotics market

Table 7.1 Metrics associated with Figure 7.1 and its information

Metrics of the report	Illustration
Market size	2018–2025
Base year	2019
Period of forecasting	2020–2025
Quantity or measurement	Value – USD (billions)
Covered segments	Industry, type, and geography
Region (geographical) covered	Europe, APAC, North America, and RoW

associated with Figure 7.1 that indicates how the quantity of the robots-based automation systems have increased during the years and also predicted its values till 2024. From Figure 7.1, we can see that from the year 2021, this has been increased, as this will reduce the human interactions as well and will be helpful to stop spreading the infection of the pandemic like COVID-19.

When compared with other types like industrial robots, articulated robots are used in a wider range of industries. Because their manufacturing processes are largely automated, the automotive industry accounts for the majority of articulated robots. Despite the fact that the COVID-19 epidemic has damaged almost every industry, flexible robots are likely to hold the maximum share of the industrial robotics market over the projection period. Therefore, for the whole projected period, articulated robots will continue to retain the highest share in the market.

AI in healthcare 5.0 addresses the concepts of realistic and mechanized disease prediction, remote patient observations, monitoring and its analysis, robotics surgical systems, AI therapy, and is for persons suffering from social apprehension, collecting and evaluating results, and so forth [6]. AI is a concept that has no bounds when it comes to smart innovations. It is the capacity of computers program or algorithm to estimate the final results for machines without the need for human involvement. The term "smart healthcare system," which refers to the growth and progress of methodology in healthcare with the fifth-generation wireless touch [7]. Pretty much every single research facility, from pathology labs to disease observation research facilities and blood repositories, enables sophisticated, intelligent, interactive advanced technologies that incorporate powerful and highly effective AI algorithms for automated matching and detection of blood groups of any blood samples collected, etc. When it comes to decision-making, AI techniques are the primary choice for developing or designing any intelligent model. As AI-based technologies and applications in healthcare may assist to help understand each patient's care plan, medical decisions, and the emergence of digital pharmaceuticals, it has become a main research focus for researchers and many companies [8]. Tests, X-rays, CT scans, information input, and other tedious activities are all analyzed faster and more precisely by the AI robots. It is based on the same idea as genomics and proteomics, which hunt for mutations and ways to forecast disease using data taken from DNA samples. Based on the patient's DNA, body scan reports can detect malignant cells and vascular illnesses sooner and predict the likely health issues that would influence the patient's healthcare [9]. An AI-enabled smartphone webcam automatically validates that patients are taking their medications and assist them in managing their illness. Users with significant medical illnesses, individuals who tend to ignore doctor advice, and clinical trial participants all accept the digitized smart wearable devices. A digital nurse is one of the brilliant AI applications [10]. What this does is that, between medical visits, it monitors and records the patient's status and follows up with prescriptions. The software employs AI to assist patients with chronic diseases, as well as provide basic health advice and guidelines to parents of sick children. Based on the past records of the registered user's medical documents and standard health knowledge, the app or AI-powered web portals can answer inquiries regarding prescriptions and also decide whether or not symptoms demand a doctor visit [11]. These details are put into an AI-powered application that utilizes speech recognition to match the information to a database of ailments and provide a query result. Radiology and cardiology are two disciplines in which large amounts of data must be analyzed, which takes time. The following two disciplines will be the emerging research areas of healthcare 5.0 in the future, and they should only be used in the most complex circumstances where human supervision is necessary [12].

In the recent decade, there is a significant rise in research and commercial interest in the field of healthcare robots [13]. Robotic rehabilitation technologies are actively offering physical or occupational treatment, allowing for a higher frequency of nursing that is constantly adjustable to a patient's demands and requirements. A wide range of disorders and diseases can be treated using modern AI-based robots [14]. This increases the standard care and improvises the health diagnosis and management of the patients. While robots are already starting to affect human health through clinical use, further investigations and commercial successes will be facilitated by carefully considering society's drivers for improving healthcare systems [15]. The specific capacity for robotic systems to influence healthcare scenarios and the basic technological improvements requires significant improvements. Among the numerous fields of service robots, healthcare for the old and disabled is the one with the most promising prospects. The present trend toward lengthy care at home rather than in hospitals and nursing homes, as well as the loss of staff in all medical associations, makes the creation of new amazing and advanced technical aid tools especially in a timely manner. Academia and businesses are actively working on a number of medical robotics research programs. The first commercial version of a hospital delivery robot is detailed herein. It consists of a sensor-based platform with beacons and visual cues that can navigate unattended in medical situations [16]. A prototype of a Mobile Robotic Unit for disabled assistance is detailed, it is a transportable station with a dexterous controller, a stereo camera, and a sonar ring that makes up this gadget [17]. A control center attached to an interface (graphical) is in charge of the control. This technology is now being transformed into a prototype that may be tested in real-world scenarios. In another research work, a notion of autonomous medical support is presented where the robotic equipment is in the hospital room itself, which is outfitted with sensors and actuators for recognizing, expounding, and working on

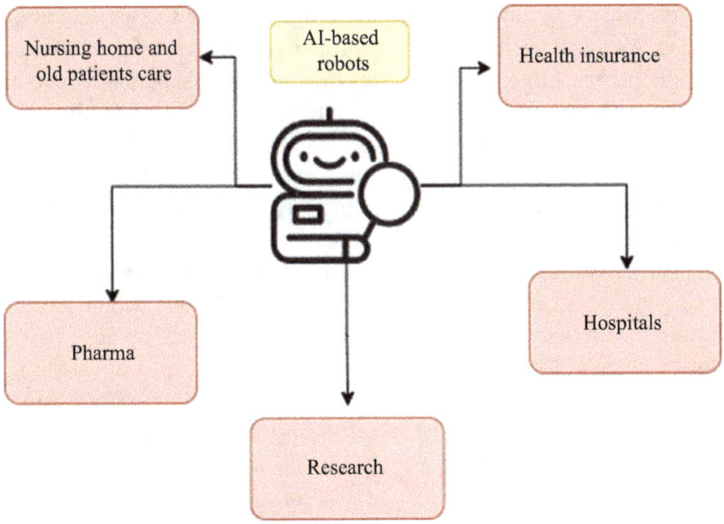

Figure 7.2 Flow chart depicting the process undergoing inside the hospital

the aided person's demands [18]. The three important concepts for the design of supporting robotic infrastructure are mentioned and these are: the utilization of motions and verbal commands for the robotic system, a dispersed architecture of simple and robust devices associated with the robotics, and cognition to tolerate unpredictability and cooperate with the client. It is widely agreed that a small mobile system comprising a manipulating arm, a computer vision-based mechanism, and an interaction for robotic medical is necessary. Figure 7.2 shows the applications of the AI-based robots in the healthcare industries. The robots can be used to assist the nursing centers, health insurance centers, hospitals, pharma, and research centers [19].

The framework is developed with a variety of capabilities based on the task and the type of user. This also leads to the personalizing of the robot to the type of user who is receiving the services.

The purpose of this chapter is to outline some of the most essential capabilities and technical breakthroughs in medical and healthcare robotics that are required to improve human health. We go over application domains, societal factors, motivating scenarios, desirable system capabilities, and foundational research areas to think about while designing medical and healthcare-automated robots [20].

AI systems that can learn so much that they can surpass the doctors by synthesizing all accessible knowledge from all medical sources and will be the wish granting of the twenty-first-century robo–doctor combinations. Most experts believe, however, that AI will not replace skilled medical personnel; rather, it will improve their efficiency in various areas, including, increasing precision: the robots are used in various prestigious hospitals, as they work without tiredness and are not similar to human surgeons [21]. This could do tasks with great precision, tremendous confidence, and no blade quivering, bridging the gap among humans and machines. As much as the software is properly configured for the process being performed, the human surgeon serves as a supplementary, supervising persona [22].

Accuracy in the diagnosis: The true potential of AI lies in identifying patterns in health data and other data to describe distinct disorders. Machine can work throughout the 24 × 7 on thousands of the parameters related to the medical scenarios. These robots outperform and compete many greatest healthcare surgeons and physicians [23].

Treatment from afar: The robots for the medical utility for remote locations were initially proposed by the DARPA in the late 1990s. Currently, the fourth-generation and fifth-generation technologies are adding advantages to the communication systems of the robotic environment. They have a great role to solve the issues which have occurred before. However, it appears that robotic surgery still necessitates the use of human assistance for hygiene/cleanliness and other activities, which complicates matters and makes it financially unviable [24].

Concerns about malpractice: The effect of malpractice is an issue for the humanoid machines and are called as robots. The technology is merely a tool used by the doctor and may lead to hazards if properly not trained. When the computer program or an AI program is mature enough to make judgments without human input, the scenario changes. It is whose responsibility if the decision taken leads to a failure. This may be the doctor who failed to turn off the equipment at the appropriate time? Who was the coder who did not see that coming? There is currently no apparent response due to the novelty of this topic. With time, and as the AI-based healthcare industry becomes more standardized, there will be a more exact manner of addressing these issues. One method through which the AR and AI can assist surgeons through producing a real-time tailored overlay while surgery, indicating blood arteries and other vulnerable locations. On the other hand, if we talk about the robotic arm, then the expertise library can recommend different tools based on existing guiding principles. Another sort of remote medical robot is a basic bot-pill, which performs an endoscopy much more conveniently than before. This "miracle pill' transmits photographs of your intestines, and you naturally remove it [25].

Enhancing human capabilities: Certain medical robots, in addition to medical professionals, assist patients. Like exoskeleton, robots assist paralyzed people regain its ability to assist through walking and to become self-sufficient. Another example of technology in action is a smart prosthesis. These bionic limbs have sensors that make them more reactive and precise than real body parts, and they can be shielded with a bionic skin and linked and associated with the user's muscles operations. Supporting psychological health as well as day-to-day tasks: the service robots can assist sick elderly patients with loneliness. These patients must be monitored with their regular check-ups and BP measurements. The robots can provide companionship and communicate with these patients and also remind them to take medicines and check their body temperatures at regular intervals. These robots also help patients to maintain a happy environment.

There is another terminology called a personal assistant. The personal assistance embeds itself with the person and analyses their temperament. These are helpful for assisting the depressed people.

Robots that assist the main robot: Due to so much work to be done in the healthcare sector, and existing doctors cannot do much work, so robotic assistance can be beneficial to nurses and hospital workers. Robot handles refilling, providing goods, and sanitizing, allowing health workers to spend a lot more time with patients. A UV light disinfectant robot is another fantastic auxiliary robot that enters a hospital room and does not depart until it is bacteria/germ-free. Many consumers use social networking, Netflix, Siri, Alexa, and other "smart" technologies without realizing those machine learning powers many of them [26].

7.1.1 Motivation

The automated robot is motivated to increase the efficiency of an operation by combining data and actions in the medical scenario or intervention and by transcending human physical limitations in operation and other intervention procedures, but still allowing the human to control. The present-day robots used in surgery are controlled directly by a surgeon frequently in a telescope where the operator manipulates one master input device, and the input is followed by the patient-side robot. In contrast to typical minimally invasive operations, robotics enables the neurosurgeon to manipulate the body with expertise and to scale down the operator from ordinary dimensions of man to very small distances. When the necessary data is available, many medical operations can be identified and carried out in a reasonable way in advance with human supervision of the robot.

7.1.2 Objectives

- To study and examine the rising role of robotics in healthcare;
- To identify current and probable hurdles to gain the potential of robotics and related convergence innovations;
- And, to focus on the health robotics categorization and outline how to offer high-quality patient care, successful medical treatments, and a relaxing work environment for medical professionals.

7.1.3 Organization of chapter

The rest of the chapter is organized as: Section 7.2 discusses about various methodologies used in healthcare field and robotics and section 7.3 concludes the chapter with future scope.

7.2 Methodologies in healthcare and robotics

In the previous section, we have seen how the medical robots facilitate operations, simplify hospital operations, and allow healthcare providers to focus more directly

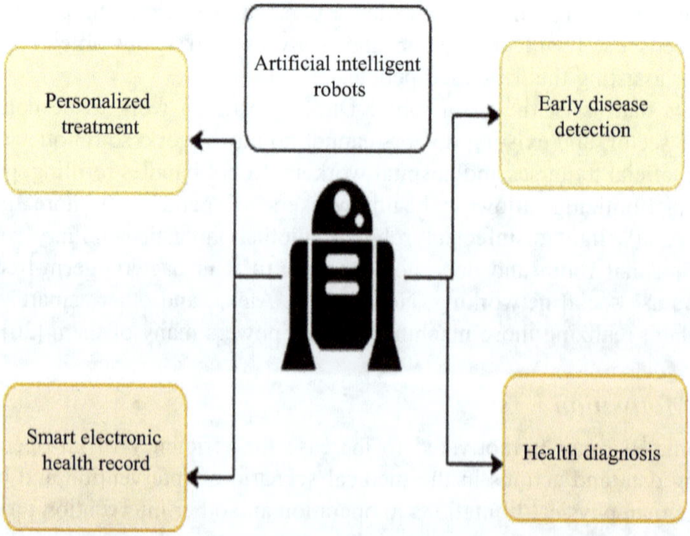

Figure 7.3 AI robots and healthcare operations

on patients. The introduction section also mentions about how the robots are revolutionizing surgeries, which are carried out, speeding supply delivery and cleaning, and giving medical professionals more time to interact with the patients. In this section, we stress on the discussion towards methodologies in healthcare and robotics.

Figure 7.3 indicates the use of the AI robots in the medical sector such as personalized treatment, to maintain the smart electronic health records, early disease detection based on the previous healthcare records and health diagnosis.

One of the key bodies of contemporary society is hospitals. Socio-demographic characteristics, such as the increase in the elderly, raise the need to revise health services to maximize the efficiency of hospital operations. Figure 7.4 indicates the scenario of the working of the AI robot-based systems in the healthcare organizations. This also depicts how the communication takes place between the robots and the medical healthcare stakeholders. The robots are made intelligent using the AI technology and act as a central decision-making controller and interact with the IoT systems installed at the healthcare organizations, data management systems, and usages deep learning concepts to make the decisions [27]. The knowledge is then used for the various applications just as smart blood bank management, for intelligent pathology lab testing and disease prediction, intelligent waste management of the hospital's residuals, ambulance services, to generate the emergency alert systems and also to monitor the smart and intelligent devices attached to the patients' body parts. This health information is then transferred to the base stations that are connected to the central cloud services. When this data is accumulated at the cloud, then the registered users such as doctors, patients, and

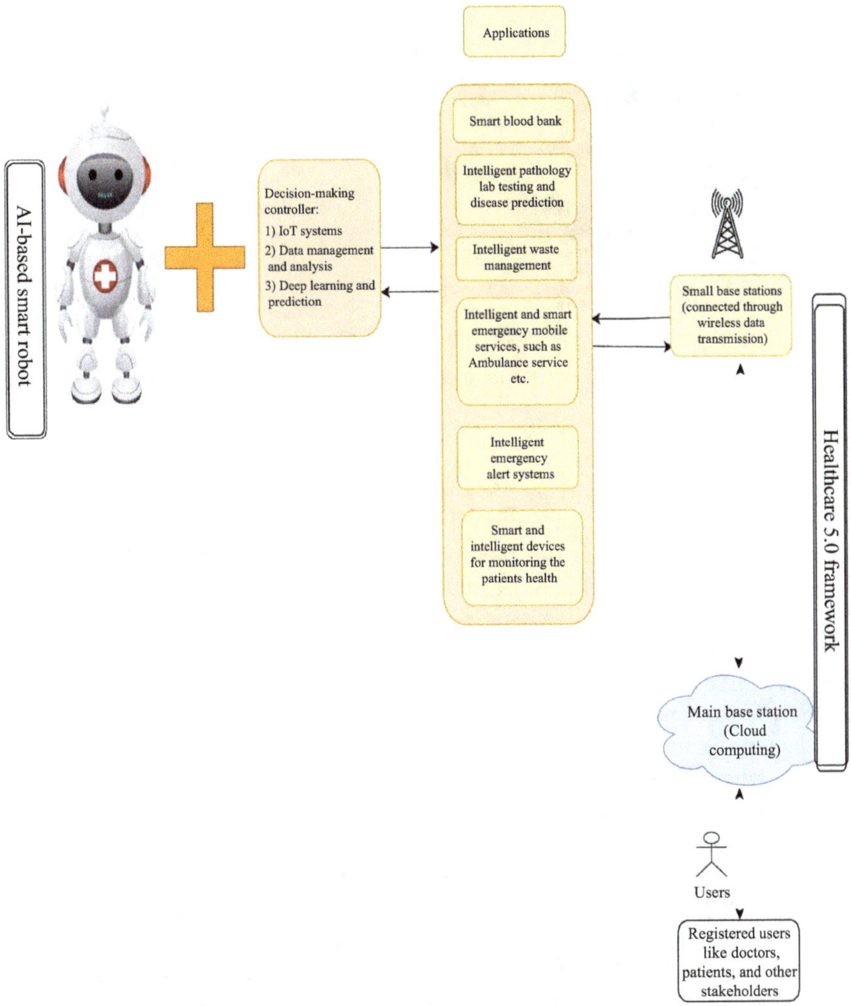

Figure 7.4 Flow chart depicting the process undergoing inside the hospital

other stakeholders of the healthcare systems can connect to this information and act accordingly. Figure 7.5 depicts how the hardware and software are used for the research-based AI robots. The hardware components are essential for the design of the robots. Apart from the hardware, to make the decision, software is also essential for the robot design. The software is required to make the robot operational as per the logic of the healthcare systems.

Figures 7.6 and 7.7 illustrates that the installing the app on one's smartphone is required before using the healthcare app, and the following paragraph shows how the interaction between the robot and the healthcare stakeholder takes place.

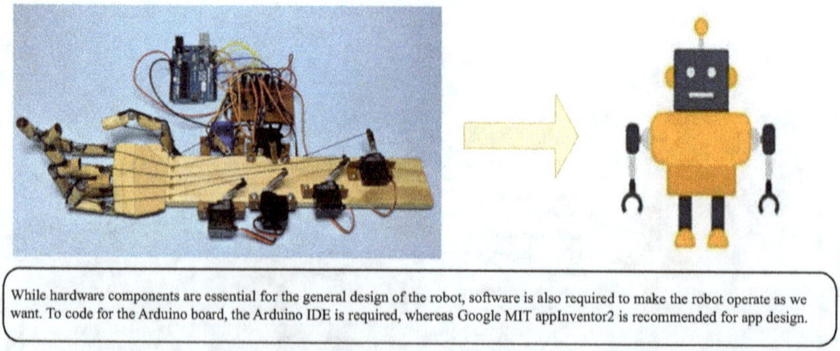

Figure 7.5 How the hardware and software are used for the research-based AI robots

The client or a family member must first enroll on the App by entering the client's id as well as other information, including the identity of the attending physician, the house number or bed numbers or ID, as well as other information for security reasons. After registering, the users can utilize the app for emergencies and other purposes only when medical personnel are not nearby. Then they must first log in; while logging their credentials in, they must just type the patient's ID and push the log in button. The login is not required every time. They can log in for the first time right after registering, or anytime they wish. As a result, when you log in, the app will validate the user using data that was previously submitted during registering. If it is inaccurate, it will prompt you to supply the right data; if the number is accurate, you will be directed to the app's final screen. There are two buttons available, one for START and another for STOP. When people touch the touchpad of the mobile, the robot will quickly move by validating the path you have chosen. If any difficulties are encountered along the way, or if a failure type occurs, the hurdles or problems are addressed first. Then gain the press, the start button needs to be touched. Until and unless the fault is corrected, the robot may not proceed to its destination. When everything is in order and there are no more errors or roadblocks in the path. The robot will come to the destination prescribed. Where the stakeholders find the assistance that they require.

Healthcare robots are no longer limited to laboratory work and assisting patients, but with technological advancements, they are also being employed for small diagnostic and therapeutic services as well. The robots are intended to save human resources while simultaneously boosting healthcare service [28].

The fully accessible Arduino Software (IDE) simplifies code writing process and uploads the same to the ARDUINO board. It is compatible with Windows, Mac OS X, and Linux. The ecosystem is implemented in Java and is built on open-source software such as Processing. This model is equipped with any Arduino board. The MIT App Inventor2 is a cloud-based application that allows apps to be

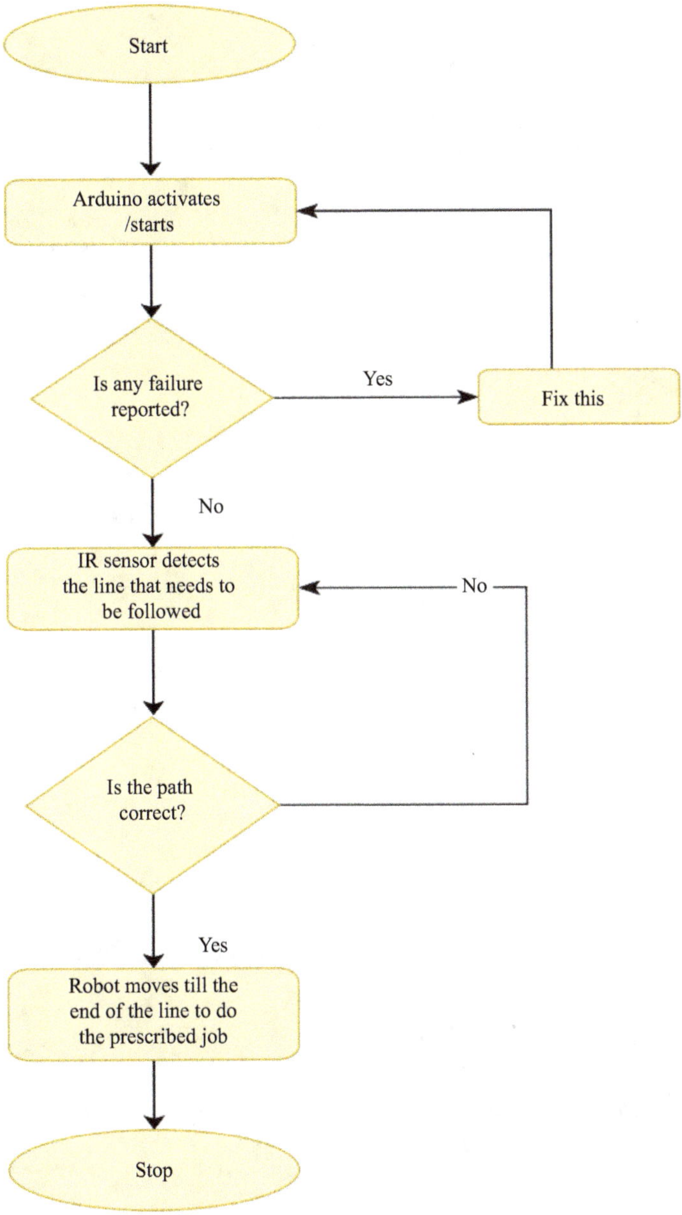

Figure 7.6 Flow chart of installing the app on smartphone

produced directly in the web page or a web browser. This website provides all of the assistance required to create your own apps. The App Inventor program, often known as the "service," can be found at ai2.appinventor.mit.edu. Robots are now utilized in clinical practice to assist health personnel and improve quality of care, in

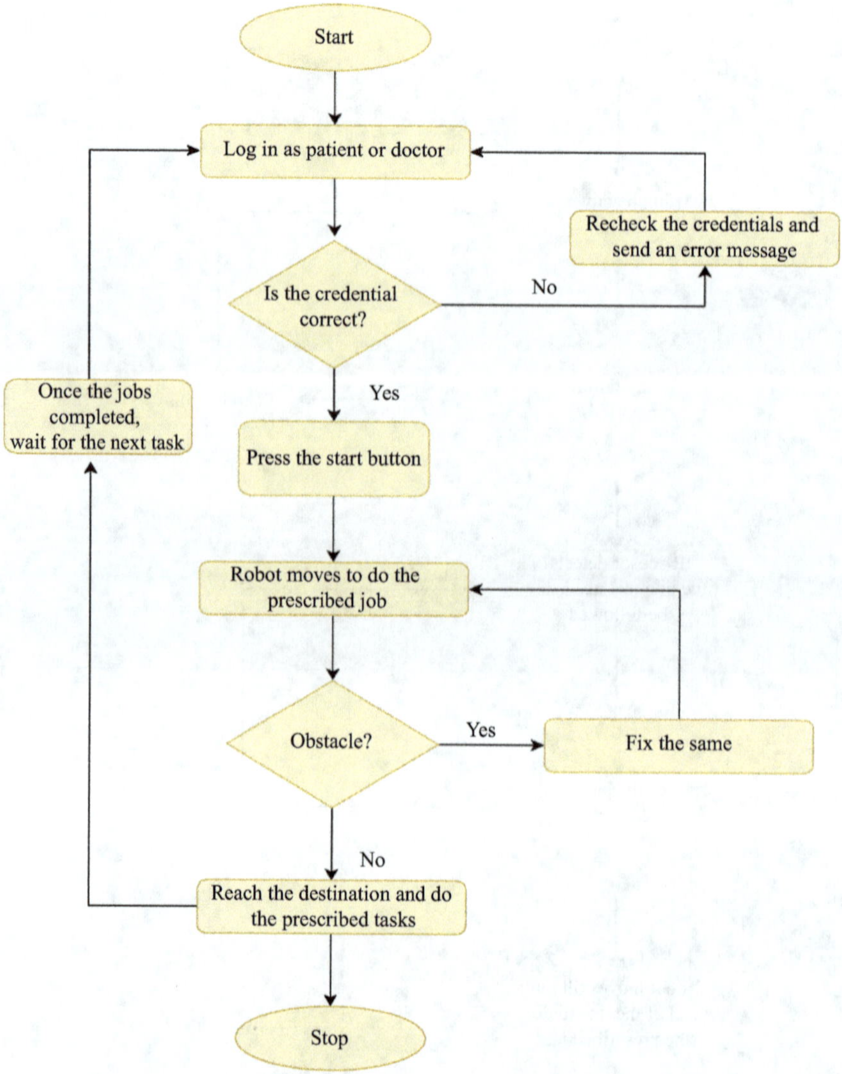

Figure 7.7 Flow chart of installing the app on smartphone required before using the healthcare app

addition to operating rooms. During the COVID-19 outbreak, clinics and hospitals began using robots for a far broader range of jobs in order to help decrease infection exposure. This is evident that health robotics' administrative savings and risk minimization give benefit in a variety of areas [29]. Infected wards can benefit from robots that clean and organize patient rooms on their own, reducing individual to individual interaction. Algorithm 7.1 depicts how the robots maintain the beds in the hospitals.

7.2.1 Algorithm regarding how ROBOTS handles the patients in the hospitals

---

1. Input: The expected value of the patients (P).
2. Output: To allocate beds to affected patients (CB).

/* Initialization and Declaration of the Variables*/
Get the COVID-19 patients count = A.
Identify the empty hospital beds information = V.

3: Get the information of the COVID-19 patients of the specified city.
4: The number of patients are denoted as A.
5: Finding the relation between patient details and available vacant beds to allocate them in defined regions hospital
  if Patients(A) < Available vacant Beds(V) **then**
    These patients can be admitted to the hospitals of that region and as per the availability of the vacant beds. A <CB.
  else
    Healthcare data (of the vacant beds)(V) need to identified and these patients(A) can be directed to the nearby located hospitals. V < A.
6: Repeat Steps - 3 to 5 until all the Patients not got Discharged.

7.2.2 Healthcare benefits of using robotics

Medical robots provide effective clinical procedures, best patient safety, and a secure work conditions for patients and medical personnel. The following points highlight the benefits of utilization of Robotics in healthcare:

1. Patient care with the highest quality: Medical robots help with non-invasive operations, personalized, and regular observation for people with chronic conditions, cognitive medicines, and social interaction for the elderly. This is due to the robots with reduced responsibilities, nurses and other caregivers may provide more empathy and human connection to patients, which can improve healthcare use-cases.

With the enhancement in the DL, big data, image processing, and other technologies, healthcare robotics are enhanced with better operations. All kinds of the robots will continue to get better their utilities for jobs which are to be done independently, quickly, and precisely [30].

2. A secure working environment: In healthcare scenarios where infection exposure is a problematic, service robots move equipment, towels, and prescriptions for the safety of the medical workers. By minimizing microbial transmission, sweeping and sanitizing robots help decrease healthcare-associated illnesses. Already, hundreds of healthcare institutions are using them. Social robots can aid with lifting, such as moving bedding or individuals, relieving the physical burden on medical workers [31].

3. Surgical-assistance robots: Surgical robots, the name itself is indicating its meaning and this works under the concept of the technology called a motion controller. Careful incisions are done to carry out complex operations, which are usually micro-operations. While doing the incision, the robots use the technology of computer vision. This is also to be noted that while doing the operations, the barriers such as nerves and other obstacles are to be carefully moved before doing any cuts, etc. Some robots perform the operations under the monitoring and supervision of a doctor and can be classified into two different types and are mentioned below [32].

For the torso, minimally invasive operations are available and discussed here, robotic hysterectomy, bariatric surgery, robotic prostatectomy, and other soft tissue surgeries are among the procedures available. After being placed through a small incision, these robots lock themselves, offering a dependable platform for remote control procedures. Previously, most inside surgeries required open surgery with large scars. Reaction times were much slower with additional indications of contaminants as well as other issues. Although for the highly experienced surgeon a button-size incision could be problematic. Robotic surgery, like Intuitive's da Vinci robot, makes these operations easier and more accurate in order to reduce pathogens and complications [33].

4. Interventions in the orthopedics field: Common orthopedics' procedures, such as knee and hip surgeries, can be integrated into machines such as Stryker's Mako robot. These robots assist the surgeon by using spatially defined limits and combining sophisticated robotic arms, 3D imaging, and data analytic to offer more precise data. Using AI programming, the Mako robot may be taught in specialized orthopedics operations, with exact guidance on where to go and how to perform the therapies [34].

Surgical robots are expanding with deep learning-based technologies. Surgical robots make use of the image recognition to detect a variety of tissue with various view and range. Surgical robots, for example, can now assist doctors for the operations to be done on muscles and nerves during operations. During surgeries, high-definition 3D image processing and computer vision can offer surgeons with specific information and enhanced efficiency. Robots can perform the required activities and movements with the assistance of surgeons. As well as in many cases, the trained robots educate and provide training to the newly joined surgeons. The mimic development environment, the virtual reality, and DL are used to teach new physicians how to operate robotic surgery. By adopting robotic controllers, physicians will practice procedures and refine their expertise in a virtual world [35].

5. Robots that are self-aware and autonomous: To deal with the patients in the healthcare organization and support the patients with room cleaning services and communicating with the remotely located patients, the self-decision-based robots with the support of LiDAR, map-skills, and visual computing will be an advantageous option. These robots also take care of the self-charging energies and when they detect a fall in the energy they come to the facility of the recharge and charge themselves and these robots also disinfect and cleanse the areas under their supervision and where the patients are present. Akara is a new start-up in this technological era and uses UV lights to clean and sanitize the surfaces that are polluted. This is also an advantage for COVID-19-like situations.

6. Robots on the move or mobile robot: The mobile robots track and work with the backbone of the predefined pathways in and around the medical facilities. This includes sanitization and cleaning of the patient's surroundings, the movement of patients, and for the movement of heavy machinery from one area of the healthcare center to another. Even the cleaning of the robots would not be a problem, as disinfect and sanitizing mobile robots can utilize air filtration, ultraviolet (UV) radiation, hydrogen peroxide vapors, or other cleaning agents to consistently clean accessible areas or assist avoid contamination [36].

7. Robots that are social or social robots: Humans interact directly with social robots. These "friendly" robots can provide human contact and surveillance in lengthy care facilities. They may foster active learning or encourage patients to adhere for treatment plans, keeping them focused and upbeat. They also are employed to guide both patients and visitors around the facility. Generally speaking, social robots help caretakers by easing burdens and enhancing patients' mental well-being [37].

8. Robots that provide services and are social: The major stress for healthcare professionals is related to the logistical duties. These routine duties related to the logistical can be very well handled by the service robots. These robots are self-contained and will report the action, which indicates that the work has been done by the robot. Apart from the aforementioned task, the robots can work for searching and setting up rooms for the hospitals, also keep track of medical supplies, maintain the orders related to the filing purchase, for transporting the linens of the beds, refilling of the healthcare supply cabinets, and washing facilities as well. Similarly, the service robots are also used for performing certain regular day-to-day activities such as any essential clinical preferences [38].

9. Robots with modular components: Modular robots can be programmed for various healthcare tasks. Examples of such robotic will be therapeutic exoskeletons and prostheses of humanoid legs and arms related to the healthcare sector. Therapeutic robots are useful for rehabilitation followed by any neurological damage, strokes, paralysis, and neurodegenerative disorders. The patients can be monitored by these AI-enabled robots for the prescribed workouts, positions with respect to the various degree of the motions, and charting improvements more accurately than the human eye (Table 7.2). The same operations can be used for communications with the patients for training purposes and for motivating them to do the workouts in a better manner [39].

10. Self-sufficient robots in hospital transport systems: In hospitals, logistics is one of the most important and sometimes undervalued services and supports. Hospital logistics service is primarily committed to organizing and maintaining the flow of commodities in the hospital. In a healthcare center, a wide range of materials is important, which leads often to complex transportation networks and a complicated material flow. Logistics automation in hospitals is increasingly a concern. The major motive is to cut operating costs, which are the consequence of logistical efforts around 30% of healthcare costs. Statistics also reveal that in

Table 7.2 Different types of robots and its description

Types of AI robots	Description
Intel Movidius VPUs	These robots help doctors in their daily routine work and also for sanitizing the medical devices
Intel Real Sense technology	These robots are used for physiotherapy kind of verification and to observe the movements of the patients
Intel Distribution of OpenVINO toolkit3	These robots are used for surgical assistance and monitoring of the healthcare premises
Intel Atom processors and Intel Core processors	Used for disinfecting and surgical purpose
Scalable processors: Intel Xeon	Used for the edge computing-based systems
Intel supported 5G networks	Used in AR/VR-based systems and video-based examinations

western countries where the number of hospital patients is increasing, significantly increasing the load of logistical support tasks. Finally, there is an increase in the range of materials and equipment utilized in hospitals. New equipment and materials are progressively being produced and throwaway products are becoming frequent for one-time usage. Consequently, the volume of transport is significantly expanding; more commodities need to be transported and often transferred. Automation can also enhance transport capabilities. Hand-held transportation duties in hospitals are frequently limited by the human strength available. The available transport modes and personnel will be used to plan routes, amounts, loads, and intervals. On the other hand, an automated system can produce a considerably more adaptable transit scheme. Routes can be adjusted and regular deliveries both at night and during the day can be scheduled. If the tracking and stock systems and transport systems are closely integrated, the time of responding to requests can also be increased.

The healthcare transit system is quite sophisticated. Some of the reasons for complexity are highlighted as varied material flow sources, time limitations on distribution routines, employee working regulations and transportation equipment. Based on the findings, the autonomous robots in transport systems in hospitals are developed in three choices.

Existing system adjustment: In the global arena, the transport system continues the same. Robots are employed to perform existing transport activities in some types of transport. As they are, routines endure [40].

Systems that are partly reconfigured: Some new tasks gets discovered and robots are created for the best performance of these jobs. Transport protocols based on robotic capabilities are streamlined. Existing warehousing systems are used for 24-h operation as well as deliveries routines are extended.

System restructuring: A central stock system is implemented to assist system-wide optimization. The central stock can contain many types of supplies. The fundamental advantage of this approach is that various types of commodities can be combined and shipped together at the central stock, thus reducing the number of consignments. Various sorts of robots are thus introduced and routes are designed to be operated for 24 h.

Most robotic systems are simplest solution when it comes to overhead development; since there are little changes to routines, the robot carriers just take place where possible from the human counterparts [42,43]. There are still not very frequent service robots in hospitals, and existing robots are under developing stage. It is therefore common to note that they just attempt to adapt to current transit infrastructures. However, the possibility of deploying independent mobile robots and automation is not fully exploited in this alternative. A partial reconfiguration or reconstruction is necessary to make the most of an autonomous robotic transport system. It is evident that various units are needed in the current scenario for different workers to be serviced several times. Support can instead be integrated and supplied together in a single freight unit from sterile central, sterile manufacturing, central warehouse, pharmacy, and the core laundry. The inbound load can then be sorted and delivered inside the unit. Likewise, output materials can be mounted to the unit and packaged in a single freight unit. There are also many collections and distributions that you can run day and night. Even though it is more difficult to implement than the first, it is considered that the second option is far better at exploiting the benefits of deploying mobile robots for transporting. Existing healthcare facilities should explore this alternative instead of the first if they are renovating or redesigning. The third option is the most efficient approach to construct an automated hospital transit system. The goal is to rebuild the entire system and replace the different decentralized shipping containers with a logistics center. Because this method would necessitate a significant number of environmental changes, it is not a viable choice for existing hospitals. However, it should be the primary option to explore when new healthcare systems are being developed. In terms of infection risk, transportation equipment as well as routinely touched surfaces such as doors and elevators are vulnerable. A robot system that is easy to clean, as well as automatic doors and elevators that are required for robot operation, can help to ameliorate the issue.

Finally, it is worth noting that integrating IT technology for buying supplies and resource planning into transportation systems that use robots is another major issue that can improve effectiveness and efficiency. This combination still has a lot of room for development, and more advancement is predicted in this area. The necessity for mobile robots for the transit of hospital means of transport jobs is apparent, but their use must be based on a comprehensive examination of the entire logistics system and must be designed accordingly. Although existing robotic systems demonstrate the possibility of future applications, major advances can be realized by recreating the transport frameworks wholly or partially [41].

11. Interaction of more than one robot in the healthcare field: The objective is to create networked, diverse multifunctional robot platforms that can integrate with healthcare organizations and are easy to modify as required by changing operational demands. A receptionist robot system, a nurse assistance robot design, and a medical server are the three subsystems of the prototype solution. The nurse and receptionist assistant robots' jobs are to assist the human receptionist and nurse with tasks. The healthcare robots use a web interface to upload and download patient data from a medical server and send data summaries to human caregivers. Multiple robots of each type could be used in a robot healthcare system. Speech recognition and generation, as well as facial recognition, are used by the receptionist robot to interact with the patient. Without the assistance of a human caregiver, the nursing assistant robot system assists assigned patients in measuring vital signs and reporting the results to the medical server. Personal information and medical history of registered patients are stored on the medical server. Medical personnel can access the medical server web site from any web-enabled device to check the measured health data [42]. These automated robot systems ought to be capable of a wide range of intellectual interactions, including identification and response mechanisms [43], they should have a variety of intelligent functions for interaction [44]. The same has been depicted in Figure 7.8. Hence, in the medical field, robots are revolutionizing surgery-based applications, expediting supply delivery, used for decontamination, and freeing up time for clinicians to interact with patients. Different vendors also provide a variety of healthcare robotics technology including surgical assistance, modular, service, social, portable robotics. and many more to come in future [45].

7.3 Conclusion and future scope

One of the key bodies of contemporary society is hospitals. Socio-demographic characteristics, such as the increase in the elderly, raise the need to revise health services to maximize the efficiency of hospital operations. Along with improvements in deep learning, business intelligence, image recognition, and other innovations, health robots will evolve over time. Robots of all kinds will continue improving their ability to accomplish jobs independently, swiftly, and effectively. Because of rising technology capabilities, lower costs, and increased need to save costs, robot installations in healthcare environments are forecast to expand. However, because robots have the potential to be extremely disruptive technologies, it is essential to understand the socio-technical difficulties that will be encountered as robots are implemented in order to develop mitigating measures. Complex adaptive issues related to efficient robotic system implementation in healthcare facilities, especially in patient-facing roles, may indeed be important. It is necessary to identify and address these for efficient acceptance and development. The system design requires a low-cost robot support system for easy home care and chosen duties in medical institutions to meet the

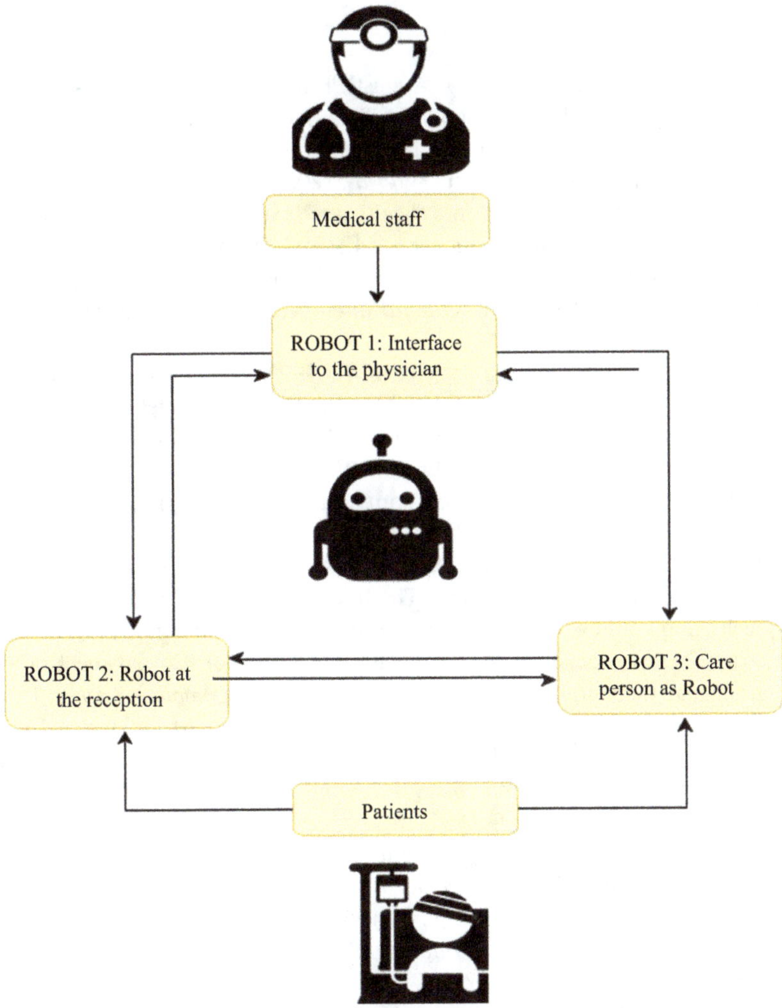

Figure 7.8 Flow chart depicting the process undergoing inside the hospital

healthcare community's requirements. These issues must be foreseen and, if feasible, addressed in a proactive manner. The provocation is to preserve, enhance, or increase care labor in an efficient technological and cultural forces. This can only be accomplished if we foresee and handle the issues that come with emerging innovations and when they are integrated into current social structures. The findings should be viewed as a foundation for more in-depth talks about these issues. It can also assist healthcare organizations and policymakers in making crucial strategic decisions about robotic application purchase, development, and deployment.

References

[1] Wang, W. and Siau, K. (2018, August). Ethical and moral issues with AI: a case study on healthcare robots. In *Twenty-fourth Americas Conference on Information Systems*. Retrieved from July (vol. 16, p. 2019).

[2] Foster, M. E., Ali, S., Litwin, S., *et al.* (2020). "Using AI-enhanced social robots to improve children's healthcare experiences." In *International Conference on Social Robotics*, pp. 542–553. Springer, Cham.

[3] Bohr, A. and Memarzadeh, K. (2020). The rise of artificial intelligence in healthcare applications. *In Artificial Intelligence in Healthcare* (pp. 25–60). Academic Press.

[4] Mohanta, B., Das, P., and Patnaik, S. (2019, May). Healthcare 5.0: a paradigm shift in digital healthcare system using artificial intelligence, IOT and 5G communication. In *2019 International Conference on Applied Machine Learning (ICAML)* (pp. 191–196). IEEE.

[5] https://www.marketsandmarkets.com/Market-Reports/impact-on-industrial-robotics-market-158051867.html

[6] Özdemir, V. and Hekim, N. (2018). Birth of industry 5.0: making sense of big data with artificial intelligence, "The Internet of Things" and next-generation technology policy. *Omics: A Journal of Integrative Biology*, 22(1), 65–76.

[7] Islam, M. M., Rahaman, A., and Islam, M. R. (2020). Development of smart healthcare monitoring system in IoT environment. *SN Computer Science*, 1, 1–11.

[8] Budida, D. A. M. and Mangrulkar, R. S. (2017, March). Design and implementation of smart HealthCare system using IoT. In *2017 International Conference on Innovations in Information, Embedded and Communication Systems (ICIIECS)* (pp. 1–7). IEEE.

[9] Zhu, H., Wu, C. K., Koo, C. H., *et al.* (2019). Smart healthcare in the era of internet-of-things. *IEEE Consumer Electronics Magazine*, 8(5), 26–30.

[10] Mohapatra, S. and Swarnkar, T. Artificial intelligence for smart healthcare management: brief study. In *Intelligent and Cloud Computing*, pp. 365–373. Springer, Singapore, 2021.

[11] Kamruzzaman, M. M. (2020). Architecture of smart health care system using artificial intelligence. In *2020 IEEE International Conference on Multimedia Expo Workshops (ICMEW)*, pp. 1–6. IEEE.

[12] Prasad, V. K., Tanwar, S., and Bhavsar, M. D. (2021). Advance cloud data analytics for 5G enabled IoT. In Tanwar, S. (eds), *Blockchain for 5G-Enabled IoT*. Springer, Cham. https://doi.org/10.1007/978-3-030-67490-87

[13] Konwar, A. N. and Borse, V. (2020). Current status of point-of-care diagnostic devices in the Indian healthcare system with an update on COVID-19 pandemic. *Sensors International*, 1, 100015.

[14] Hager, G., Kumar, V., Murphy, R., Rus, D., and Taylor, R. (2020). The role of robotics in infectious disease crises. *arXiv preprint arXiv:2010.09909*.

[15] Qureshi, D., Salim, M., Singh, P., Swarnkar, P., and Goud, H. (2020). Robotics solutions to combat novel Corona virus disease-2019 (COVID-19).

In *Pushpendra and Swarnkar, Pankaj and Goud, Harsh, Robotics Solutions to Combat Novel Corona Virus Disease-2019 (COVID- 19).*

[16] Mehta, H., Prasad, V. K., and Bhavsar, M. (2017). Efficient resource scheduling in cloud computing. *International Journal of Advanced Research in Computer Science,* 8(3), 809–815.

[17] Lobo, A., Kadam, R., Shajahan, S., Malegam, K., Wagle, K., and Surve, S. (2014). Localization and tracking of indoor mobile robot with beacons and dead reckoning sensors. In *2014 IEEE Students' Conference on Electrical, Electronics and Computer Science*, pp. 1–4. IEEE.

[18] Scioscia, F., Loseto, G., Ieva, S., *et al.* (2018). The physical semantic web for knowledge-based M2M interactions in precision agriculture. In *I-CiTies 2018-CINI Annual Workshop on ICT for Smart Cities and Communities.*

[19] Patel, A. R., Patel, R. S., Singh, N. M., and Kazi, F. S. (2017). Vitality of robotics in healthcare industry: an Internet of Things (IoT) perspective. In *Internet of Things and Big Data Technologies for Next Generation Healthcare*, pp. 91–109. Springer, Cham.

[20] Kim, J., Min Gu, G., and Heo, P. (2016). Robotics for healthcare. In *Biomedical Engineering: Frontier Research and Converging Technologies*, pp. 489–509. Springer, Cham.

[21] Fang, B., Guo, X., Wang, Z., Li, Y., Elhoseny, M, and Yuan, X. (2019). Collaborative task assignment of interconnected, affective robots towards autonomous healthcare assistant. *Future Generation Computer Systems* 92, 241–251.

[22] Nejat, G., Sun, Y., and Nies, M. (2009). Assistive robots in health care settings. *Home Health Care Management Practice* 21(3), 177–187.

[23] Seo, J., Cho, J., Woo, H., and Lee, Y. (2015, October). Development of prototype system for robot-assisted ultrasound diagnosis. In *2015 15th International Conference on Control, Automation and Systems (ICCAS)* (pp. 1285–1288). IEEE.

[24] Okamura, A. M. (2004). Methods for haptic feedback in teleoperated robot-assisted surgery. *Industrial Robot: An International Journal.*

[25] Evans, A. J., Chetty, R., Clarke, B. A., *et al.* (2009). Primary frozen section diagnosis by robotic microscopy and virtual slide telepathology: the University Health Network experience. In *Seminars in Diagnostic Pathology*, 26(4), 165–176. WB Saunders.

[26] Candelas, F. A., García, G. J., Puente, S., *et al.* (2015). Experiences on using Arduino for laboratory experiments of automatic control and robotics. *IFAC-PapersOnLine* 48(29), 105–110.

[27] Patel, K., Mehta, D., Mistry, C., *et al.* (2020). Facial sentiment analysis using AI techniques: state-of-the-art, taxonomies, and challenges. In *IEEE Access*, 8, 90495–90519.

[28] Prasad, V. K. and Bhavsar, M. D. (2021). SLAMMP framework for cloud resource management and its impact on healthcare computational techniques. *International Journal of E-Health and Medical Communications (IJEHMC)*, 12(2), 1–31. http://doi.org/10.4018/IJEHMC.2021030101.

[29] Wang, M., Pan, C., and Ray, P. K. (2021). Technology entrepreneurship in developing countries: role of telepresence robots in healthcare. *IEEE Engineering Management Review*, 49(1), 20–26.

[30] Gombolay, M., Yang, X. J., Hayes, B., *et al.* (2018). Robotic assistance in the coordination of patient care. *The International Journal of Robotics Research*, 37(10), 1300–1316.

[31] Higgins, F., Tomlinson, A., and Martin, K. M. (2009). Survey on security challenges for swarm robotics. In *2009 Fifth International Conference on Autonomic and Autonomous Systems*, pp. 307–312. IEEE.

[32] Hager, G. D., Okamura, A. M., Kazanzides, P., Whitcomb, L. L., Fichtinger, G., and Taylor, R.H. (2008). Surgical and interventional robotics: part III [Tutorial]. *IEEE Robotics Automation Magazine*, 15(4), 84–93.

[33] Zamora-Ortiz, P., Carral-Alvaro, J., Valera, Á., Pulloquinga, J. L., Escarabajal, R. J., and Mata, V. (2021). Identification of inertial parameters for position and force control of surgical assistance robots. *Mathematics*, 9(7), 773.

[34] Cleary, K. and Nguyen, C. (2001). State of the art in surgical robotics: clinical applications and technology challenges. *Computer Aided Surgery*, 6 (6), 312–328.

[35] Brandt, G., Zimolong, A., Carrat, L., *et al.* (1999). CRIGOS: a compact robot for image-guided orthopedic surgery. *IEEE Transactions on Information Technology in Biomedicine*, 3(4), 252–260.

[36] Horvat, T., Karakasiliotis, K., Melo, K., Fleury, L., Thandiackal, R., and Ijspeert, A. J. (2015, September). Inverse kinematics and reflex based controller for body-limb coordination of a salamander-like robot walking on uneven terrain. In *2015 IEEE/RSJ International Conference on Intelligent Robots and Systems (IROS)* (pp. 195–201). IEEE.

[37] Belpaeme, T., Kennedy, J., Ramachandran, A., Scassellati, B., and Tanaka, F. (2018). Social robots for education: a review. *Science Robotics*, 3(21), eaat5954–21.

[38] Niemelä, M. and Melkas, H. (2019). Robots as social and physical assistants in elderly care. In *Human-Centered Digitalization and Services* (pp. 177–197). Springer, Singapore.

[39] Dashevskiy, V., Budkov, V., and Ronzhin, A. (2017, September). Survey of modular robots and developed embedded devices for constructive and computing components. In *International Conference on Interactive Collaborative Robotics* (pp. 50–58). Springer, Cham.

[40] Thrun, S., Fox, D., Burgard, W., and Dellaert, F. (2001). Robust monte carlo localization for mobile robots. *Artificial Intelligence*, 128(1–2), 99–141.

[41] Rubenstein, M., Cabrera, A., Werfel, J., Habibi, G., McLurkin, J., and Nagpal, R. (2013, May). Collective transport of complex objects by simple robots: theory and experiments. In *Proceedings of the 2013 International Conference on Autonomous Agents And Multi-Agent Systems* (pp. 47–54).

[42] Ahn, H. S., Kuo, I. H., Datta, C., *et al.* (2014, October). Design of a kiosk type healthcare robot system for older people in private and public places. In

International Conference on Simulation, Modeling, and Programming for Autonomous Robots (pp. 578–589). Springer, Cham.

[43] Hu, Y., Wang, Y., and Wang, S. (2020). Whole scene reconstruction of medical service robots fused with key frame focus stack images. In *Proceedings of the 2020 International Symposium on Artificial Intelligence in Medical Sciences* (pp. 33–37).

[44] Fong, T., Nourbakhsh, I., and Dautenhahn, K. (2003). A survey of socially interactive robots. *Robotics and Autonomous Systems*, 42(3–4), 143–166.

[45] Strengers, Y. and Kennedy, J. (2020). *The Smart Wife: Why Siri, Alexa, and Other Smart Home Devices Need a Feminist Reboot*. MIT Press.

Chapter 8

Evolving technologies: IoT and artificial intelligence for healthcare informatics

Rachana Mehta[1,2], Vivek Kumar Prasad[1], Shakti Mishra[2], Sudeep Tanwar[1] and Yash Patel[3]

Abstract

Computational advancement like Artificial Intelligence (AI) and Internet of Things (IoT) has marked their importance in a variety of domains, healthcare being one such domain. The one major need of involving such advancements in healthcare is the data explosion. The modern healthcare uses these technologies in different medical area; in medical diagnostics to detect chronic disease like cancer and tumors, automated real-time monitoring of patient using sensors, medical image analysis and processing using deep learning (DL) and machine learning (ML), medical data analytics with the help of smart devices, treatment plan decision, development of medical equipment, drug discovery, robot-assisted surgeries, personalized healthcare, early disease detection, and many more. AI and IoT-enabled healthcare is currently at the niche stage and is evolving. The efficient usage of AI and IoT can change healthcare dynamics drastically in terms of decision making, predictive modelling, time consumption, and resource management. However, the inclusion of AI and IoT also tails its challenges like the accountability and trust of decisions made by AI, assurance of data quality which is transmitted through sensors, manpower for handling the AI- and IoT-enabled devices in healthcare environment, management of AI and IoT devices especially when linked with important medical equipment, consumption of energy and power by the devices that use sensor data, storage of data generated by the medical record, real-time management of the generated data, interpretation of data by machine, privacy, legal and security concerns of health records, human intervention, comfort of IoT devices for a patient and clinical staff and much more.

With the current world scenario of pandemic and epidemics spreading around, it is of prime importance to dwell and see where these technological advancements

[1]Department of Computer Science and Engineering, Institute of Technology, Nirma University, India
[2]Department of Computer Science and Engineering, School of Technology, Pandit Deendayal Energy University, India
[3]Department of Data Science, Saint Peter's University, USA

stand in Healthcare Informatics and how they can be efficiently used for the futuristic healthcare sector. This chapter focuses on the proliferation of AI and IoT in various healthcare segments; clinical and non-clinical, along with the context of ML and DL and includes domains of healthcare benefit, AI and ML algorithms and techniques involved at various stages, use of IoT in healthcare, challenges faced, their limitations, and future scope.

Keywords: Artificial Intelligence; Internet of Things; Healthcare informatics; Machine learning; Clinical AI; Internet of Healthcare Things

8.1 Introduction

Current revolutions in technological advancements and communication in various industries have signified the proliferation of Artificial Intelligence (AI) and Internet of Things (IoT). AI and IoT have marked its onset of footprints in almost all domains covering e-commerce, machine automation, food industry, healthcare to name a few. Healthcare is one the versatile and conventional domain of study in human evolution. Healthcare is still in a niche stage when it comes to the use of AI and IoT. The technological advancement in machinery, laboratory equipment, and medical instruments has led to the generation of humongous data of various forms in the healthcare domain, especially electronic records [1]. These demands the need of technologies like AI and IoT which helps to analyze, store, transmit, and appropriate utilization of data.

AI, machine learning (ML), and deep learning (DL) have revolutionized many industries since the past few decades, healthcare being one such. Quite recently, such techniques are used for medical image recognition [2], reconstruction of medical image [3], segmentation of brain tumor [4], diseases diagnostic and managing patient health records [1] to consider a few. This will significantly help medical professionals to provide good healthcare solutions to patients. The wide acceptance and usage of wearable devices like smart watches, fitness bands, and portable devices like smart phones have given a new vision to the concept of remote health monitoring.

Under the purview of IoT, the objects and sensors communicate with one another remotely via the Internet. The remote communication and interaction of devices/sensors, transmission of data are increasing on a daily basis. In the healthcare domain, remote data access provides the benefit of remote health monitoring, cost effective health treatment and personalized healthcare to consider [5]. Such benefits of IoT-enabled healthcare can be beneficial for timely monitoring of a patient's health as compared to physical rounds by medical professionals, as well multiple patients' health statistics can be seen at one place. AI- and IoT-enabled technologies are not only limited to patient's monitoring but also in varied healthcare sub-domains like drug-discovery, robot-assisted surgeries, treatment plan decision, and much more.

AI and IoT are still at the niche stage in the healthcare domain. There is much to explore and evolve. Along the advancement, there still remains a question how such technologies can be trusted, held accountable for the decisions they have made, how much reliable and accurate they are, the level of human intervention, skilled professionals and manpower needed to operate, maintain such devices, the downtime of sensor devices, power consumption, security, and privacy of the data of patients. Such concerns need a proper addressing as well one need to evaluate the model into consideration on all such factors.

With the current scenario of rapidly spreading epidemics and pandemics around the world, it is desired to see how AI and IoT stand for futuristic healthcare, how they can be beneficial and what breakthrough they can bring. The scope of the current study is to bring an overview of how AI, ML, DL, and IoT have been used in the healthcare domain. How much they have been used and what are their further advancements. AI helps majorly in the classification and prediction of disease-related task using ML- and DL-based approaches, imaging and segmentation-based task using the concepts and notions of neural networks. Natural language processing (NLP) is used for audio and text recognition and classification. Whereas, IoT is not only concerned for the analysis of medical data, rather it is useful in collecting the generated data. Using the real-time and dynamic data, IoT helps healthcare professionals to monitor patient, see their progress, and health status. IoT sensors play a major role in data collection.

Nowadays, IoT sensors are also embedded in wearable devices and medical equipment for data monitoring. The major healthcare segments which have started using AI and IoT includes medical imaging, disease classification, electronics health records, clinical notes, telemedicine, robot-assisted surgery, and much more. In the current chapter, we have overviewed on various applications of AI and IoT in healthcare informatics. The main objective of the study is to see how AI and IoT have penetrated into the healthcare domain and what advances and improvement they can bring in the healthcare informatics.

The major contributions of the chapter are:

(i) To study of AI and IoT in healthcare domain and its applications and current use cases;
(ii) To enlist various challenges that AI- and IoT-based healthcare informatics faces;
(iii) And, to state the future research direction for AI- and IoT-based healthcare informatics.

8.1.1 Organization of chapter

The rest of the chapter is organized as: Section 8.2 gives a brief overview of AI and IoT. Section 8.3 highlights healthcare informatics with its usage in AI and IoT alongside their applications and various current use cases in different literatures. Section 8.4 brings forward the challenges of healthcare in AI and IoT. Section 8.5 concludes the chapter with the future direction.

8.2 Technology overview

The section provides a base of technological advancement taken under consideration for healthcare: AI, ML, DL, and IoT.

8.2.1 AI

AI is a domain of computing, which works on the principle of replicating the human cognitive mechanism in machines, to make them work efficiently as humans do. The goal of AI is to achieve the benefit of cognitive science in machines for the need of decision making, data analysis, image recognition, and several other purposes. The domain of AI is amalgamation of various techniques like ML, neural networks, DL, NLP, expert systems, robot automation, computer vision, and others [6]. Each of them provide a significant advantage of their own.

Out of them, ML and DL are preferred in healthcare relevant tasks. The domain of ML falls into two sub categories: supervised learning and unsupervised learning. The goal of supervised learning is to develop a system such that it has sufficiently learned from the given input–output pair, whereas the unsupervised learning model develops a system with the given input data set only. The dataset taken into consideration for model learning gets usually divided into parts: one for training and one for testing purposes. In the training phase, the model under consideration learns features from the training data, and then the model is verified against the unknown testing data based on the learned features of the training phase. If the model provides the expected result on the testing data, then it is accepted with the learned feature for the appropriate task. ML takes various metrics like precision, recall, F-score, accuracy, root mean squared error, and many others to validate the model performance.

The current AI has been largely driven by a sub-domain of ML: DL. DL using the neural network variation can help us to infer hidden features and patterns in a large dataset, which are unless not feasible [6]. Artificial neural network (ANN) consists of numerous connected computational units which are called neurons and these neurons are arranged in layers. In usual, neural networks consist of input layer, output layer, and multiple hidden layers. The input layer is the entry point of the data, then it gets passed from one hidden layer to another till it reaches the output layer, where it provides the predicted output value for the given input data. The neurons in the hidden layer get tuned until they produce the expected or likely output for the given input values. The learned features from the hidden layer can then be used on unseen or unknown data for the prediction purpose. The number of hidden layers a model can define the model complexity too.

Among the various ANN, convolutional neural network (CNN) has been significantly used in the various healthcare domains like diagnostic, medical imaging, segmentation, restoration and others image-based tasks. In CNN, the hidden layers are responsible for the task of convolution and pooling work of CNN. Restricted Boltzmann machine, deep auto encoder, and recurrent neural network are other preferred architectures of DL in healthcare.

NLP is the domain of AI that deals with the recognition and classification of human text, audio, and speech. It translates the unstructured data into the machine readable form. NLP is used in the medical domain to recognize the clinical notes of the past [7]. Expert systems have been useful since long in AI, the help to make a decision, are also used for electronic health record (EHR) management [6]. AI-assisted robots have largely revolutionized the market of healthcare; it is one of the most sought applications of AI [8]. It has given the healthcare professional the benefit of achieving the minimum invasion in the human body, which otherwise is not achievable by human hands.

8.2.2 IoT

Most of the AI techniques and models are efficient in disease diagnosis using medical imaging, clinical notes recognition and translation, prognosis, and expert system development. The task at the hand of AI is mainly concerned around the available data be it medical image, clinical note, patients health record, doctor's expertise, and all, the data forms the base for the development of AI model in healthcare. IoT, on the other hand, focuses majorly on collecting the data first and then operating on it.

The main goal of IoT lies in capturing the real-world information, which otherwise is lost by humans due to time constraints, attention or accuracy, using sensor technology [9]. The sensor-generated data at core in IoT is usually free from human intervention. IoT has drastically shifted the paradigm of current wireless technology. IoT captures the real-world information using radio frequency identification (RFID) tags, sensors, mobile phones, and actuators, they are termed as objects or Things of IoT. These devices accumulate data of the physical entities, to which they are embedded, and exchange or transfer it to other devices for the utilization purpose over the Internet. Most of the IoT sensors are miniature in size and cheap in cost, which has led them to be used in various industries. IoT sensors have transformed the automation industry significantly of manufacturing, transportation, logistics, healthcare, home, grids, and many others. It has leveraged the notion of remote control and ubiquity. This gives access to remote things/humans to be monitored from any place.

The remote health monitoring system aided by IoT has become a center of study to monitor human health at every microsecond [10]. IoT has given a new perspective of a patient's health, at prior it was limited to human–human communication, screening or report, now the real-time data accumulation gives the benefit of continuous monitoring and giving appropriate treatment. IoT has benefitted all the stakeholders of healthcare: patients, doctors, hospitals, families, and insurance organizations [11]. Wearable devices like smart watches, fitness belts, and glucometers are prevalent now in health enthusiasts. Not only for humans IoT sensors are also embedded in X-ray machines, oxygen pumps, wheelchairs, nebulizers, and several other medical equipment [11]. IoT-generated data can be further transferred directly to the doctor or physician for monitoring purposes or can be fed to AI, ML, or DL models for analysis and diagnostic purposes.

8.3 Healthcare informatics

Healthcare informatics deals with combining the healthcare with current computing technologies; in order to obtain the efficient working and better healthcare. This section provides a detailed overview on various application of healthcare that uses AI, various IoT sensors and wearable devices, and application of healthcare benefitted from IoT. Figure 8.1 shows the various applications of healthcare that uses AI and IoT for efficient working.

8.3.1 AI in healthcare

The domain of ML has largely evolved in healthcare owing to the enormous data getting generated on a daily basis. The main task of ML-based algorithms is to infer the hidden patterns and features from such heavy data. The ML model uses several supervised and unsupervised algorithms like support vector machine (SVM), logistic regression, anomaly detection model, neural networks, linear regression, Naïve Bayes, clustering, principal component analysis, and many others [6]. Here we will see how such techniques are useful in various applications and areas of healthcare considering especially CNN-based approaches.

8.3.1.1 Medical imaging: segmentation and image reconstruction

Medical diagnostic involves the task of identifying the health problem or issues a patient is having based upon his medical records or symptoms. Medical diagnostics

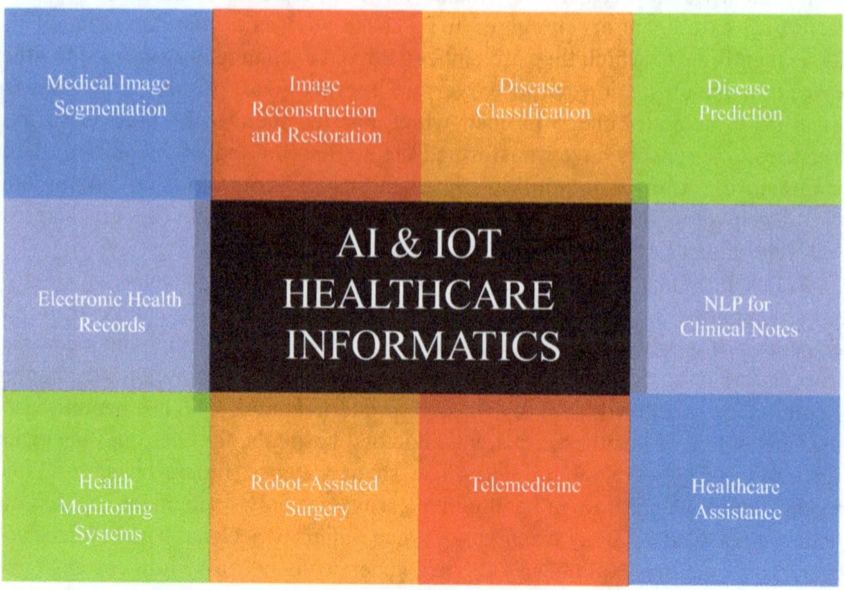

Figure 8.1 AI and IoT applications in healthcare informatics

has remained in the purview of the specific medical practitioner treating a patient. This task of manual decision can many a time be error-prone. The inclusion of AI can give the power of auto diagnosis to lead to less error in decision making and as well reduce the time [8]. Medical image diagnosis is one of the most sought AI applications in the domain of healthcare.

In order to detect a disease or diagnose the condition of a patient, medical imaging is preferred in many medical cases. It includes various radiological techniques like X-ray, computer tomography, magnetic resonance imaging (MRI), microscopy, fluoroscopy, ultrasound, nuclear, and many others [12]. Most medical cases need them in one or other ways. Medical imaging provides a high-resolution image of the body part under consideration, most of the imaging techniques are costly, require a large amount of data storage, and vary in image dimensionality to provide a detailed image information needed for diagnosis purpose. Researchers have explored various ways to reduce the dimensionality reduction using principal component analysis and singular value decomposition [13].

As most radiological tasks involve imaging, the classical image diagnostic task can be converted to ML- and DL-based diagnostic task. In the scenario of radiology, enormous data of high-end images are available, such that they can be efficiently utilized as training data for neural networks and develop a neural network model. AI systems are being used to improvise the reconstruction of medical image, noise removal, quality maintenance, object segmentation, radio genomics, computer-aided detection, and classification using medical images. Image segmentation defines the outline border over the object in the given image and classification and signifies the presence of the object in the image. Among the various DL models, CNN has outperformed in the sector of medical imaging.

MRI is a most opted technique for the brain tumor analysis and detection. It needs the timely segmentation to get proper insight on the stage a patient is. In [14], authors used CNN-based models for segmentation of brain tumors in MRI images for oncological detection of gliomas, a type of brain tumor. In order to detect gliomas, a number of MR images are required, segmenting them manually is inefficient and a time consuming task. The CNN model adopted [14] performs the automatic segmentation using the concept of kernels, building the deep architecture, and leaky rectified linear unit. In [15], authors used multilayer perceptron (MLP), a type of neural network and active contour model for segmenting the structures in MR images for lungs. The MLP generates binary images, based on the boundary and non-boundary points, which get further transferred to the contour model to detect the correct segmentation. In [16], authors used dual models of CNN for the pulmonary CT scan images. One model of CNN worked on the softmax classifier and the second one on radial basis function (RBF) with SVMs. These models are used for automatic feature extraction.

In [17], authors used the offline CNN for learning the end-to-end mapping between zero-filled MR images and fully sampled MR Images. The network was sufficient for restoring the details and minute structures of MR images. As well it was efficient for online image reconstruction algorithms. The CNN-based image restoration approach was used for filling in the details in zero-filled images. A convolutional recurrent neural network was preferred in [18] for image reconstruction of brain MR images that are under sampled. The model learns representation of MR images over time and iterations, efficiently utilizing the benefits of CNN and optimization. In the image reconstruction of ultrasound images, the CNN-based model was also used along with the adversarial attack approach [19].

8.3.1.2 Disease classification

Medical imaging of various types are used for varied purposes. Apart from the segmentation and image reconstruction, AI has also been prevalent in the classification of disease through images. In many of the medical images, the second step after segmentation is classification. Classifying whether a disease is there or not.

In [20], K-means clustering and SVM-based approach was used for the classification of leukemia, with two types of cells. K-means clustering was used for the nuclei segmentation in the initial stage. In the next stage, appropriate features were extracted and SVM was trained using those features to classify the leukemia cell in the given two classes. These models segment and classify the microscopic blood images. Further the classification results can be used by physicians for diagnostic purposes. In [21], the basic CNN model was used for the heart disease classification, using the large dataset of heart disease images.

The DL architecture-based models are not only for early stage detection of cancers but have also been used in various studies of Corona Virus Disease (COVID-19). A CNN-based approach of image classification was used in [22] to detect the positive and negative cases of COVID-19 using the chest CT scan images. The axial slices of publicly available CT scan images were used and the model was found to be more than 80% efficient in correct detection. The AI-based models were not only preferred for heart disease, cancer, COVID-19, but the image-based classification was also useful for cardiovascular disease, retinal, and fundus images.

8.3.1.3 EHRs

In any domain of the medical fraternity, health records of patients are of prime importance for disease diagnosis, treatment, future plan, and research. The health record can take any form; it can be a doctor's clinical note/prescription, medical images, varied test reports of bloods, vitamins, urine, and much more. These medical records change from person-to-person pertaining to their disease or symptoms. However, in some general medical scenarios like coughing, sneezing, or fever, many doctors follow a practice of prescribing the same set of medicines and test reports. The medicinal prescription may be common in such cases, but the test reports may differ individually. Health records relevant data are extensively increasing on a day to day basis and it is almost impossible to keep track of them in the physical world, be it to patients or to doctors. These give rise to the domain of EHRs, sometimes alternatively known as electronic medical records (EMRs).

EHR aims at storing the medical records of varied forms of patients digitally over the system. EHR leverages the benefit of digitization through which medical records can be managed, stored, tracked, and accessed efficiently. It has reduced the physical hard work of maintaining an individual's file. With the exponential advancement in healthcare, the amount of EHR has also grown exponentially. The primary purpose of EHR inclusion was not to digitize and maintain medical records rather efficiently plan the billing of a patient based upon the treatment taken [23]. Rather than just dumping the medical data on computers and servers for billing purposes, it would be more powerful to use them for analyzing and detecting patterns, trends, or similarity in medical cases, research, and teaching purposes. Several models of AI and pattern recognition are now being used along with EHR to infer the pattern analysis, define the predictive modelling systems, and define the expert systems based on the previous research.

EHR has been used with ML and DL techniques in [24] for the ophthalmic disease diagnostic, and assessment of risk in patients and the progression of disease in the patient due to varied factors. The main focus was on the ophthalmic disease which can lead to permanent blindness, such as diabetic retinopathy, macular degeneration due to age surgical complications of cataract and glaucoma. Supervised ML techniques were used for the diagnosis and prediction of glaucoma using the EHR and other clinical data. In case of surgical complications of cataract, NLP was used to extract data from textual clinical notes. Furthermore, SVM and neural network models were used for the extracting data for diagnosis and risk analysis. For macular degeneration due to age, DL techniques were used for the improvement of diagnosis. Medical images combined with EHR were used. In diabetic retinopathy, screening of patients leads to huge resource consumption. In order to reduce this concern, AI technique along with EHR data are used for risk assessment of diabetic patients. In the patient case with cardiology, echocardiography is one of the most important imaging forms. These echocardiograms give a lot of information about a patient's health status to the doctors.

As per [25], the echocardiogram infers more than 500 different factors, it is not feasible for physicians to extract the entire information single handedly with minimal time constraint. In such scenarios, ML technique can act as a boon by provisioning the auto analysis of large-scale clinical information. Random forest classifier was used to predict the survival of patients after echocardiography using the information of EHR data. As large as 19 years of EHR data was processed for the model analysis. EHR is significantly used in various medical research for the analysis of trends for diagnosing the disease, risk assessment, progression of disease, and several purposes. One of the major sources of clinical information is handwritten prescriptions in many places. EHR can be efficiently built if those clinical notes are digitized correctly. In lieu of that, NLP techniques are used for clinical note recognition.

8.3.1.4 NLP for clinical notes

The excessive and extensive use of EHRs with help of AI has opened various other domains. EHR data is majorly used to decision support systems and disease diagnosis, but it has also given rise to translational research in the medical domain. Among the challenges of EHR, the one is that the large amount of a patient's medical history is present in clinical notes, which are not accessible directly for computation purposes and in digital form. In lieu of that, technologies like NLP are leveraged for extracting structured medical information from the notes or prescription. Evidence-based medical practices have quite proliferated with the patient case research in medical history. NLP in medical domain are used to model several attributes and features like content of document for getting information of patient status and report, section of medical prescription like current details, past details or summary of discharge, named entity recognition (NER) to recognize name of patient, disease, tablets prescribed, symptoms, and other details; and semantic attributes like severity [26].

8.3.1.5 IoT in healthcare

The IoT has shifted the paradigm of personal and remote healthcare significantly. IoT connects the medical devices with a communication mechanism to transfer the live data to intelligent end-devices that are capable of processing those healthcare data. The enormous availability of medical devices has led to the generation of vast medical data and in the domain of healthcare. IoT can play a major role in the development of cost-effective and smart healthcare applications that can monitor patients in real time in order to save them in severe cases of medical emergencies like cardiac arrest, asthma attack, and others. The domain of IoT for medical applications is sometimes called Internet of Medical Things (IoMT). IoMT is a collection of sensors used in medical devices and infrastructure.

8.3.1.6 IoT sensors and wearables

IoT provides the network of interconnected medical devices to communicate and transfer data, the domain is termed as "IoMT". IoMT does so with the help of various types of sensors. The sensors of IoMT fall into two basic categories: one for clinical purpose and another for non-clinical purpose. The clinical sensors are those sensors that work along with smart devices to provide efficient patient monitoring and healthcare, while the non-clinical ones are the one used in different domains of healthcare like machineries, logistics, tracking, and much more [27]. The clinical ones are stuck to the human body in one way or another, they can be wearable devices like smart gears or they can be planted in the human body to monitor and track the working of specific organs or immune level. Some of the clinical sensors are shown in Figure 8.2.

The wearable clinical sensors are the one that humans wear on the outer body, these sensors keep track of varied human vitals. These sensors are non-invasive and non-intrusive, they have as well now become an indispensable part of healthcare monitoring in the current world. The data collected from these wearable devices is now directly available on mobile devices and is easily transferred to physicians. These sensors are capable of monitoring physiological signs, activities, movement,

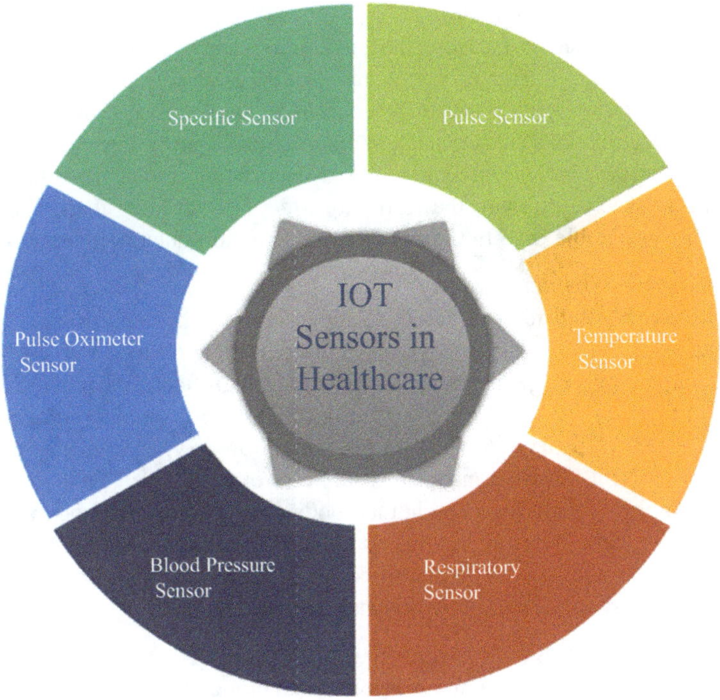

Figure 8.2 AI and IoT applications in healthcare informatics

and fluctuation in body parameters. Most of these sensors use Bluetooth, RFID technologies, and Long Term Evolution (LTE).

8.3.1.6.1 Pulse sensor

Pulse sensor reads the pulses of human beings. The fluctuation in pulses is used to monitor change in pulmonary embolism, indicate signs of cardiac arrest, to indicate the reflex syncope that occurs due to heart rate drop. The readings of the pulse sensors can be obtained by attaching the sensor to human parts like ear, wrist, chest, or fingertip. Among all, the wrist is a preferable, comfortable, and highly adaptable in industries. Pressure, ultrasonic, RF, and photoplethysmographic are some other sensors available for monitoring pulses [28].

8.3.1.6.2 Temperature sensor

Temperature sensors are used to measure the human body temperature. These sensors are primarily used to measure the fever, heatstroke, and drop in body temperature (hypothermia). Among the various types of temperature sensors, thermistors are preferable and in demand. Temperature sensors are more efficient when they are in close proximity to the human body. Smart textile and soft patch are preferable options to keep the sensor attached to the human body.

8.3.1.6.3 Respiratory sensor

These sensors are used to measure the breathing rate and number of breaths per minute by a human. Respiratory rate monitoring is useful in scenarios where patients are suffering from diseases or conditions like asthma, tuberculosis, lung cancer, disorders of sleep like apnea, and others. Nasal sensor, microphone, stretch sensor, fiber optic sensor, and ECG-derived respiration sensor are some sensors which can be used to measure the respiratory rate [28]. Among them, the stretch sensor is the preferable one. In them, there is fluctuation due to tensile force.

8.3.1.6.4 Pulse oximeter sensor

Pulse oximeter is used to measure the oxygen level of humans in blood. These devices measure the slightest fluctuation in the oxygen and blood pressure level.

The importance of pulse oximeter is highly significant during the current scenario of COVID-19. The drop in oxygen level has been one of the indicators for the health status of patients in COVID-19, where low oxygen indicates a critical state of the patient. These sensors measure the oxygen level using the photoplethysmographic signals from the human body [28]. Photoplethysmographic signals using the light emitting diodes to measure the light that gets absorbed, according to that, the oxygen is measured. In general, the fingertip sensor is the more efficient way to measure the oxygen level.

8.3.1.6.5 Blood pressure sensor

As the name suggests, it is used to measure human blood pressure. The main reason to sense the blood pressure is that the fluctuation in it can lead to severe issues like heart attack. In order to measure the blood pressure, several researchers have proposed to use the pulse transit time, however, it is not too accurate. A good option would be to use the photoplethysmographic [28]. Devising the blood pressure sensor is still an open challenge.

8.3.1.6.6 Specific sensors

Echocardiogram sensors are used to measure the heart condition. Wearable devices are preferred for using this sensor. Electroencephalogram-based sensors are used to measure the activity of the human brain along with the disorders of sleep, seizures, and head injury progress. Several other sensors are available in the market for different disease and monitoring purposes. Some of these sensors are directly usable while some need to be embedded within wearable devices to operate on a daily basis.

Some of the commercially available different types of wearable products that include different types of sensors are given in Table 8.1. In general, many companies develop smart watches and fitness bands, almost all of them provide similar functionalities [29]. So, here different types are covered rather than similar types of wearable devices from different manufacturers. Figure 8.3 shows the different form of wearable devices available in the market [28,30–33]. Table 8.1 covers the different wearable devices commercially available, places it can be worn or attached, activity carried out by them and groups they are preferable for. Different types of wearable devices are covered; however, the list is more which can be further

Table 8.1 Commercially available wearable devices

Sr. no.	Wearable device name	Wearable type	Task	Preferable for
1	Fitbit Fitness Tracker	Watch, wrist band	Heart rate, daily step counter, sleep monitor, exercise tracking, health metrics, oxygen saturation, skin temperature	Adults
2	TempTraq	Soft patch	Bluetooth wearable temperature monitor, fever alert; mostly used for small children	Children
3	Ava fertility Tracker	Wrist band	Skin temperature, heart rate, breathing rate, pulse rate, perfusion, menstrual and fertility tracker	Women
4	Motiv Ring	Ring	Monitor sleep, fitness, heart rate	Adults
5	My Skin Track UV	Wearable sensor (Clip)	Ultraviolet rays exposure, pollution, pollen and humidity, skin sensor	Adults
6	Owlet Smart Sock	Socks	Oxygen level, heart rate, sleep trends, activity change	Toddlers
7	Omron Platinum	Upper arm band/ D-ring cuff	Blood pressure monitor	Adults
8	Philips Biosensor	Chest monitor patch	Vital signs, posture, environmental signs	Adults
9	Hexoskin Biometric Shirt	Shirt	ECG and heart rate, respiratory, activity, steps, sleep	All

explored [30]. Some of which are smart glass, smart shoes, ECG monitor, blood pressure monitor, smart jewellery, and much more.

8.3.1.6.7 Implantable sensor

Implantable sensors are those sensors which need to be implanted in the human body to track and monitor various vital signs. The major market of implantable sensors is the neurological disorder in humans. In such cases, patients are not supportive of wearable devices. Nowadays implantable sensors are used in varied health conditions. Some of the implantable sensors used are given below.

8.3.1.6.8 Pacemaker

Pacemaker is used to measure the fluctuation in the heart rate. The pacemaker is used when a patient is suffering from Dysrhythmia. Dysrhythmia occurs when the heart beats are either too fast or too slow or at an irregular pace. Pacemaker is used to send the electrical pulses to the heart in order to make it beat at normal human rate [34]. Pacemaker are also used to help the heart chamber beat in synchronous manner so that it can pump the blood efficiently.

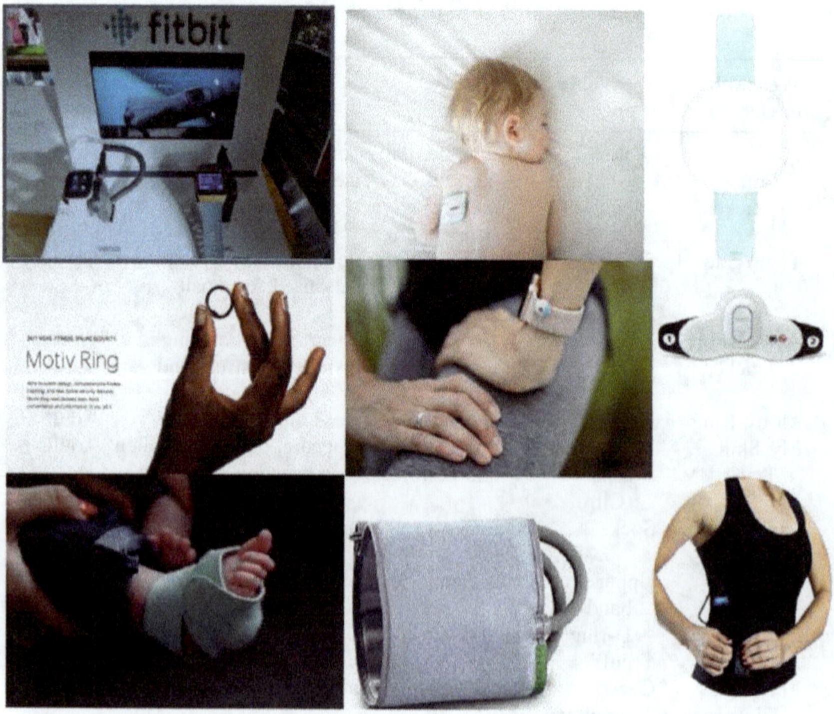

Figure 8.3 Different types of wearable devices

8.3.1.7 IoT applications in healthcare

IoT sensors generate large amounts of data on a daily basis. The data generated by those sensors is of prime importance to utilize it efficiently for the well-being of patient and futuristic research. Thus can be done by transferring the data from IoT sensors and wearables to mobile devices or computers for analysis purposes using various communication technologies like Bluetooth, ZigBee, RFID, and other wireless technologies. Using them, the data from the sensor gets transferred and effectively stored, monitored, and analyzed. In general, the architecture of IoT for healthcare follows a three-layer hierarchy starting from the sensing layer, fog layer, and cloud layer [27,35]. In the sensing layer, the sensor and wearable device perform the monitoring task, then the data gets transferred to fog later where it gets communicated within the local system for alert indication and at last it goes to the cloud layer where it is stored and analyzed to infer the findings, trends, and patterns. Most of the IoT applications follow a similar architecture. This section focusses on various IoT applications in the healthcare domain.

8.3.1.7.1 Health monitoring systems

Health monitoring is one of the major tasks in healthcare especially for the patients suffering from chronic disease or undergoing medical treatment due to accidents or organ failure. It is important to keep track of health parameters in order to maintain the patient's health in good condition. In the physical scenario, it is not possible to measure all body signs at every minute or microsecond, as the task is resource consuming and not feasible. In order to provide the real-time monitoring of patients, IoT sensors are used to measure vital signs. The data generated and gathered from these sensors can be useful to design the treatment plan and track the recovery progress of patients. IoT sensors have also leveraged the remote health monitoring of patients. In [36], the monitoring system uses the Phidgets interface kit which provides various sensors, ECG sensor, temperature sensor, and Arduino UNO. The sensing layer senses the various parameters using the sensors and passes this using the Arduino. The data then gets displayed on the desktop or web interface. In [37], a monitoring system was developed for measuring the temperature, ECG, heart rate, and foot pressure. The generated data is gathered and sent to the cloud with the help of IoT through Arduino UNO. Furthermore it uses the Global Positioning System to keep track of patients, as well as uses the RFID for security purposes. Real-time monitoring of health is just not limited to wearable devices; rather in the scenario of critical illness, where a patient is hospitalized, IoT sensors are also embedded with the medical equipment to obtain the data fluctuation and alert the doctor.

8.3.1.7.2 Disease prediction

The health monitoring system stores the data in servers or over clouds, and this data can be further analyzed for the prediction of disease or diagnosing the disease. In [38], an IoT-based heart disease prediction model is developed with the help of a variant of CNN. Wearable devices, smart watch and heart monitor device are used to measure the electrocardiogram and blood pressure. The data generated using these IoT sensors is transferred to long-range cloud and provided to medical professionals. The same data from a long-range cloud server is used for the prediction of heart disease using the modified deep CNN model. It classifies the sensory data into normal or abnormal states. Diabetes prediction model was developed by Kumar *et al.* [39] using the IoT and cloud computing. The model proposed uses the wearable IoT to fetch the real-time data of the patient, then it makes use of 5G network to transfer the data to the cloud server. At cloud, along with the sensor data, normal data for marking and hospital records are present. The data is then fed to the fuzzy neural classifier for the prediction of diabetes. In the fuzzy neural classifier, a severity module is present to define the severity level, which is used for knowledge base generation.

Most of the IoT-based disease prediction applications involve taking the data generated from the IoT sensors of the wearable devices transferring them to the cloud server for the storage purpose. Then the stored data undergoes the analysis module which involves using the various ML and DL algorithms to identify the presence of disease based upon the collected sensor data and the standard marker data [40]. Sensors used for this purpose can be wearable or are implantable based upon the patient's condition and medical needs.

8.3.1.7.3 Telemedicine

Telemedicine in healthcare falls into various domains like a consultation of a doctor over a call or e-mail or message, booking an appointment with the physician, or monitoring the patients' health remotely using the sensored devices and appropriately managing the healthcare. The major benefit of this is that any patient can remotely achieve the best healthcare from their choice of doctor. Quiet recently there has been expansion in telemedicine due to the ongoing surge in COVID-19. Many patients are consulting the doctor from their place of isolation. Abdellatif and Mohamed [41] proposed a system for telemedicine. Wearable devices are used by patients for sensing of heart rate, humidity, blood pressure, temperature, and alcohol. The data generated from the sensor is sent using Bluetooth, GSM, or Wi-Fi. The real-time data is provided to the doctor, the doctor reads the data online and provides the patient with an appropriate plan of treatment.

8.3.1.7.4 Robot-assisted surgery

IoT applications in healthcare vary from use of sensors in patients or wearables to acquire the real-time health metric to assisting the doctors in surgery using the robots. Robots are more powerful for minute invasion in the human body as well as assisting the doctor using the controllable robotic arms and much more. The use of robot-assisted surgeries has increased owing to the high-end machineries being available. Skilled robots have also fulfilled the shortage of experienced nurses and staff. In [42], authors used the robotic arm to assist the doctor in surgery and replace the unavailability of nurses, The robotic arm provides the movement and grip to acquire the tools for assisting the doctor and helping the patient to acquire objects nearby. An IoT-based model was developed using the Raspberry Pi and smartphone to control the robotic arm. Raspberry Pi was responsible for controlling the robotic arm based on the instructions received from the android-based smartphone [43]. The arm can be controlled from long distance also, since it is embedded with the video streaming module. Assistance of robots in rehabilitation is also proliferating nowadays especially in limb rehabilitation, in minimally invasive surgeries, prosthetics, and much more.

8.3.1.7.5 Healthcare assistance using IoT for patients

The monitoring of health of patients is not only limited to the hospital premises or in purview of specific doctors. Evolution of IoT has seen significant benefits to the old age people at home. In many cases, it is not feasible to keep them in long observation in hospitals in normal to mild treatment. It will be more feasible for them to be monitored regularly in their home environment. IoT not only involves monitoring with the help of wearable and implantable sensors it can also leverage the sensor technology on the medical equipment that is being used or things used for medication. An alert-based system was developed to remind the elderly patient to take the medicine in [44]. The system was embedded with a weight sensor, if there is a reduction in weight then the medicine is considered to be taken else it will generate an alarm to remind the patient. In [45], an IoT-based healthcare system was developed to help the patients who need the assistance of a wheelchair. A mobile healthcare system was developed using the IoT technologies like wireless sensors for heart rate, electrocardiogram, environment sensing, pressure, and actuator. The system also embedded motion sensor to keep track to keep track of the movement of the patients.

The IoT has proliferated heavily in many domains of healthcare including child healthcare, smart medical infrastructure, logistics of medicine and their smart transportation, medical imaging, assistance of physicians, and much more. The IoT combined with AI is very powerful in providing remote healthcare on a timely basis, disease prediction, and diagnosis on the live data. Amalgamation of both technologies can work wonders for the upcoming health sector, especially in the data analysis domain of disease.

8.4 Challenges in AI and IoT healthcare

The evolving technologies and their inclusion in healthcare, also bring new set of challenges with needs to be catered to in the appropriate functioning of the system. Here the limitations and challenges pertaining to AI and IoT in healthcare are explored.

AI gives the benefit of analysis of the medical data to infer new trends and patterns, diagnose a disease based upon health records and medical imaging, image reconstruction and image segmentation, and much more. In order to apply the AI on medical data, skilled personnel are required who can understand both the modalities of AI and terminologies of healthcare soundly.

In the disease classification and medical imaging tasks of AI, the conventional computing architecture is not powerful and efficient enough. This requires the high-end processors like graphical processing units (GPUs) to extract the underlying information or restore the image or analyze the information from the image. GPUs are not cost effective in all the healthcare domains, for small clinics and middle scale hospitals they are not affordable since they come with their own challenge of purchase, maintenance, resource consumption and requirement of skilled staff. At times, the result generation is delayed due to system failure or performance degradation, they may lead to their inefficiency and underutilization.

ML- and DL-based models have potential to extract meaningful information from the data. In healthcare, the clinical or medical data can take varied forms like text, image or audio, the data preprocessing is important as not all AI architectures are efficient working with varied data forms. One system may be good with text-based data while the other might be efficient with image-based data. This requires the development of new models and architectures for every other healthcare task. It is also important to check and validate the model accuracy. A wrong prediction by the AI model can lead to wrong treatment of a patient and may result in severe health conditions. Unlike other applications of AI, which reduce the dimension or remove some features for model training, in healthcare each feature and information needs to be maintained. This may lead to delayed performance of the AI model.

AI models also come along with the security, ethical, and privacy concerns. In many cases, the analysis of the data is not carried out by the health professionals themselves rather it is outsourced to the AI-skilled personnel. In such cases, ethical breach may occur, as some patients are reluctant to share their data to anyone else except the doctors. The security and privacy should be adhered to, if the data leaks it might affect the reputation of the medical professional. These may also occur if the AI personnel use them for their own benefit.

IoT-based healthcare systems are dependent on the sensor technology and communication among them and the cloud server. It is of prime importance to check that the sensors used in healthcare are working correctly and in an efficient manner. The IoT architecture developed should be fault tolerant in a scenario when the sensor stops working or stops sensing or is not able to communicate with the cloud server properly. An alternate mechanism should be considered as a backup for such scenarios, as well as regular conditioning and validating of the sensed data should be done.

In IoT, the communication technologies used for the purpose of data transfer are the most important middleware between the sensed data and the place of data storage. If the communication technology fails to deliver or receive the data, the sensing would not be of any use. This should be checked on a regular basis. The communication technology used also comes along with other set of challenges like latency, range issues, quality issues, and other network issues. The network should be available all time in the scenario of critical patients and in chronic disease, the lack of availability may lead to patients' death in severe cases. The data should be delivered on time to the health professionals, a slightest of delay is not tolerable. The communication technology taken into consideration should have appropriate range. The data transmitted over the network should be free from noise and attacks [46] .

IoT sensors used have a dedicated battery life. Before using the sensor or implanting them, the battery life of the sensor should be appropriately checked, so that it does not run out of battery while being used. In emergency healthcare, it is of prime importance to see that they do not shut down, backup nodes should be kept ready for the smooth functioning of the system [27].

IoT sensors used and the communication technology used should be interoperable on both sides as well as data storage servers. If the system is not interoperable it may fail to interpret the data correctly, losing the main gist of remote healthcare.

IoT-based wearable and implantable sensors are at center for the remote monitoring of health parameters of the patients. Sensors used should be comfortable to the patient's body, if not then some patients might remove or not even use them [47].

IoT devices require the transmission of data generated from sensor to cloud server or any other database server. The amount of data generated is large, an appropriate and dedicated storage should be available as well on regular basis data clearance should be done, to avoid the wastage of space. Scalability of the system also should be taken into consideration while developing the IoT systems.

One of the major issues in IoT-based devices is privacy and the security of the data being transferred. Most of the IoT sensor devices operate on low efficiency, appropriate security, and encryption mechanisms are needed in order to protect them from the malicious attacks. Authentication and authorization of the data and patient should be done, one way to do so is the RFID-based sensors. The real-time examples and case study usage pertaining to AI and IoT in the healthcare as discussed in previous also gives insights on challenges and future direction.

The major research direction in healthcare informatics includes: designing of cost-efficient processing units for image segmentation and medical image analysis, such that it is useful for small-scale and medium-scale clinics. Most of image reconstruction [14–19] is done using neural network-based models which are efficient, but resource intensive. An offline models or service models should be designed for reducing the resource consumption. Healthcare informatics in the field of classification and prediction works on the limited set of data or data pertaining to specific region or hospital, or for specific disease, a wide applicability of the model should be validated and standardized in future. EHR are used for managing the health records, their security and privacy over global domain is major concern and research area. As well, the current usage of EHR data is limited, an efficient usage needs to be defined. Currently the amount of data being generated is vast, leading to big data and its management, how AI and IoT are useful for big data in healthcare is also an important research direction. This will form a good platform for large-scale medical analysis and healthcare IoT. Furthermore, this large-scale EHR is also an issue and appropriate mechanisms needs to be developed to cater to such data and their analysis. In the scenario of IoT, the sensor, devices, and actuators are battery operated, the consumption of battery is their major concerns. Energy consumption should be done efficiently and option such as its harvesting can be thought of. The quality of service of the networking and communication technologies should be properly adhered as well their security and privacy should be well taken care of. Also, as in [36], the sensor placement for remote monitoring of health is very important, if the sensors are placed in wrong position, it may produce wrong results. Further the communication via Internet is dependent on the connectivity and bandwidth, in such scenario, alternate mechanism should be available. In [37], an EHR system for monitoring of the patients was designed using IoT. A low cost, portable HER has been developed. It provides the dynamic and real-time data for patient's ECG, temperature, and other medical data [48]. Here, GPS plays a major role, so the connectivity over the locations is important factor for such IoT devices.

8.5 Conclusion and future scope

AI and IoT have set their foot in many domains of healthcare to aid the doctors, patients, hospitals, and all other stakeholders. In this chapter, major focus was on healthcare informatics and clinical applications using AI and IoT. However, the AI and IoT have emerged significantly in vast domains of healthcare. The combination of AI and IoT have shown tremendous results in new edge healthcare. Some of the future directions of healthcare which needs an eye are covered in this section.

AI requires a model having high resource utilization and is not cost effective in many cases. Focus should be on building an efficient AI model that is cost effective.

The conventional medical sector relies on the health parameters and human touch to give appropriate medication and treatment. Human intervention is must in the AI and IoT models, system, and protocol should be developed to draw a boundary between both. At times, the results generated from AI are not explainable or understandable, an effort should be made in the direction to make the AI-based models and evidence-based models. The models should be made as transparent as possible.

AI- and IoT-based healthcare models have emerged more since last decade. It is a need of time to devise a standard system or rules to make health professionals aware about their benefits and usage. Many rigid healthcare companies are not even trying to adapt these technologies because of lack of knowledge and its potential. Comfortability of IoT devices will result in surge in their demands and usage. Industrial focus should be on developing comfortable wearable devices for all ages and all types of patients. AI and IoT models are being made accurate and accurate. However, they also come with a bit of error. In such cases, one cannot make the model accountable for the mistake; rather in those cases, the medical professional intervention is mandatory.

Communication technologies are changing drastically over the years since the invention of the Internet. The IoT-based healthcare system so developed should always look for the upward and downward compatibility. Non-clinical domains like smart infrastructure, smart logistics, AI in drug discovery and development, and new edge technologies like blockchain should be leveraged in healthcare.

AI and IoT are used widely in many healthcare applications. AI and IoT have given the benefit of data analysis, remote monitoring for better healthcare. This chapter provides an overview on the AI and IoT domain, as a technological advancement. Then the various applications of AI are covered followed by that of the IoT. It also covers the basic clinical sensors used extensively for monitoring health parameters, various commercially available wearable devices. Subsequently the challenges and the future scope are covered. The computing technologies are evolving at a very fast rate and have signified their mark prominently during the current time of pandemic, where digitization and remote facility are of very much need. This has also impacted healthcare informatics drastically. The major futuristic focus should be on how they impacted healthcare in COVID times. This will pave the new road for the healthcare industry in general.

References

[1] Wiens, J. and Shenoy, E. S. (2017). Machine learning for healthcare: on the verge of a major shift in healthcare epidemiology. *Clinical Infectious Diseases*, 66(1), 149–153.

[2] Yan, Z., Zhan, Y., Peng, Z., *et al.* (2016). Multi-instance deep learning: discover discriminative local anatomies for body part recognition. *IEEE Transactions on Medical Imaging, 35*(5), 1332–1343. https://doi.org/10.1109/tmi.2016.2524985

[3] Schlemper, J., Caballero, J., Hajnal, J. V., Price, A. N., and Rueckert, D. (2018). A deep cascade of convolutional neural networks for dynamic MR image reconstruction. *IEEE Transactions on Medical Imaging*, 37(2), 491–503. https://doi.org/10.1109/tmi.2017.2760978

[4] Havaei, M., Davy, A., Warde-Farley, D., *et al.* (2017). Brain tumor segmentation with deep neural networks. *Medical Image Analysis*, 35, 18–31. doi: 10.1016/j.media.2016.05.004

[5] Kodali, R. K., Swamy, G., and Lakshmi, B. (2015, December). An implementation of IoT for healthcare. In *2015 IEEE Recent Advances in Intelligent Computational Systems (RAICS)* (pp. 411–416). IEEE.

[6] Davenport, T. and Kalakota, R. (2019). The potential for artificial intelligence in healthcare. *Future Healthcare Journal*, 6(2), 94.

[7] Jiang, F., Jiang, Y., Zhi, H., *et al.* (2017). Artificial intelligence in healthcare: past, present and future. *Stroke and Vascular Neurology*, 2(4), 230–243.

[8] Kaur, S., Singla, J., Nkenyereye, L., *et al.* (2020). Medical diagnostic systems using artificial intelligence (AI) algorithms: principles and perspectives. *IEEE Access*, 8, 228049–228069.

[9] Ashton, K. (2009). That 'internet of things' thing. *RFID Journal*, 22(7), 97–114.

[10] Darshan, K. R. and Anandakumar, K. R. (2015, December). A comprehensive review on usage of Internet of Things (IoT) in healthcare system. In *2015 International Conference on Emerging Research in Electronics, Computer Science and Technology (ICERECT)* (pp. 132–136). IEEE.

[11] Wipro.com. (2021). IoT in Healthcare Industry. *IoT Applications in Healthcare – Wipro*. Retrieved 14 July 2021 from https://www.wipro.com/business-process/what-can-iot-do-for-health care

[12] Panayides, A. S., Amini, A., Filipovic, N. D., *et al.* (2020). AI in medical imaging informatics: current challenges and future directions. *IEEE Journal of Biomedical and Health Informatics*, 24(7), 1837–1857.

[13] Tanwar, S., Ramani, T., and Tyagi, S. (2017, August). Dimensionality reduction using PCA and SVD in big data: a comparative case study. In *International Conference on Future Internet Technologies and Trends* (pp. 116–125). Springer, Cham.

[14] Pereira, S., Pinto, A., Alves, V., and Silva, C. A. (2016). Brain tumor segmentation using convolutional neural networks in MRI images. *IEEE Transactions on Medical Imaging*, 35(5), 1240–1251.

[15] Middleton, I. and Damper, R. I. (2004). Segmentation of magnetic resonance images using a combination of neural networks and active contour models. *Medical Engineering & Physics*, 26(1), 71–86.

[16] Polat, H. and Danaei Mehr, H. (2019). Classification of pulmonary CT images by using hybrid 3D-deep convolutional neural network architecture. *Applied Sciences*, 9(5), 940.

[17] Wang, S., Su, Z., Ying, L., *et al.* (2016, April). Accelerating magnetic resonance imaging via deep learning. In *2016 IEEE 13th International Symposium on Biomedical Imaging (ISBI)* (pp. 514–517). IEEE.

[18] Schlemper, J., Caballero, J., Hajnal, J. V., Price, A. N., and Rueckert, D. (2017). A deep cascade of convolutional neural networks for dynamic MR image reconstruction. *IEEE Transactions on Medical Imaging*, 37(2), 491–503.

[19] Byra, M., Styczynski, G., Szmigielski, C., *et al.* (2020, September). Adversarial attacks on deep learning models for fatty liver disease classification by modification of ultrasound image reconstruction method. In *2020 IEEE International Ultrasonics Symposium (IUS)* (pp. 1–4). IEEE.

[20] Laosai, J. and Chamnongthai, K. (2014, March). Acute leukemia classification by using SVM and K-means clustering. In *2014 International Electrical Engineering Congress (iEECON)* (pp. 1–4). IEEE.

[21] Saito, K., Zhao, Y., and Zhong, J. (2019, December). Heart diseases image classification based on convolutional neural network. In *2019 International Conference on Computational Science and Computational Intelligence (CSCI)* (pp. 930–935). IEEE.

[22] Mishra, A. K., Das, S. K., Roy, P., and Bandyopadhyay, S. (2020). Identifying COVID19 from chest CT images: a deep convolutional neural networks based approach. *Journal of Healthcare Engineering*, 2020.

[23] Kim, E., Rubinstein, S. M., Nead, K. T., Wojcieszynski, A. P., Gabriel, P. E., and Warner, J. L. (2019, October). The evolving use of electronic health records (EHR) for research. In *Seminars in Radiation Oncology* (vol. 29, no. 4, pp. 354–361). WB Saunders.

[24] Lin, W. C., Chen, J. S., Chiang, M. F., and Hribar, M. R. (2020). Applications of artificial intelligence to electronic health record data in ophthalmology. *Translational Vision Science & Technology*, 9(2), 13–13.

[25] Samad, M. D., Ulloa, A., Wehner, G. J., *et al.* (2019). Predicting survival from large echocardiography and electronic health record datasets: optimization with machine learning. *JACC: Cardiovascular Imaging*, 12(4), 681–689.

[26] Velupillai, S., Suominen, H., Liakata, M., *et al.* (2018). Using clinical natural language processing for health outcomes research: overview and actionable suggestions for future advances. *Journal of Biomedical Informatics*, 88, 11–19.

[27] Baker, S. B., Xiang, W., and Atkinson, I. (2017). Internet of things for smart healthcare: technologies, challenges, and opportunities. *IEEE Access*, 5, 26521–26544.

[28] A. W. S.L, "Ava Fertility Tracker," AvaWomen. https://www.avawomen.com/wo_en.

[29] Majumder, S., Mondal, T., and Deen, M. J. (2017). Wearable sensors for remote health monitoring. *Sensors*, 17(1), 130.

[30] Built In. (2021). *Care Worn: 11 Companies Using Wearable Technology in Healthcare*. Retrieved 14 July 2021 from https://builtin.com/healthcare-technology/wearable-technology-in-healthcare

[31] Living with the Owlet Smart Sock 2: The Health Monitoring Accessory for Babies, Wearable (2021). Retrieved 15 July 2021 from https://www.wareable.com/smart-clothing/owlet-smart-sock-2-review.

[32] OMRON Platinum Wireless Upper Arm Blood Pressure Monitor, Healthcare Wellness & Healthcare Products (2021). Retrieved 15 July 2021 from https://omronhealthcare.com/products/platinum-wireless-upper-arm-blood-pressure-monitor-bp5450/.

[33] Philips Launches Next Generation Wearable Biosensor for Early Patient Deterioration Detection, Including Clinical Surveillance for COVID-19, Philips (2021). Retrieved 15 July 2021 from https://www.philips.com/

[34] Koydemir, H. C. and Ozcan, A. (2018). Wearable and implantable sensors for biomedical applications. *Annual Review of Analytical Chemistry*, 11(1), 127–146.

[35] Vora, J., Kaneriya, S., Tanwar, S., Tyagi, S., Kumar, N., and Obaidat, M. S. (2019). TILAA: tactile internet-based ambient assistant living in fog environment. *Future Generation Computer Systems*, 98, 635–649.

[36] Ghosh, A. M., Halder, D., and Hossain, S. A. (2016, May). Remote health monitoring system through IoT. In *2016 5th International Conference on Informatics, Electronics and Vision (ICIEV)* (pp. 921–926). IEEE.

[37] Shanin, F., Das, H. A., Krishnan, G. A., *et al.* (2018, July). Portable and centralised e-health record system for patient monitoring using internet of things (IoT). In *2018 International CET Conference on Control, Communication, and Computing (IC4)* (pp. 165–170). IEEE.

[38] Khan, M. A. (2020). An IoT framework for heart disease prediction based on MDCNN classifier. *IEEE Access*, 8, 34717–34727.

[39] Kumar, P. M., Lokesh, S., Varatharajan, R., Babu, G. C., and Parthasarathy, P. (2018). Cloud and IoT based disease prediction and diagnosis system for healthcare using Fuzzy neural classifier. *Future Generation Computer Systems*, 86, 527–534.

[40] Chauhan, K., Jani, S., Thakka, D., *et al.* (2020). Automated machine learning: the new wave of machine learning. In *2020 2nd International Conference on Innovative Mechanisms for Industry Applications (ICIMIA)*, Bangalore, India, pp. 205–212.

[41] Abdellatif, M. M. and Mohamed, W. (2020). Telemedicine: an IoT based remote healthcare system. *International Journal of Online & Biomedical Engineering*, 16(6), 72–81.

[42] Ishak, M. K. and Kit, N. M. (2017, November). Design and implementation of robot assisted surgery based on Internet of Things (IoT). In *2017 International Conference on Advanced Computing and Applications (ACOMP)* (pp. 65–70). IEEE.

[43] Gupta, R., Shukla, A., and Tanwar, S. (2021). BATS: a blockchain and AI-empowered drone-assisted telesurgery system towards 6G. *IEEE Transactions on Network Science and Engineering*, 8(4), 2958–2967.

[44] Sohn, S. Y., Bae, M., Lee, D. K. R., and Kim, H. (2015, October). Alarm system for elder patients medication with IoT-enabled pill bottle. In *2015 International Conference on Information and Communication Technology Convergence (ICTC)* (pp. 59–61). IEEE.

[45] Yang, L., Ge, Y., Li, W., Rao, W., and Shen, W. (2014, May). A home mobile healthcare system for wheelchair users. In *Proceedings of the 2014 IEEE 18th International Conference on Computer Supported Cooperative Work in Design (CSCWD)* (pp. 609–614). IEEE.

[46] Anand, P., Singh, Y., Selwal, *et al.* (2020). IoT vulnerability assessment for sustainable computing: threats, current solutions, and open challenges. *IEEE Access*, 8, 168825–168853.

[47] Selvaraj, S. and Sundaravaradhan, S. (2020). Challenges and opportunities in IoT healthcare systems: a systematic review. *SN Applied Sciences*, 2(1), 1–8.

[48] Tanwar, S., Parekh, K., and Evans, R. (2020). Blockchain-based electronic healthcare record system for healthcare 4.0 applications. *Journal of Information Security and Applications*, 50, 102407.

Chapter 9

Confidentiality, integrity, authentication (CIA) model for secured storage and processing of personal health records using Merkle hash tree and recurrent neural networks (RNN)

Veeralakshmi Ponnuramu[1], *Latha Tamilselvan*[2], *X. Mercilin Raajini*[3], *Kavitha Karthikeyan*[3] *and Syed Bilal Hussain Shah*[4]

Abstract

Healthcare is a major sector generating huge amount of data rapidly. It requires a large storage space for storing the personal health records (PHR), thereby insisting the use of the storage services of third-party cloud providers. But in cloud, the security of PHR is compromised since the user is not having direct control over the data. To handle this issue, we aim at proposing novel Confidentiality, Integrity, Authentication model (CIA model) to ensure the security of stored data in cloud. Here, confidentiality of data is achieved by encrypting the PHR using the 2-Keys Symmetric Encryption Algorithm. Data integrity is ensured by using Merkle hash tree (MHT). Authentication is done by using fingerprint because of its consistency and uniqueness. PHR can be retrieved only by the authenticated users. For authenticating the user, a signature is generated from the hash values of the user's fingerprint by using the MHT and is submitted to the cloud providers during registration. And for accessing the data, the user needs to submit one leaf of MHT. From the leaf, the provider is able to generate the signature of the user. So, it is not required to submit the fingerprint in its entirety, thereby reducing the misuse of fingerprint of the user. For data representation and retrieval, recurrent neural network (RNN) deep learning technique is used. The cloud environment is set up with OpenNebula cloud and the algorithms for confidentiality, integrity, and authentication algorithms are implemented in java. Cryptanalysis against different cryptographic attacks prove that the proposed method is contributing a secure and

[1]Government Polytechnic College, Sevvapet, TamilNadu, India
[2]B.S. Abdur Rahman Crescent Institute of Science and Technology, India
[3]Prince Shri Venkateshwara Padmavathy Engineering College, India
[4]School of Computing and Mathematics Manchester Metropolitan University, UK

proficient mechanism for storing health information. After comparing the encryption time of our system with the standard encryption algorithms, it is inferred that this CIA model incurs less overhead for encryption. And also, better performance has been observed on the factors like error rate of 2% and accuracy rate of 98%, minimum overhead on computation, communication, and storage costs.

Keywords: Merkle hash tree; Confidentiality; Integrity; Authentication; Health informatics; Recurrent neural network

9.1 Introduction

Now-a-days, huge volume of data is generated relating to the personal health records (PHRs). PHRs are digitally represented as electronic health records (EHR). EHR is the electronic version of patients' health records kept by the healthcare providers. With the propagation of immense volume of data on healthcare, it is strongly understood the storage requirement for storing healthcare data is more prominent and is well known that cloud is one of the major solutions for huge storage requirements. But, the problem with the cloud is that the data may be stored in data servers maintained at different locations and the user does not exercise direct control over the data stored in cloud servers, thereby compromising the security of the health records. So, the data needs to be protected with administrative, physical, organizational safeguards, policies and procedures. These safeguards can protect the technology, people and information of the healthcare providers. But due to the abundant use of Internet, there is a possibility of cyber-attack and loss of sensitive healthcare data stored in cloud servers. Protection of this sensitive data depends on three important security goals like confidentiality, integrity and authentication.

Confidentiality is nothing but hiding data from unauthorized individuals. This can be ensured by encrypting the data either by using symmetric or asymmetric encryption techniques. The standard encryption algorithms like DES, 3-DES, AES, etc., are not appropriate for the encryption of massive data since these algorithms incur more computation complexity. Moreover, if the same data is repeated in plaintext, the same set of ciphertext is generated which may lead to the prediction of encryption key. So, we have used 2-keys symmetric encryption algorithm that is suitable for the encryption of a huge amount of data with less computation overhead. This algorithm is applied on variable sized blocks with variable sized keys. By applying this algorithm, PHR is split into different blocks each of size *n*. Then a recurrence relation has been used for generating a binary sequence. The input for recurrence relation can be chosen by the user. Then 2-keys symmetric encryption function is carried out either by using key1 or key2. Choice of key1 or key2 depends on the next bit in the binary sequence generated by applying the recurrence relation. One of the important features of this encryption function is that the encryption process depends on the keys as well as the position of text in the data blocks.

Integrity is ensuring the consistency of data by allowing only the authorized users to modify the data. If the data is modified by unauthorized users, it should be identified by the data owners. The methods like Message Authentication Code (MAC), checksum, Provable Data Possession (PDP), precomputed tokens, and Proof of Retrievability (POR) can be used to assure integrity. Even with the existence of these algorithms for checking integrity, it is unrealistic to use these techniques for ensuring integrity of huge collection of data due to high computation overhead. So, to ensure integrity of PHRs, we proposed an integrity assurance scheme which uses secure hash algorithm (SHA) and Merkle hash tree (MHT).

Authentication is the required procedure or action in order to prove the genuineness of user or data. While storing sensitive PHR data in cloud, it needs a proper authentication system allowing only the intended users to access the data. There are different types of authentications like biometric authentication, certificate-based authentication, password-based authentication, multifactor- and token-based authentication. Among these, biometric is one of the universally accepted authentication systems using human's physical or behavioral traits like fingerprint, face, eye, voice, etc. as discussed by Dargan and Kumar [1].

Now-a-days, artificial intelligence has the wide range of applications in various fields like health, multimedia processing, stock market, banking sector, intelligent systems (smart homes and automatic cars), bioinformatics, etc. These applications are in need of extensive analysis of huge volume of data, that can be accomplished by machine learning methods. These methods can extract the implicit date patterns and these patterns can be used for future prediction. Mostly training and testing phases are involved in building a machine learning model for analysis and prediction. In the training phase, training data is analyzed to train the machine learning model. Then the trained model is used in predicting the feature of new data in the test phase. Deep learning methods like convolution neural networks (CNNs) as well as recurrent neural networks (RNN) gained significant progress in recent years and solve the problems in different domains. CNN methods are used for multimedia data analysis, while RNN can be applied for the analysis of sequential data and time series data. Here RNN-non-linear auto-regressive with exogenous inputs (RNN-NARX) model has been used for predicting the patient visit date to the hospital.

Although PHR is processed by RNN for analysis and prediction, it does not guarantee the security of this sensitive information. Stealing PHR storage devices is interesting to thieves because of the availability of large amount of sensitive data in one place. Moreover, if healthcare data is stored in hard disks, it may be difficult for the patient/hospital to access it from remote location. In order to avail this feature of remote access, health information can be stored in cloud servers. Since, in cloud the users' data is stored in data servers maintained in various locations and also the users are not fully aware of the about the exact storage location of their data, there is possibility of security breach. In order to thwart such security breaches, it is highly required a mechanism for ensuring data security in cloud. Data security involves three major components like confidentiality, integrity, and authentication. So, we proposed a Confidentiality, Integrity and Authentication (CIA model) for secured storage of PHRs in cloud.

Confidentiality can be achieved by encryption algorithms. We have used 2-keys symmetric encryption algorithm and compared the performance of this algorithm with the standard encryption algorithm. Then for ensuring integrity, MHT has been used here. Authentication of data access is provided here by using fingerprint of the patient/doctor as the authentication structure. If the fingerprint is submitted as a whole, it may lead to the misuse of fingerprint by the cloud server. In order to avoid this, fingerprint is not submitted as a whole to the cloud server. Rather, a signature of the patient/doctor is derived from the fingerprint using MHT and the signature is submitted as the authentication structure to the cloud server. This avoids the misuse of fingerprint by the cloud server.

9.1.1 Organization of chapter

The rest of the chapter is organized as: Section 9.2 offers the details of related works in confidentiality, integrity, authentication issues, and also about RNN. Section 9.3 consolidates the problem definition. Section 9.4 elaborates about different kinds of deep learning methods for data processing and representation of healthcare data. Sections 9.5–9.8 discuss about the algorithm and approaches of CIA model. Section 9.9 explains about the data representation and processing using RNNs. Experimental results and analysis of proposed CIA model is highlighted in Section 9.10 and Section 9.11 concludes the chapter with future scope.

9.2 Related works

Machine learning algorithms are utilized in recognizing complex data patterns in various domains for classification, regression, and clustering of data. And also, deep learning approaches such as RNNs [2–6] are generally more precise and more efficient than the heuristics approaches. By using the input, inner state, and output, RNN attempts to model the system. In medical storage management system, $x(t)$ – the input data denotes the access request at time t and $y(t)$ – the output denotes suitable mechanism for data retrieval. The inner state $s(t)$ acts like the memory and sums up the set of the past events ($x(1)$, ..., $x(t-1)$). RNN is recursive in implementing a scenario: (1) read the new input data, (2) verify the previous state, (3) combine the new input data and previous state together to build the current state, and (4) generate the corresponding output by deploying the present state [7]. Miotto *et al.* [8] used unsupervised deep learning method for predicting patients' future based on EHR. Mohana *et al.* [9] use Internet of Things for healthcare data processing through convolutional neural network.

Data confidentiality is the protection of sensitive information like PHRs against unauthorized access and is the extent of the ability of the system to secure data from malicious insiders or outsiders as said by Yan *et al.* [10]. There are two kinds of cryptography: Symmetric and Asymmetric cryptography. Since symmetric cryptography is efficient and fast, a large volume of data are encrypted using only symmetric cryptography [11]. Different encryption methods like Data

Encryption Standard, 3-DES, Advanced Encryption Standard, Vernam Cipher, Blowfish were discussed in [12,13]. Tek, 2017 [14] reported that 81% of Protected Health Information (PHI) records were breached due to hacking attacks in 2016. Abouelmehdi *et al.* [15] discussed the importance of security of healthcare information and also surveyed about how security is compromised. Keshta and Odeh [16] reviewed the privacy and security issues on the electronic medical records (EMR) and also suggested some solutions. Chen *et al.* [17] discussed about protecting EHRs that are stored in cloud. Haque *et al.* [18] elaborated safeguarding EHR data using blockchain technology. Hambouz *et al.* [19] proposed a mechanism of hybrid cryptography for ensuring confidentiality by combining Advanced Encryption Standard (AES) and Rivest–Shamir–Adleman (RSA) algorithms [20]. But this method can be used for encrypting small messages. Hussain *et al.* [21] proposed an algorithm for the encryption of IEC 61850 GOOSE (Generic Object-Oriented Substation Events) messages which can be applied to minor data. Alkhudhayr *et al.* [22] discussed the different data security issues when the data is stored remotely. Aijaz *et al.* [23] elaborated the mechanism to protect the data servers from Distributed Denial of Service (DDoS) and Denial of Service attacks (DoS). Joshy *et al.* (2021) [24] used machine learning techniques and proposed a method to provide cyber security. Vijayalakshmi *et al.* [25] elaborated a medical record storage system that uses blockchain technology. Akkaoui *et al.* [26], Tao and Ling [27], Hylock and Zeng [28] and Gupta *et al.* [29] discussed a framework using block chain technology that facilitates the secure exchange of patient health records.

In 1981, Lamport [30] recommended a remote user authentication method, in that the hash values of the user's password are stored in the password table. However, Hwang and Li [31] in 2000 found that the method of storing password become invalid if the security of the password table is compromised. So, authors proposed a new scheme for remote user authentication that uses ElGamal's public key cryptosystem. This scheme did not demand the system to maintain a password table. But this scheme become vulnerable to impersonate attack of one user impersonating another valid user by using identity and password. In 2002, Chien *et al.* [32] suggested a method for remote user authentication that provides mutual authentication. In 2004, Ku and Chen [33] found out that Chien *et al.*'s scheme was susceptible to reflection and insider attacks. In 2010, Chen and Huang [34] proposed an authentication technique based on visual secret sharing and CAPTCHA. Then, Li and Lee. [35] claimed that masquerading attack would happen on Chen *et al.*'s scheme for authentication.

Choudhury *et al.* [36] suggested another remote user authentication framework providing authenticity through the two-steps: smart card and password. But the scheme was vulnerable to the attacks like password change flaw, out of band attack and masquerading attack. Kim *et al.* [37] proposed another method by using both fingerprint authentication and smart card in which the table of password is not maintained at the server, but providing security against replay attacks through the use of nonce technology. However, Scott [38] proved that this technique is vulnerable to imitation attack.

Wenyi Liu *et al.* [39] introduced the concept of multi-factor authentication that uses the user behavior and the password as the authentication factors. Another multi-factor authentication system was developed by Banyal *et al.* [40] using the concept of Cloud Access Management (CAM) pattern. Then, Bhattasali and Saeed [41] proposed a 2-factor remote authentication system using behavioral biometrics on keystroke analysis and the secret PIN mechanism. Masrom *et al.* [42] developed the mechanism of using graphical password. Then Castiglione *et al.* [43] proposed the scheme of one-time authentication that uses Hash Message Authentication Code (HMAC). Nimmy and Senthumadhavan [44] developed a model for mutual authentication using the concept of steganography.

Fan *et al.* [45] developed remote three-factor authentication mechanism that uses password, biometric template, and a smart card as the authentication factors. Here, the security of the biometric template was compromised. Harakannanavar [46] discussed about the various biometric authentication techniques. Win *et al.* [47] discussed about fingerprint classification algorithm and its usage in identifying criminals. Venkatachalam *et al.* [48] proposed a mechanism of biometric authentication for giving the users the permission to access to the cloud services.

Even with the emergence of these many researches in the area of data security, still security in whole is not achieved in many sectors. The hackers and malicious users still find the path to achieve security breach. So, in this chapter, we have proposed a new mechanism for protecting the medical data stored in cloud servers.

9.3 Problem definition

This chapter aims at providing a CIA model for ensuring secured storage of PHRs and data processing using RNNs. It has the following objectives.

- *Providing confidentiality*: 2-Keys encryption algorithm is developed for the encryption of PHRs with minimum computation and storage overhead.
- *Ensuring integrity*: Integrity is ensured here by generating the hash value of PHR data using SHA and signature generation using MHT.
- *Authentication*: Fingerprint is used as remote user authentication structure allowing only the authorized users gaining access to the data stored in the cloud.
- *Prediction using RNNs*: Patient health information history is analyzed in order to predict the date of visit of the patient to hospital.

9.4. Deep learning methods for data representation and data processing

Deep learning involves set of computational models with multiple layers of processing to study about the data representations that consists of multiple levels of abstraction. These deep learning methods have been used in image classification, image captioning, video processing, text processing, healthcare data, etc. Health informatics is a significant area that intersects with domains of computer science, machine learning, deep learning, healthcare, etc. It deals with the required resources and strategies in order to enhance the data storage and retrieval process. Deep learning methods are used to solve different problems in healthcare informatics that uses medical images, pervasive sensing, previous diagnosis, immunization details, and public healthcare information [49]. Healthcare data can contain a variety of data that can be structured, unstructured biomedical, incomplete, continuous, discrete, homogeneous, heterogeneous, high dimensional, irregular, etc. There are many deep learning methods for health informatics in order to overcome challenges in medical data processing. But these existing deep learning methods are sufficient for processing new kinds of data that are generated every day. So, it is required for the in-depth exploration of these methods in order to address the new challenges. Table 9.1 lists the different types, functions, and applications of deep learning methods.

Neural networks are similar to the connection of different computing devices that are working parallelly simulating human brain functions. The interconnecting devices are known as neurons working together to solve the problems. CNN is a special class of neural network consisting of several layers with the grid-like arrangements. CNNs apply filters to process the data and tunes the filters to extract the features of images. RNN is a special type of artificial neural networks (ANN)

Table 9.1 Deep learning methods

Type	**Functions**	**Application**
Classical Neural Network	Linear function, non-linear function, Sigmoid curve, Rectified Linear Unit (ReLU), Hyperbolic tangent (tanh)	Any dataset represented as table with rows and columns
CNN	Convolution, max-pooling, flattening, full connection	Image datasets that contain OCR document analysis
RNN	Gated RNNs, LSTM	Image classification, sentiment analysis, image captioning, video classification
Generative adversarial networks	Generator and discriminator	Image and text generation, image enhancement
Self-organizing maps	Best matching units (BMUs)	Projects in text, music, and video
Boltzmann machines	Produce model parameters	System monitoring, developing binary recommendation system
Deep reinforcement learning	Predict the future reward of each action	Self-drive cars, robotics, board games like chess
Autoencoders	Sparse, denoising, stacked, and contractive	Feature detection, recommendation model
Backpropagation	Learn about any errors in data prediction	Data debugging
Gradient descent	To find the optimum solution	Updating parameters in a given model

for processing natural language and time-series data [50,51]. Gated recurrent units (GRUs) and long short-term memory (LSTM) networks are the variants of RNN used to extract the features by applying filters on the medical data.

9.5 CIA model for securing data stored in server

CIA model shown in Figure 9.1 protects sensitive PHR from unauthorized access. Initially for authentication, the patient has to acquire the fingerprint, apply the image enhancement operations, split the fingerprint into eight equal shares, generate the hash values of the fingerprint shares, and generate the root signature of the user using MHT. Then the patient needs to register to the cloud server with the signature [52]. To access data, the user needs to submit the signature and a part of fingerprint. Using the fingerprint share, the server is generating the whole signature. The server allows the user for data if the generated signature matches with the stored signature of the user.

Confidentiality is achieved here by encrypting the data by using 2-keys symmetric encryption algorithm. The user selects 2 keywords of variable size and generates a binary sequence. The record is either encrypted with keyword 1 or keyword 2 depending on the bit present in the binary sequence. If the bit in the binary sequence is 1, then the record is encrypted using the keyword. Otherwise, the data is encrypted using the keyword 2. For ensuring integrity, hash value of the record is generated using SHA. Hash value of the individual records are considered

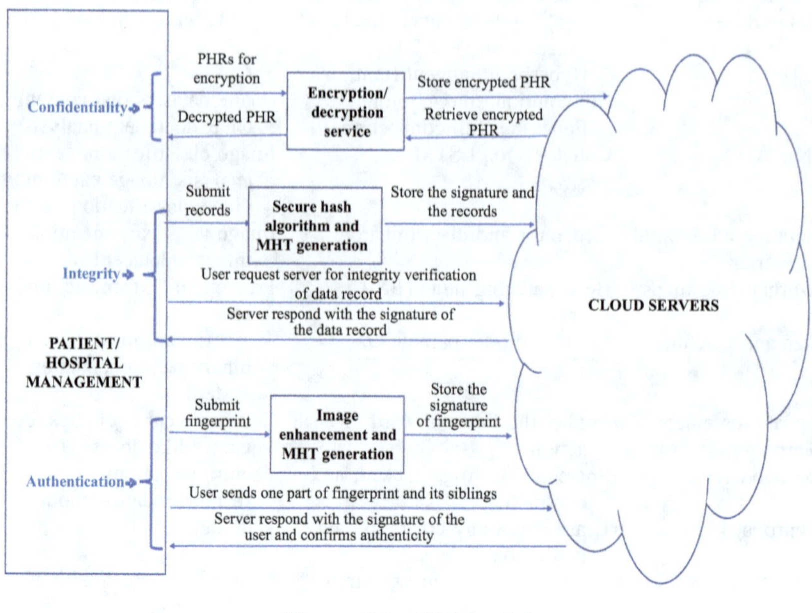

Figure 9.1 CIA model

as the leaves of the MHT [53,54]. The adjacent 2 values are combined together to form the upper-level node of the MHT. The procedure is repeated until the root of the tree is formed. The root is the tree is the signature of the data. This signature can be used by the data owner for data verification by the data owner.

9.6 Confidentiality by 2-keys symmetric encryption algorithm

PHRs are represented in table structure. This is encrypted by applying the algorithm of 2-keys symmetric encryption [55] that involves binary sequence generation, key generation, and encryption.

Step 1: Generation of binary sequence:
This binary sequence is used to select the key1 or key2 to be used for encryption. If the bit in the binary sequence is 1, then the block can be encrypted with the key1. Otherwise, the block should be encrypted with the key2.

The sequence of 0s and 1s is produced by applying the recurrence relation. The general form of recurrence relation is shown in (9.1):

$$X_{n+m} = (a_0X_n + a_1X_{n+1} + a_2X_{n+2} + a_3X_n + .. + a_{m-1}X_{n+m-1}) \bmod 2 \quad (9.1)$$

The initial value of $m = 5$, initial vector $(x1, x2, x3, x4, x5)$ of (0, 1, 0, 0, 0) and the initial coefficients $(a1, a2, a3, a4, a5)$ of (1, 0, 0, 0, 0) generate the recurrence equation of the form as shown in (9.2):

$$X_{n+5} = X_n + X_{n+2} \quad (9.2)$$

It generates the binary sequence of 0100001001011001111100011011101010000, . . .

Step 2: Key generation:
The user can choose a random number and 2-keywords of variable length and generate the encryption keys based on (9.3). The initial value of $i = 0$ and key = 0:

$$\text{key} = \text{key} + \text{ASCII}(\text{keyword}[i] * (i + 1)) \quad (9.3)$$

Step 3: Encryption:
The PHRs are stored in the cloud as the table consisting of rows and columns. Each row has the record of 1 patient. The column contains various attributes of medical information of the patients. Here, each row is encrypted separately either using key1 or key2. The choice of selection of key is depending upon binary sequence. The encryption procedure is as follows:

1. Read through the binary sequence for a bit.
2. If (bit=1) then

Encrypt the record using the following encryption function with the key1:
$\text{PHRCipher}[i,k,j] = (\text{key1} + \text{PHR}[i,k,j] + \text{r and}^{(i+k+j)}) \bmod 256$

3. Else

   Encrypt the record using the following encryption function with the key2.
   $PHRCipher[i,k,j] = (key2 + PHR[i,k,j] + r \text{ and}^{(i+k+j)}) \bmod 256$
   i—> record number of the patient
   k—> column number of the corresponding patient i
   j—> Position of the character of the particular column k

9.7 Integrity assurance by MHT

MHT is a tree data structure in which the hash of a block of data is stored in leaf node, and the hash of its children is stored in non-leaf node. Each non-leaf node in MHT has two child nodes.

Here, in order to assure the integrity of the PHR, MHT is used. It involves two different stages.

9.7.1 Signature generation

MHT is generated by the data owner. Hash value for each record is generated using secure hash function (SHA-256). This hash value is treated as the leaf of MHT. Each adjacent leaves are combined; hash is generated and a non-leaf node (parent node) is generated. Then two non-leaf nodes are combined, hash value is generated. This process is repeated until generating the root node of the MHT as shown in Figure 9.2. The value of the root node is known as signature. The data owner retains the signature for verification.

9.7.2 Integrity verification

To verify the data's integrity, the data owner stores the signature, hash values of any two adjacent combined nodes, and their siblings at the upper level. For example, the

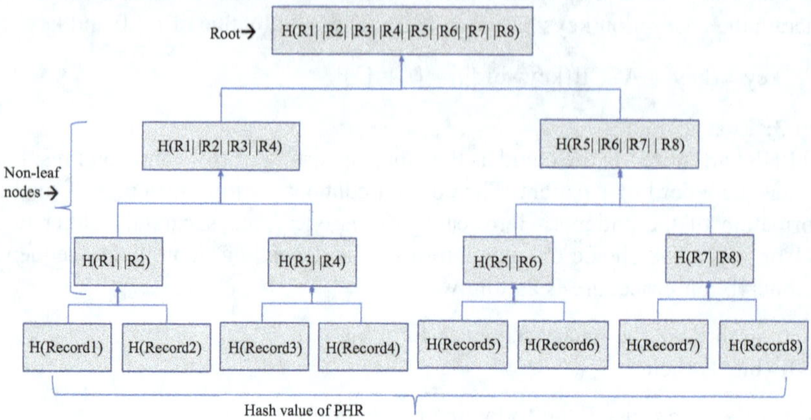

Figure 9.2 Hash signature generation using MHT

data owner should retain Hash(Record1), Hash(Record2), Hash(R3 || R4), and Hash (R5 || R6 || R7 || R8). From this, the data owner can able to generate the signature:

- Hash(R1 || R2) = Hash(Record1) || Hash(Record2)
- Hash(R1 || R2 || R3 || R4) = Hash(R1 || R2) || H(R3 || R4)
- Generated signature = Hash(R1 || R2 || R3 || R4) || Hash(R5 || R6 || R7 || R8)

If the generated signature matches with the signature stored by the data owner, the data integrity is assured. Otherwise, the data is modified by the unauthorized individuals.

9.8 Biometric authentication using fingerprint for data access

The proposed system uses the fingerprint as an authentication structure because of its uniqueness and consistency. Here, rather than submitting the whole fingerprint to the server, the user submits only the hash value of part of the fingerprint. The server stores the signature of the user that is generated from the hash value of the fingerprint using MHT. This signature is used as an authentication structure in order to prove the authenticity of the user. This system for remote user authentication involves three phases as shown in Figure 9.3:

1. Initialization
2. Registration
3. Authentication

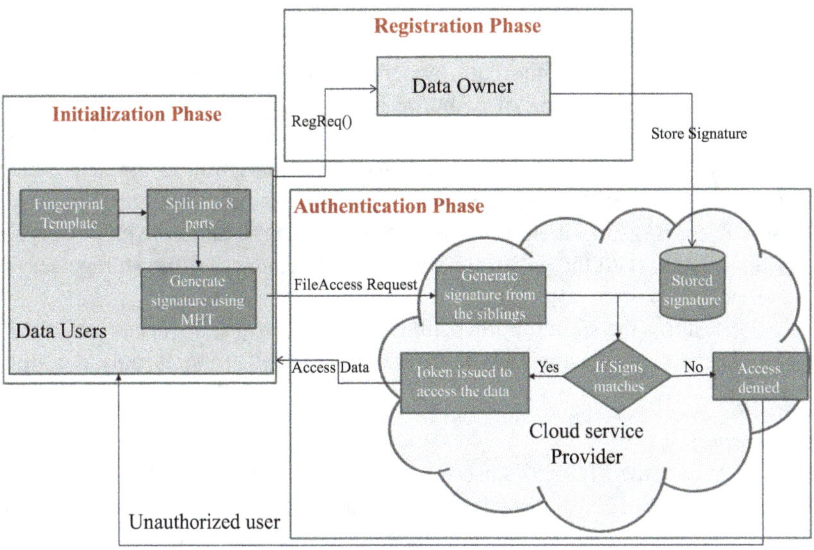

Figure 9.3 Remote user authentication system in cloud

Output image	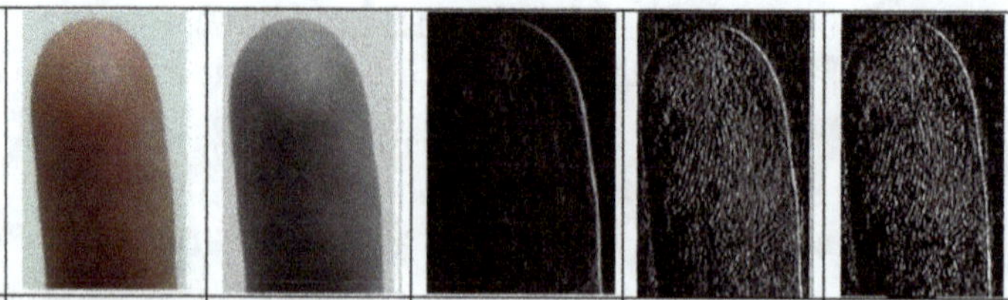				
Operations	Original image	After grayscale conversion	After edge enhancement	After filtering	After binarization

Figure 9.4 Image enhancement operations

Initialization

In this phase, the patient and the doctor acquire their fingerprint. The fingerprint is applied with the image enhancement operations of grayscale conversion, edge enrichment, filtering, and binarization. The result of image enhancement operation is shown in Figure 9.4.

An enhanced image of fingerprint template image is obtained as the result of image enhancement operations. This template is split into eight equal Finger Print Shares (FPS) (FPS1, FPS2.FPS8) as four rows and two columns. Hash values of these shares are generated using SHA-256. The hash values of the fingerprint shares are called as authenticated values (AV). MHT is generated by considering the hash values of the fingerprint shares as the leaves. Root of the Merkle hash tree is the signature of the user that is shown in Figure 9.5b. These fingerprint shares and the signature are maintained by the data user. The eight equivalent shares fingerprint template are shown in Figure 9.5a.

Registration

The user (patient) registers the data owner (hospital) by sending the registration request (RegReq()) consisting of name, password, AV. The data owner sends the RegUser (Name, AV, signature) message to the CSP in order to provide access to the storage services. The DO also adds this DU's information in the data owner's local database. The CSP stores the details, including name, signature, and AV of data user. Then, RegAck (identity) message is sent from CSP to the patient through the hospital. Now, both the patient and the hospital can access the storage services of the cloud.

CSP maintains the sibling information of the data user (patient) as in Table 9.2. Here, CSP is not storing the fingerprint shares, instead it stores only the sibling information.

Authentication

In order to access the PHRs, the user sends the AccessReq() message that consists of the hash values of 2 adjacent fingerprint shares to the server. The server generates the signature – Rootsig of the user using the algorithm1. If this Rootsig matches with the stored signature of the user, the user is given access to the PHRs. It is demonstrated in Algorithm 9.1.

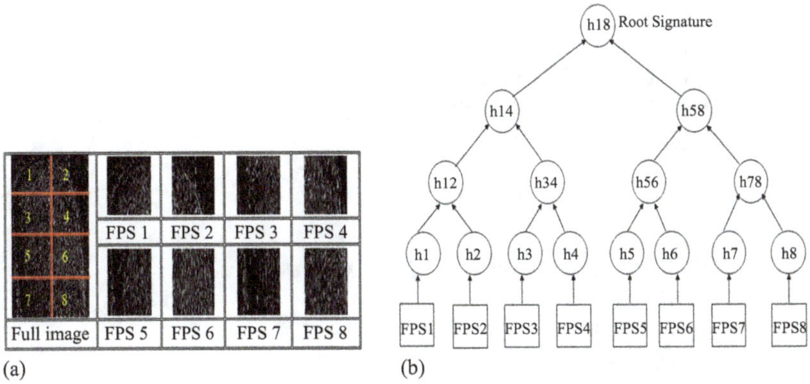

Figure 9.5 (a) Fingerprint shares. (b) Root signature generation from fingerprint shares.

Table 9.2 Sibling information required for generating signature

For the fingerprint templates from the DU by CSP	Values maintained by CSP	
	Sibling1	**Sibling2**
FPS1 and FPS2	h34	h58
FPS3 and FPS4	h12	h58
FPS5 and FPS6	h78	h14
FPS7 and FPS8	h56	h14

Algorithm 9.1: Signature Generation at the server

Input: Hash value of any 2 fingerprint shares, Eg.H(FPS1), H(FPS2)
Output: Root signature (Ω)

1. The server receives the hash values of the 2 adjacent finger print shares from the user and assign them as

```
        HASH-1=H(FPS1);
        HASH-2=H(FPS2);
```

2. Retrieves the corresponding sibling information HASH-1 and HASH-2 from the Table 9.2 and uses the following procedure.

```
    If (HASH-1==h1 and HASH-2==h2)
    {
        HASH-12=HASH-1||HASH-2;
        HASH-14=HASH-12||h34;
        HASH-18=HASH-14||h58
        Rootsig=HASH-18;
    }
    else if (HASH-1=h3 and HASH-2=h4)
```

```
{
        HASH-34=HASH-1||HASH-2;
        HASH-14=h12||HASH-34;
        HASH-18=HASH-14||h58
        Rootsig=HASH-18;
}
else if (HASH-1=h5 and HASH-2=h6)
{
        HASH-56=HASH-1||HASH-2;
        HASH-58=HASH-56||h78;
        HASH-18=h14||HASH-58
        Rootsig=HASH-18;
}
else
{
        HASH-78=HASH-1||HASH-2;
        HASH-58=h56||HASH-78;
        HASH-18= h14||HASH-58
        Rootsig=HASH-18;
}
        Return Rootsig;
```

9.9 Data representation and processing using RNNs

RNN-NARX (RNN-Non-linear Auto-Regressive with eXogeneous inputs) model in Figure 9.6 is proposed here in order to predict the requirement of immediate treatment for the patient. While the PHR is retrieved from the cloud, RNN model is

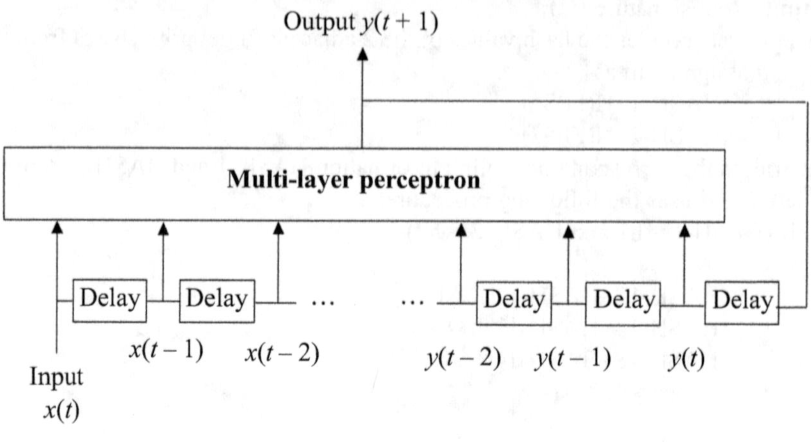

Figure 9.6 RNN-NARX model for prediction

activated. It analyzes the history of patient records and the last date of hospital visit. From this, we predict visit date for the patient. RNN contains at least one feed-back connection, in order to allow the flow of activations in a loop. It consists of multilayer perceptron (MLP) with additional loops. It enables the network to work with temporal processing and learning the sequences.

GRUs are used here rather than using LSTM. It is because that GRUs can achieve the same performance of LSTM with simpler architecture. GRUs have more persistent memory in order to allow RNN thereby to capture dependencies and uses h_{t-1} and x_t to produce the next hidden state h_t by using the following equations:

$$
\begin{aligned}
z_t &= \sigma(W^{(z)}x_t + U^{(z)}h_t - 1) \\
r_t &= \sigma(W^{(r)}x_t + U^{(r)}h_t - 1) \\
\widetilde{h}_t &= \tanh\ r_t \circ Uh_{t-1} + Wx_t \\
h_t &= (1 - z_t) \circ \widetilde{h}_t + z_t \circ h_{t-1}
\end{aligned}
$$

We strongly believe that this RNN-NARX model predict the correct visit date because of the following reasons: (1) RNN is using GRUS for learning and estimating the visit date based on PHRs. (2) Patient visits are correlated with poor health.

9.10 Experimental results and analysis

Experimental setup: We established a private cloud environment using OpenNebula cloud computing software. In this OpenNebula cloud environment, we set up a front end and three virtual machines in order to demonstrate the significance of the CIA model. OpenNebula consists of three components: front-end: used to access the services of cloud server; Hosts: provide the resources for creating virtual machine; Datastores: store the images of virtual machine.

The cloud environment is configured with two server machines of 32-bit Ubuntu server edition and one client machine of 32-bit desktop version. The servers and client machines are configured with 1 GHz dual core processor, 300 GB hard disk, and 2 GB RAM. The algorithms for key generation, encryption, decryption, hash generation, signature generation, and signature verification have been implemented in Java. The client and server machines are installed with the Ubuntu desktop version. All the algorithms, including preprocessing, key generation, encryption, and decryption, have been implemented in Java. Moreover, the communication between the client and server has been accomplished through Java remote method invocation (RMI) concepts.

Experimental parameters:

Encryption and decryption time

We compared the execution time of 2-keys symmetric encryption algorithm with the other standard existing algorithm like DES, 3-DES, AES, RC2, and RC4 and is

Table 9.3 Encryption time of RC2, RC4, DES, 3-DES, AES, and 2-keys symmetric encryption

Data size (MB)	**Encryption time (ms)**					
	RC2	**RC4**	**DES**	**3-DES**	**AES**	**2-Keys symmetric encryption**
2	377	543	1,438	892	1,757	211
4	407	600	1,544	1,003	1,930	264
6	433	703	1,752	1,251	2,167	311
8	489	791	1,931	1,350	2,353	369
10	569	912	2,145	1,529	2,452	412
12	572	990	2,204	1,682	2,483	444
14	596	1,023	2,292	1,720	2,502	495

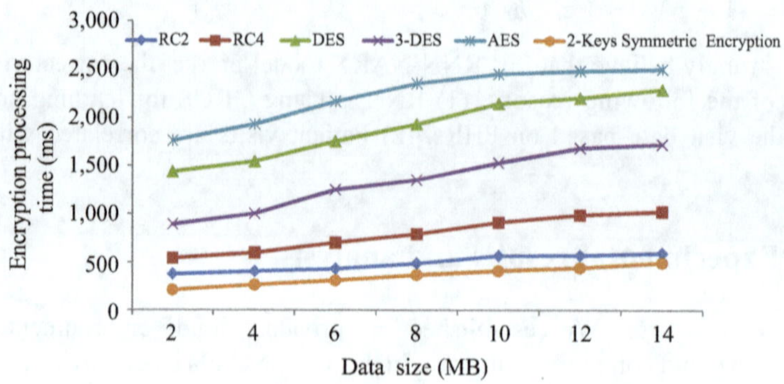

Figure 9.7 Encryption time of RC2, RC4, DES, 3-DES, AES, and 2-keys symmetric encryption

shown in Table 9.3 and Figure 9.7. Among these algorithms, the 2-keys symmetric encryption algorithm consumes lesser time for encryption and decryption.

Integrity verification

Since SHA is used here for the generation of the hash value of PHRs, possibility of collision attack is very lesser. Moreover, if there is a 1-bit change of the original PHR, it results in a significant change in hash value, thereby assuring data integrity. Generation of signature using MHT is more than the integrity verification time. But here signature generation is one time operation. Only integrity checking is done whenever required to check the integrity of data. It is shown in Figure 9.8 and Table 9.4.

In authentication, communication cost is measured based on the number of messages exchanged between the user and the cloud server, both in the registration phase and the authentication phase. From the Figure 9.9 and Table 9.5, it is inferred that the number of exchanges increases when the number of user increases.

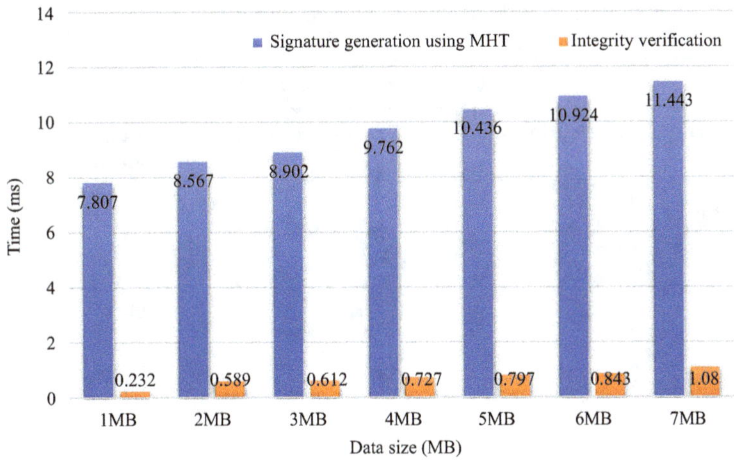

Figure 9.8 Signature generation and integrity verification time

Table 9.4 Signature generation and integrity verification time

Data size (MB)	Signature generation time using MHT (ms)	Integrity verification (ms)
1	7.807	0.232
2	8.567	0.589
3	8.902	0.612
4	9.762	0.727
5	10.436	0.797
6	10.924	0.843
7	11.443	1.08

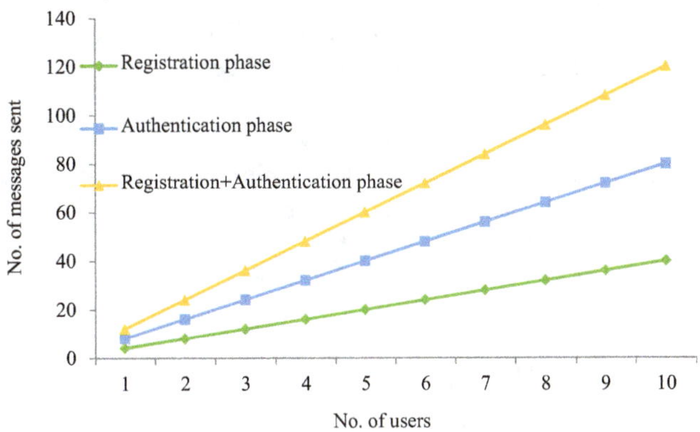

Figure 9.9 Communication between the patient and the cloud server

Table 9.5 Communication between the patient and cloud server

	No. of messages communicated		
No. of users	**Registration phase**	**Authentication phase**	**Registration + authentication phase**
1	4	8	12
2	8	16	24
3	12	24	36
4	16	32	48
5	20	40	60
6	24	48	72
7	28	56	84
8	32	64	96
9	36	72	108
10	40	80	120

Table 9.6 Summary of storage and computation cost of CIA model

	Storage cost		**Computation cost**	
	Patient/hospital	**Cloud server**	**Patient/hospital**	**Cloud server**
Confidentiality	O (1)	O (n)	O (1)	O (n)
Integrity	O (1)	O (n)	O (1)	O (n)
Authentication	O (1)	O (n)	O (1)	O (n)

Storage and computation cost: Table 9.6 shows the storage and computation cost of CIA model. Patient/hospital needs to store only lesser amount of data in terms of O(1) for adopting this security mechanism. And also the computation that is to be done by the user is the order of O(1). Only the server computation and storage cost of the cloud server is O(n). It is well known that computation and storage space can be easily hired from the cloud server. It incurs only less overhead in the user side.

This fingerprint authentication system was tested against the dataset of 60 patients collected from Kaggle. In this, 40 people are previous registered and 20 people are not registered their signatures to the cloud. Accuracy and error rate are measured using (9.4) and (9.5). Based on the results, it is observed that our system can identify a person with an accuracy of 98% and an error rate of 2%:

$$\text{Accuracy} = \frac{\text{TN} + \text{TP}}{\text{TN} + \text{TP} + \text{FN} + \text{FP}} \tag{9.4}$$

$$\text{Error rate} = \frac{\text{FN} + \text{FP}}{\text{TN} + \text{TP} + \text{FN} + \text{FP}} \tag{9.5}$$

Security analysis:

Our CIA model has been proved to be secure against the security attacks.

Brute force attack: Here the encryption process depends on 2-keys of variable sizes. So, the attacker cannot use brute force attack to predict the keys.

Ciphertext only attack: In this attack, the cryptanalyst is trying to find the details about the key only from ciphertext. Since, here the ciphertext depends on both the keys and the position of text, it is difficult to form the relationship between ciphertexts, thereby reducing the possibility of ciphertext attacks.

Known plaintext attack: The attacker can try to find the keys by setting up linear equations representing the relationship between plaintext and ciphertext. Here, it is difficult because of the fact that different ciphertexts can be generated for the same plaintexts at different positions.

Chosen plaintext attack: With the chosen plaintext and ciphertext pairs, the opponent generates linear equations and try to get keys. But, here the encryption depends on the keys as well as position of characters making the possibility of chosen plaintext attack lesser.

Frequency analysis attack: By counting the number of occurrences of the same plaintext ciphertext pairs, the adversary can try to extract the key. But, in this approach, the same plaintext can be converted to different ciphertext at different places due to the inclusion of position of data in the encryption process.

Collision attack: The chance of getting the same hash value for two different messages is called collision attack. Since uniqueness is the primary feature of fingerprint and hash value is generated for fingerprint, possibility of collision attack is very less.

Security of fingerprint: Since the user is not submitting the fingerprint in its entirety and only the hash value of the fingerprint is stored in server, security of the user fingerprint is ensured.

9.11 Conclusion and future scope

This chapter demonstrates the security mechanism of storing sensitive PHR in the cloud through the CIA model by using the suitable algorithms. Here, the significant portion of computation and storage are dedicated to the cloud servers, thereby releasing the patients and hospitals from storage overhead. Performance and security analysis proves the successiveness of this approach. But still 100% security cannot be achieved due to the abundant use of internet. So, in the near future, the security of this model can be increased further by incorporating the security parameters like authorization, access control, non-repudiation, etc.

References

[1] S. Dargan, and M. Kumar, (2020). "A comprehensive survey on the biometric recognition systems based on physiological and behavioral modalities", *Expert Systems with Applications*, 143, 113114.

[2] I. H. Sarker (2021). Deep learning: a comprehensive overview on techniques, taxonomy, applications and research directions. *SN Computer Science.* 2, 420. https://doi.org/10.1007/s42979-021-00815-1.

[3] P. Li, J. Zhu, L. Peng, and Y. Guo (2016, April). "RNN based Uyghur text line recognition and its training strategy", In *12th IAPR Workshop on Document Analysis Systems (DAS)*, pp. 19–24.

[4] A. Graves (2012). *Supervised Sequence Labelling with Recurrent Neural Networks*, vol. 385. Springer, New York, NY.

[5] Z. C. Lipton (2015). "A critical review of recurrent neural networks for sequence learning", CoRR. http://arxiv.org/abs/1506.00019.

[6] J. Daniel, S. Irin Sherly, V. Ponnuramu, *et al.* (2022). "Recurrent neural networks for feature extraction from dengue fever", *Evidence-Based Complementary and Alternative Medicine*, 2022, Article ID 5669580, 9. https://doi.org/10.1155/2022/5669580.

[7] S. Ebrahimi, R. Salkhordeh, S. A. Osia, A. Taheri, H. R. Rabiee, and H. Asadi (2021). "RC-RNN: reconfigurable cache architecture for storage systems using recurrent neural networks", In *IEEE Transactions on Emerging Topics in Computing*, 10(3), 1492–1506. doi:10.1109/TETC.2021.3102041.

[8] R. Miotto, L. Li, B. A. Kidd, and J. T. Dudley (2016). "Deep patient: an unsupervised representation to predict the future of patients from the electronic health records", *Science Report*, 6, 26094.

[9] J. Mohana, B. Yakkala, S. Vimalnath, *et al.* (2022). "Application of internet of things on the healthcare field using convolutional neural network processing", *Journal of Healthcare Engineering*, 2022, *Article ID 1892123*, 1–7.

[10] Z. Yan, M. Wang, Y. Li, and A. V. Vasilakos (2016, March–April). "Encrypted data management with deduplication in cloud computing", *EEE Cloud Computing*, 3(2), 28–35.

[11] A. Kakkar and P. K. Bansal (2004). "Reliable encryption algorithm used for communication", Thesis, Thapar University.

[12] S. Simar Preet and R. Maini (2011). "Comparison of data encryption algorithms", *International Journal of Computer Science and Communication*, 2 (1), 125–127.

[13] T. Chatterjee, T. Das, S. Dey, A. Nath, and J. Nath (2011). "Symmetric key cryptosystem using combined cryptographic algorithms – Generalized modified Vernam Cipher method, MSA method and NJJSAA method: TTJSA algorithm", *Information and Communication Technologies (WICT)*, 2011 World Congress on, Mumbai, pp. 1175–1180.

[14] R. C. Tek, "BREACH REPORT 2016: Protected Health Information (PHI)" 2017.

[15] K. Abouelmehdi, A. Beni-Hessane, and H. Khaloufi (2018). "Big healthcare data: preserving security and privacy", *Journal of Big Data* 5, 1. https://doi.org/10.1186/s40537-017-0110-7.

[16] I. Keshta and A. Odeh (2021). "Security and privacy of electronic health records: concerns and challenges", *Egyptian Informatics Journal*, 22(2), 177–183, ISSN 1110-8665, https://doi.org/10.1016/j.eij.2020.07.003.

[17] C.-L. Chen, P.-T. Huang, Y.-Y. Deng, H.-C. Chen, and Y.-C. Wang (2020). "A secure electronic medical record authorization system for smart device application in cloud computing environments", *Human-Centric Computing Information Science*, 10, 1–31.

[18] R. Haque, H. Sarwar, S. R. Kabir, *et al.* (2020). "Blockchain-based information security of electronic medical records (EMR) in a healthcare communication system", In *Intelligent Computing and Innovation on DataScience*, Springer, Singapore, pp. 641–650.

[19] A. Hambouz, Y. Shaheen, A. Manna, M. Al-Fayoumi, and S. Tedmori (2019). "Achieving data integrity and confidentiality using image steganography and hashing techniques", In *2019 2nd International Conference on New Trends in Computing Sciences (ICTCS)*, pp. 1–6, doi:10.1109/ICTCS.2019.8923060.

[20] M. Bhavin, S. Tanwar, N. Sharma, S. Tyagi, and N. Kumar (2021). "Blockchain and quantum blind signature-based hybrid scheme for healthcare 5.0 applications", *Journal of Information Security and Applications*, 56, 102673.

[21] S. M. S. Hussain, S. M. Farooq, and T. S. Ustun (2020, October). "A method for achieving confidentiality and integrity in IEC 61850 GOOSE messages", *IEEE Transactions on Power Delivery*, vol. 35, no. 5, 2565–2567, doi:10.1109/TPWRD.2020.2990760.

[22] F. Alkhudhayr, S. Alfarraj, B. Aljameeli, and S. Elkhdiri (2019). "Information security: a review of information security issues and techniques", In *2019 2nd International Conference on Computer Applications & Information Security (ICCAIS)*, pp. 1–6, doi: 10.1109/CAIS.2019.8769504.

[23] N. Usman Aijaz, M. Misbahuddin, S. Raziuddin (2021). "Survey on DNS-specific security issues and solution approaches", In D. S. Jat, S. Shukla, A. Unal, and D. K. Mishra (eds), *Data Science and Security Lecture Notes in Networks and Systems*, vol. 132. Springer, Singapore. https://doi.org/10.1007/978-981-15-5309-7_9.

[24] A. P. Joshy, K. Natarajan, and P. Naik (2021). "An advanced machine learning framework for cybersecurity", In D. S. Jat, S. Shukla, A. Unal, and D. K. Mishra (eds), *Data Science and Security. Lecture Notes in Networks and Systems*, vol. 132. Springer, Singapore. https://doi.org/10.1007/978-981-15-5309-7_25

[25] K. Vijayalakshmi, S. Nikkath Bushra, N. Subramanian, and V. Ponnuramu (2022). "Blockchain based medical record storage and retrieval using NFT tracking system", In *2022 6th International Conference on Trends in Electronics and Informatics (ICOEI)*, pp. 1–8, doi:10.1109/ICOEI53556.2022.9776833.

[26] R. Akkaoui, X. Hei, and W. Cheng (2020). "EdgeMediChain: a hybrid edge blockchain-based framework for health data exchange", *IEEE Access*, 8, 113467–113486.

[27] J. Tao and L. Ling (2021). "Practical medical files sharing scheme based on block chain and decentralized attribute-based encryption", *IEEE Access*, 9, 118771–118781.

[28] R. H. Hylock and X. Zeng (2019). "A blockchain framework for patient centered health records and exchange (healthchain): evaluation and proof-of-concept study", *Journal of Medical Internet Research*, 21(8), e13592.

[29] R. Gupta, A. Nair, S. Tanwar, and N. Kumar (2021). "Blockchain-assisted secure UAV communication in 6G environment: architecture, opportunities, and challenges", *IET Communications*, 15(10), 1352–1367.

[30] L. Lamport, 1981, "Password authentication with insecure communication", *Communications of the ACM*, 24 (11), 770–771.

[31] M. S. Hwang, and L. H. Li, 2000, "A new remote user authentication scheme using smart cards", *IEEE Transactions on Consumer Electronics*, 46 (1), 28–30.

[32] H. Y. Chien, J. K. Jan, and Y. M. Tseng (2002), "An efficient and practical solution to remote authentication smart card", *Computers & Security*, 21(4), 372–375.

[33] W. C. Ku and S. M. Chen (2004), "Weaknesses and improvements of an efficient password based remote user authentication scheme using smart cards", *IEEE Transactions on Consumer Electronics*, 50(1), 204–207.

[34] T. H. Chen and J. C. Huang (2010), "A novel user-participating authentication scheme", *The Journal of Systems and Software*, 83(5), 861–867.

[35] C.-T. Li and C.-C. Lee (2011), "A robust remote user authentication scheme using smart card", *Information Technology and Control*, 40(3), 236–245.

[36] A. J. Choudhury, P. Kumar, M. Sain, H. Lim, and H. J. Lee (2011), "A strong user authentication framework for cloud computing", In *Proceedings of 2011 IEEE Asia-Pacific Services Computing Conference*, Jeju, South Korea, December 12–15, pp. 110–115.

[37] H. S. Kim, S. W. Lee, and K. Y. Yoo (2003), "ID-based password authentication scheme using smart cards and fingerprints", *ACM SIGOPS Operating Systems Review*, 37(2), 32–41, ISSN 0163-5980.

[38] M. Scott (2004), "Cryptanalysis of an ID-based password authentication scheme using smart cards and fingerprints", *ACM SIGOPS Operating Systems Review*, 38(2), 73–75, ISSN:0163-5980.

[39] A. Wenyi Liu, S. Uluagac, and R. Beyah (2014), "MACA: a privacy-preserving multi-factor cloud authentication system utilizing big data", In *IEEE INFOCOM Workshop on Security and Privacy in Big Data*, pp. 518–523.

[40] R. K. Banyal, P. Jain, and V. K. Jain (2013), "Multi factor authentication framework for cloud computing", In *Fifth International Conference on Computational Intelligence, Modelling and Simulation*, pp. 105–110.

[41] T. Bhattasali and K. Saeed (2014). "Two factor remote authentication in healthcare", In *International Conference on Advances in Computing, Communications and Informatics*, pp. 380–386.

[42] M. Masrom, F. Towhidi, and A. H. Lashkari (2009). "Pure and cued recall-based graphical user authentication", In *2009 International Conference on Application of Information and Communication Technologies*, pp.40–45.

[43] A. Castiglione, A. D. Santis, and F. Palmieri (2014). "An efficient and transparent one time authentication protocol with non-interactive key scheduling and update", In *IEEE 28th International Conference on Advanced Information Networking and Applications*, pp. 351–358.
[44] K. Nimmy and M. Sethumadhavan (2014). "Novel mutual authentication protocol for cloud computing using secret sharing and steganography", *Applications of Digital Information and Web Technologies (ICADIWT), Fifth International Conference on the Bangalore*, pp. 101—106.
[45] C. I. Fan, Y. H. Lin, and R. H. Hsu (2006, November) "Remote password authentication scheme with smart cards and biometrics", In *Proceedings of 49th Annual IEEE Global Telecommunications Conference (GLOBECOM)*, San Francisco, CA, USA, pp. 1–5.
[46] S. S. Harakannanavar (2019), "Comprehensive study of biometric authentication systems, challenges and future trends", *International Journal Advanced Networking and Applications*, 10(4), 3958–3968.
[47] K. N. Win, K. Li, J. Chen, P. F. Viger, and K. Li, (2020). "Fingerprint classification and identification algorithms for criminal investigation: a survey", *Future Generation Computer Systems*, 110, 758–771.
[48] K. Venkatachalam, P. Prabu, A. Almutairi, and M. Abouhawwash (2021). Secure biometric authentication with de-duplication on distributed cloud storage. *PeerJ Computer Science.* 7, e569. Published 2021 Jul 30. doi:10.7717/peerj-cs.569.
[49] P. Bhattacharya, S. Tanwar, U. Bodkhe, S. Tyagi, and N. Kumar (2021). "BinDaaS: blockchain-based deep-learning as-a-service in *healthcare 4.0 applications", IEEE Transactions on Network Science and Engineering*, 8 (2), 1242–1255.
[50] K. Greff, R. K. Srivastava, J. Koutník, B. R. Steunebrink, and J. Schmidhuber (2016). "LSTM: a search space odyssey", *IEEE Transactions on Neural Networks and Learning Systems*, 28(10), 2222–2232.
[51] A. N. Jagannatha and H. Yu (2016, November). "Structured prediction models for RNN based sequence labeling in clinical text", In *Proceedings of the Conference on Empirical Methods in Natural Language Processing* (vol. 2016, p. 856). NIH Public Access.
[52] V. K. Prasad, S. Tanwar, and M. Bhavsar (2021). "C2B-SCHMS: cloud computing and bots security for COVID-19 data and healthcare management systems", In *Proceedings of Second International Conference on Computing, Communications, and Cyber-Security* (pp. 787–797). Springer, Singapore.
[53] P. Bhattacharya, P. Mehta, S. Tanwar, M. S. Obaidat, and K.-F. Hsiao (2020). "HeaL: a blockchain-envisioned signcryption scheme for healthcare IoT ecosystems", In *2020 International Conference on Communications, Computing, Cybersecurity, and Informatics (CCCI)*, Sharjah, United Arab Emirates, pp. 1–6.
[54] C. Mistry, U. Thakker, R. Gupta, *et al.* (2021). "MedBlock: an AI-enabled and blockchain-driven medical healthcare system for COVID-19", In *ICC*

2021 – IEEE International Conference on Communications, Montreal, QC, Canada, pp. 1–6.

[55] V. Ponnuramu and L. Tamilselvan (2015) Encryption for massive data storage in cloud. In: L. Jain, H. Behera, J. Mandal, and D. Mohapatra (eds), *Computational Intelligence in Data Mining – Volume 2*, vol. 32, Smart Innovation, Systems and Technologies. Springer, New Delhi. https://doi.org/10.1007/978-81-322-2208-8_4.

Chapter 10

Blockchain technology integrity in health informatics using hyper ledger platform

Rajani Reddy Gorrepati[1], *Prathiba Jonnala*[2], *Sitaramanjaneya Reddy Guntur*[2] *and Do-Hyeun Kim*[3]

Abstract

Blockchain (BC) is an unchangeable ledger that stores data in blocks in a decentralized manner. This technology can benefit any industry that generates and transfers data. The advanced hyper ledger fabric is made up on the BC network and infrastructure/platform, offering decentralized management, robustness, security, and privacy. The hyper ledger explorer is a block-chain module that displays blocks, transactions, as well as any other relevant data that has been stored in the ledger. This state-of-the-art technology has the potential to upend a variety of data-driven industries, including healthcare. This work aims to evaluate the existing technology and proposed the novel frame work using blockchain technology to improve processes in health informatics. However, BC is potentially viable in health informatics applications. Currently, BC is not widely used in the healthcare business. This chapter focuses on the proposed conceptual frameworks and implementation of BC technologies with IoT using hyper ledger on potential applications in healthcare. A novel framework based on the most popular Ethereum and hyper ledger of fabric platforms/frameworks in this domain was proposed. In addition, the present status of IoT-based healthcare models in the domain, as well as an analysis of IoT and BC applications in healthcare is also discussed. IoT and BCT are used to improve medical image quality and storage or transmission for healthcare application challenges and issues. Medical image diagnosis benefits from the current future direction in this domain. Overall, the use of BCT in the healthcare domain is growing at an exponential rate. There are some areas of the healthcare domain where BCT has the potential to have a significant impact.

Keywords: Blockchain; Hyper ledger; Ethereum; Health information; Interoperability; m-health

[1]Department of Computer Science and Engineering, Koneru Lakshmaiah Education Foundation, Guntur, India
[2]Department of Biomedical Engineering, Vignan's Foundation for Science, Technology, Research, India
[3]Department of Computer Engineering, Jeju National University, Republic of Korea

10.1 Introduction

Blockchain technology (BCT) attempted to assess and analyze data from health informatics. In particular, the blockchain (BC) system is an implementation in the healthcare domain. Primarily, physicians could be linked to a BC system, which allows pharmaceutical companies increased access to patient information and secure patient monitoring, increasing the marketability of their products through individual product tracking, and allowing the insurance industry to rely on all client data. In such a field, smart devices continue to face critical security threats/malware from both patients and physicians; any type of attack poses a significant risk and may result in disastrous life risks, data exposure, reputation loss, and financial loss. The clinical trial network transfers various transactions and a private key participant from one host to another, estimates the benefits of the BC to improve clinical traits and patient personal data security and privacy, and stores medical records in BC via mobile apps for patient outreach. Smart IoT objects remotely managed healthcare services by collecting, reporting, and analyzing patient data in real-time. The way diseases are diagnosed and treated has changed dramatically. According to the findings, the most targeted areas for BCT are patient electronic health records and personal health records. In this field, BCT is intended to improve control of access, interoperability, provenance, and data integrity. The multidisciplinary nature of biomedical diagnosis-based applications has piqued the interest of researchers in diagnosis processing and health informatics. The main objective of this work is to examine existing technology and propose conceptual novel frameworks and implementation of block-chain and IoT on the hyper ledger of fabric platforms to improve processes in health informatics.

10.1.1 Organization of chapter

The rest of the chapter is organized as: Section 10.2 summarizes the relevant works, Section 10.3 presents the methodology consisting of system architecture, digital hyper ledger hardware and software tool kits, and smart devices for security, Section 10.4 describes the implementation of BCT using hyper ledger and analysis, Section 10.5 discusses several BCT applications in healthcare, whereas Section 10.6 discusses the benefits of BCT, Section 10.7 discusses challenges, limitations and future perspectives and Section 10.8 concludes the chapter.

10.1.2 Overview of BC technology integrity in health informatics

BC is an emerging technology that has been identified as one of the most promising captures in the last decade by several academic researchers and the health care. BCT was first proposed by Nakamoto *et al.* [1] as an unchangeable decentralized distributed public hyper ledger that is used in a peer-to-peer computer network for safely logging transactions and track assets from many industries without involving a third party. Every data block includes a hash key, a distinct digital fingerprint or unique identity, batches of recent transactions that have been time-stamped, and a hash of the block before it [2]. In this architecture, each block is linked in chronological sequence, and the collection of linked blocks is referred to as a BC. Modifying only one of the blocks in the middle chain is nearly impossible because all of them must be altered concurrently. The data blocks on the BC network are unchangeable in this mechanism. Smart contracts are computer protocols that operate automatically when a company splits from the initial BCT. This smart contract was implemented as a major feature of BC2.0, in addition to cryptocurrencies such as Bitcoin from BC 1.0. However, the recently released BC 3.0 has the potential to significantly improve the throughput of BCT that is one of the most popular technologies at present [3]. Figure 10.1 shows the interaction of BCT with healthcare and health informatics.

The unchangeable data stored in a block of the chain is one of several features of BCT. The healthcare industries were interested in the prospects of many available cases and wanted to increase the handling of medical records and the insurance claim procedure, drug supply management, diagnosis, and radiology. Clinical and advanced biomedical and healthcare data research will be accelerated with the help of hyper ledger [3]. Decentralized administration, an unchangeable audit trail, data source, resilience, and enhanced safety and privacy are some of the fundamental characteristics of BCT that can be achieved.

However, data sharing, data dependability, data safety and stability, and comforts are all critical criteria for human and animal healthcare [4]. These are the most important criteria for healthcare data when dealing with EMR systems. Existing EMR systems are inefficient in addressing these critical concerns for a smooth process because they lack a standard for data security and dependability. It

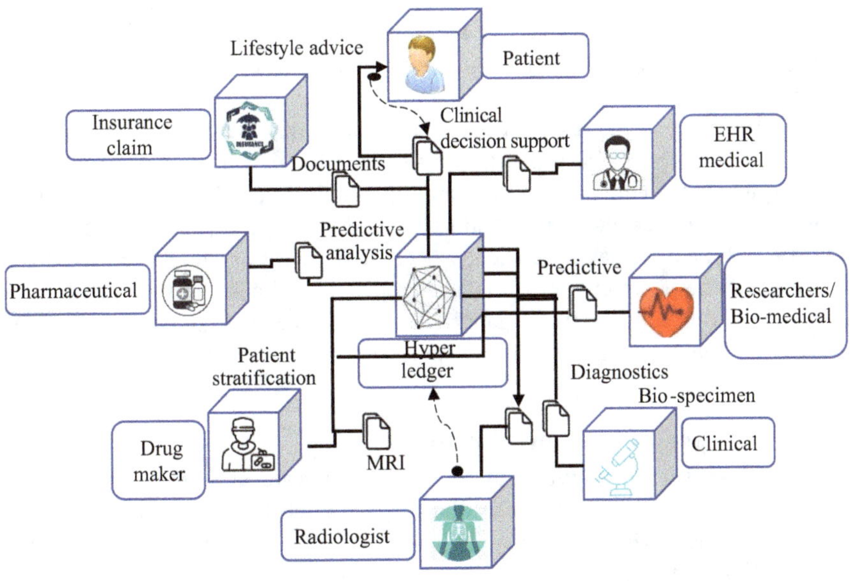

Figure 10.1 Overview of BCT interaction with healthcare

is a safe and dependable platform for improving medical information security and authoritative access [5,6].

BCT is a secure and distributed super ledger that stores and maintains immutable patient health records of online financial transactions. It is also known as distributed hyper ledger technology (DLHT) since it allows two parties to sign medical records using decentralization and cryptographic hashing methods [7,8]. BCT has potential applications; effective solution design is still in its early stages. BCT has the potential to make fundamental changes in a variety of healthcare industries, including supply chain management, patient records, radiology images, and clinical research. Lower transaction costs and complexity increased security, improved transparency, and regulation are among the enhancements.

The research and systematic inflow of healthcare discoveries are deprecatory to ensuring that association rising medical demands are properly fulfilled. According to ongoing studies, the cost of medication development is increasing at a rate of 9% per year [9]. However, the speed, cost, and success rates of drug development and working conditions worsen. As a result, the market success rate for novel medical products entering clinical testing has dropped to an all-time low of 11%. Pfizer, Amgen, and Sanofi have teamed up to figure out the best ways to use blockchain for data security to clinical trial speed and, eventually, reduction of drug development costs. However, the daily generation of massive data from clinical trials, as well as meeting the demands of commercial and regulatory organizations, necessitates the use of systems for managing data that are insufficiently strong enough to analyze and maintain all existing research study data, raising concerns about patient privacy and data integrity.

Several clinical systems have evolved over the last decade intending to streamline trial procedures and data management. Ironically, a large percentage of these function-specific solutions are incompatible with one another based on the most current industry-wide survey, and most clinical researchers struggle to keep tracking the clinical operations to determine the status of documents and procedures [10]. Despite the existence of completely integrated and all-inclusive eClinical systems are not primarily available to major pharmaceutical manufacturers due to high costs. However, approximately researchers from academics, hospitals, and smaller industries conduct 65% of the studies [11]. Regulators that audit research data are having difficulty because there is critical and protected way to assess the complex network of data interchange, offline access to the obtained outcome, and difficult to retrieve data back to its source [12]. As a result, the FDA has identified one of the most serious data issues as a lack of traceability in clinical trials [13].

10.1.2.1 Hyper ledger works in BC

Figure 10.2 shows an overview of BCT for healthcare as implemented on the digital hyper ledger platform that is divided into three layers such as representational state transfer (REST) server, BC network, and IoMT network. The hyper ledger platform includes the hardware required to run the BC. It is made up of network participants referred to as nodes 'A' node can carry out one or more of the

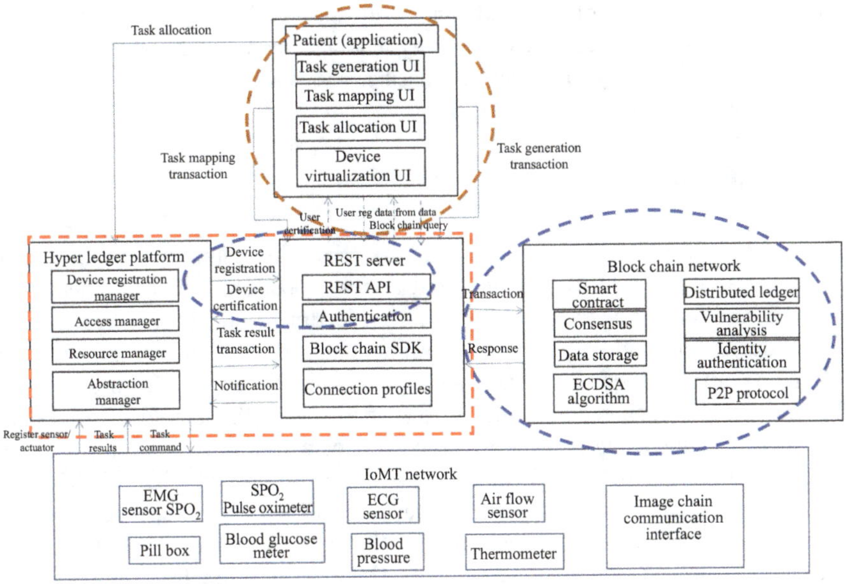

Figure 10.2 Process of BC works in the hyper ledger

functions such as initiation of transactions, validation of transactions and blocks, generation of blocks, and maintenance of blocks of a ledger copy. A ledger participant can serve as a customer, patient, and hospital, as well as a BC manager, or top BC manager.

The storage modules of the hyper ledger platform hold the network's transaction ledger. The REST server provides communication features within a BC or between multiple BC. APIs for client–server communication use REST. To ensure fault tolerance and immutability, the transaction ledger is duplicated among dispersed nodes connected by a companion network. The BC network employs a consensus algorithm to update the ledger and decide on the next block generation.

10.1.2.2 Role of BCT in health informatics

The significance of BCT for development is growing, resulting in higher-quality healthcare facilities supported by cutting-edge technology. The importance of BC in the healthcare sector cannot be overstated. In a patient-centered approach, those priorities are two essential elements as always-accessible services and adequate resources for healthcare. The BC enables a healthcare organization to provide appropriate health care with excellent medical facilities. With the use of this technology, a further tedious and time-consuming task that raises the expense of healthcare may be easily handled, enabling participants in health research projects. Data exchange, interoperability, and protection are the three most important challenges in population health management. However, BCT may resolve these issues such as safety, data exchange, interchange, probity, and real-time updating and access. Data safety is another significant issue, specifically in the areas of individualized medicine and wearable technology. Patients and medical personnel need a safe and convenient way of gathering, sharing, and referencing data via networks and these issues are addressed using BCT.

10.2 Related works

The goal of eClinical solutions is to simplify trial administration and data management. Based on a current industry-led review, most clinical researchers have problems and keep track of the state of the document and procedures in their clinical operations, and the most of these functions and solutions are incompatible [10]. Many different specialty stakeholders are involved in clinical trials, which frequently cross-organizational and national boundaries. Stakeholders in clinical trials work in silos, each using their processes, data formats, software programs, and organizational structures [12]. Zhuang *et al.* [14] proposed a hybrid BC-based test platform that includes patient recruitment as well as clinical report monitoring. All trial venues and sponsors would need to have BC accounts. The clinical trial smart contract would include these accounts. Once the sponsors had established the clinical trial criteria and get the Investigational New Drug (IND) permission, the recruiting data would be transmitted to the smart contract. Other nodes with artificial intelligence (AI) and machine learning components would handle data integrity and analytics. Nugent *et al.* [15] proposes a private BC network-based clinical trial system.

The system was built on a hierarchical structure of regular and trial smart contracts and those are logical and successfully enforced the aspects of trial protocol. Benchoufi *et al.* [16] proposed a BC-based patient consent collection system. A Proof-of-Concept protocol was created using BC, which time-stamped each stage of the patient's permission process is collected, archived, and historicized using cryptographic validation in a safe, transparent, and unfalsifiable way. Shae and Tsai [17] proposed a clinical trial BC platform. The proposed BC platform is intended to research peer-verifiable data integrity, data sharing, and working together with collaboration. The platform will work with national institutes of health to collect clinical trial data using its integrated biomedical informatics system. Smart contracts are being investigated to ensure clinical trial data integrity and to eliminate the possibility of human manipulation. Booz Allen Hamilton created and implemented a pilot platform based on BC for implementation in significant hospitals that are managing data access using the Ethereum platform.

10.3 Methodology

The researcher's study conducted a comprehensive literature review of academic and business articles published in 2008 and 2019 to assess the potential of BC healthcare based on its applicability, and identify implementation challenges. Healthcare information data is a technique of preserving and maintaining a patient's medical record to improve patient care, precisely track illnesses and their causes, use medical information for research and the creation of new medications, and effectively prevent disease. EHRs are often used by hospitals and healthcare professionals to handle patient medical data utilizing a client–server model [18–27]. Azaria *et al.* [28] proposed MedRec, a BC-based data sharing tool that connects with physicians existing data storage systems and that enables them to determine the condition of fair trade by supplying ethnicity tracking of its origin. Dagher *et al.* [29] and Li *et al.* [30] proposed a smart contract-based framework for accessing medical data in an Ethereum-based BC network. The supply chain's transactions are permanently recorded and monitored in a secure and transparent manner using BC technology. Additionally, BC technology can be used to track expenses, labor, waste, and emissions throughout the supply chain [31]. By providing ethnicity and tracing the origin of a fair-trade commodity, distributed ledger technology allows for the determination of its condition.

In this study, we proposed a BC hyper ledger fabric and developed a system for effective data exchange, management of healthcare records, and systematic access control. This proposes a health chain for EHRs, a permission patient data BC, by utilizing a distributed hyper ledger platform. The health chain framework addresses healthcare interoperability issues by preserving patient history through synchronization of records in multiple formats and retrieving data via REST server API via the framework's smart contracts that self-govern and execute by granting access and mechanisms, the patient is fully controlled over their healthcare, as well as access control authorization restrictions. Immutability is also accomplishing by cryptographic data storage, i.e. keeping hash values in the BC and encode medical documents information in the off-chain interplanetary file system database. The health chain is a decentralized architecture in which no one can modify with the documents because the data transfers are connected and adding data to the network by a consensus of stakeholders. Our approach benefits health care by tackling the vast majority of data privacy, safety, interchange, ascendable, assurance, stability, and data integrity concerns. The methodology of the proposed work consists of three main parts: (1) design and implementation of digital hyper ledger hardware and software tool kits using BC technology, (2) design smart devices for security purposes by utilizing BC technology, (3) various algorithms are applied in medical image diagnosis techniques to identify risks in medical images, and (4) test and analyze security risk, threats, and malware/vulnerable attacks in health care smart devices using BC.

10.3.1 Proposed BC healthcare system architecture

The healthcare industry is in a unique position to benefit from BCT. Figure 10.3 shows the proposed model of the healthcare system, in which the healthcare BC keeps a comprehensive, current history of all medical information, converting the EMR, appointments, medication, billing, and healthcare smart devices data that would always be available to a certain user. A self-contained data repository is a

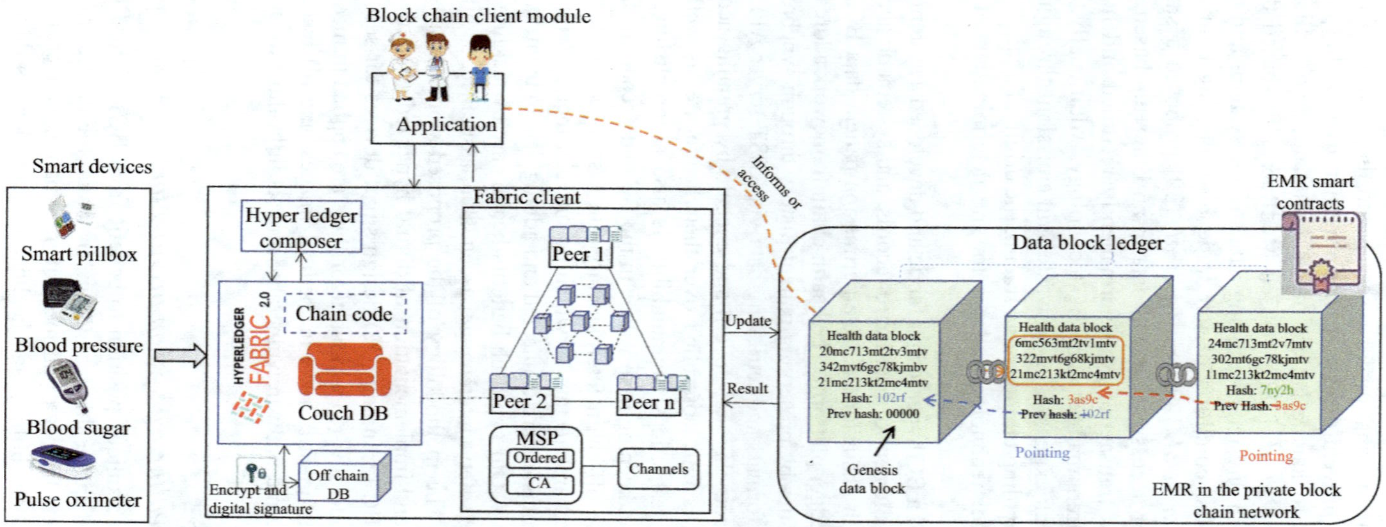

Figure 10.3 Proposed healthcare BC platform for electronic medical records (EMR)

data block ledger, also known as a stored off-chain DB. It would be a useful tool for a wide range of analyses, including health insurance, illness prevention, and research, and would not be limited to hospitals. To be more specific, the health care BC network is comprised of each trusted validated peer, and each peer maintains a copy of the network's ledger to ensure the distributed ledger's consistency. The ledger is divided into two parts: a BC, which stores the immutable, sequential transaction record in blocks, and a data block ledger, which stores various medical data.

The BC is all changes that are recorded in a transaction log to the data block ledger, whereas the information is an off-chain states database containing the current value of a collection of data. Medical equipment can transfer data continuously to linked smart gadgets for healthcare. These data can be used for data analytics, resulting in a range of services, including preventative care and emergency response. With the instant sharing of smart device health data, healthcare professionals can help their clients more quickly and accurately.

Users in various roles, such as the administrator, who can manage all of the hospital's resources, are included in the BC client module. The doctor can evaluate all of the patient's information and prescribe therapy. The pharmacist advises and educates patients on the appropriate use of prescription drugs and substances and their negative effects. The patient can access their medical data from any network peer, where it is stored. Furthermore, the patient has the option of giving other provider in the network access to their medical records.

Smart contact is accomplished by configuring an access control policy, which is distributed across the BC network to protect patient confidentiality and data security, and interactions between the end user's digital signatures to assure the system's security healthcare BC are encrypted. Every healthcare transaction is validated by ensuring the integrity of BC events for creation, access, modification, or updating the hash value of health records. If the record has been tampered with, a hash that differs from the previous hash is generated.

The functioning prototype is based on a permissioned source BC and uses a hyper ledger Composer to generate web apps for a single peer node that is decentralized. This system includes several peer nodes that have been configured to use matching Couch DB as the database and IPFS as the distributed database, a certificate authority, a member, and smart contracts that are used to connect the BC. The healthcare application is communicating with peer nodes, which execute smart contracts to keep the ledger up to date. We connect several hospital organizations to different peers, which are enabled as peer 0 (P0), peer 1 (P1), and peer n (Pn) via Channel N. When a record transaction is executed, the chain code is installed on the peers. To query or change the ledger, healthcare applications connect with peers and employ chain codes.

The transactions can be recorded to hash values within the blocks of BCT, allowing the history of changes contributed to the health chain structure can be tracked. A patient's health record is referred to as a ledger block and is made up primarily of the patient visit record to the hospital for that transaction $V_t(i)$, hash of the prior transaction $V_{P\#}(i)$, and the present transaction hash $V_{\#}(i)$. The total number of patients seen in that block's hospitals may be estimated as

$$V_{\mathrm{Tot}(i)} = V_{t(i)} + V_{P\#(i)} + V_{\#(i)} \tag{10.1}$$

The system architecture of proposed BC-based clinical trial platform as shown in Figure 10.4 is made up of several isolated channels that carry out various studies. Each channel in a clinical trial is made up of four related organizations: home, company, hospital, and contract research organization (CRO). A duplicate of the ledger is kept by each organization to ensure consistency. Smart contract defines the BC network's logic and automates the operations known as transactions, which are submitted by the users. The current ledger states are stored to be the key-value database. Moreover, the system is designed with additional cyber security policies to protect it from attack, damage, or unauthorized access. To implement the BC application, the hyper ledger composer is used to develop a tool and framework. It is compatible with the hyper ledger fabric infrastructure and runtime.

To perform operations on the BC network, participants are provided with controlled access points provided by the web server, as well as additional cyber security policies such as RESTAPI access security policies. The database stores the current state data, which represents the most recent values for all transaction log keys. The ledger viewing module is a web application that lets you see, launch, deploy, or query blocks, transactions and related data, network information, and other pertinent data contained in the ledger.

The administrator is in charge of the proposed BC-based clinical trial platform's user registration and enrollment service scenario. Profiles can be managed by the PI/CRC/CRA (read, update). Subject profiles can be managed by the PI/CRC (read, create, update, delete). The PI/CRC can manage device profiles (pillbox/bgm) (read, create, update). The device configures itself accordingly (read). Subjects' eCRF data is collected by the device (pillbox/bgm) (create).

The device's eCRF data can be viewed by the PI/CRC (read). The PI/CRC can manage eCRF PI data (read, create, update). The PI/CRC can manage eCRF lab test data (read, create, update). The CRA can handle audit queries (read, create, update). The PI/CRC can read and update the PI consults data/lab test data by the CRA audit (read, update).

10.3.2 Digital hyper ledger hardware and software tool kits using BC technology

In a medical BC platform, the main design elements of hyper ledger fabric are used to build and construct the network architecture for clinical research/trial management. Clinical trial-related data contains sensitive information that should be kept private. The peer is the central component of the BC network, hosting a ledger and smart contracts that execute basic ledger actions. A channel is a secure BC that is used to provide secrecy and order services from nodes in charge of organizing transactions into blocks.

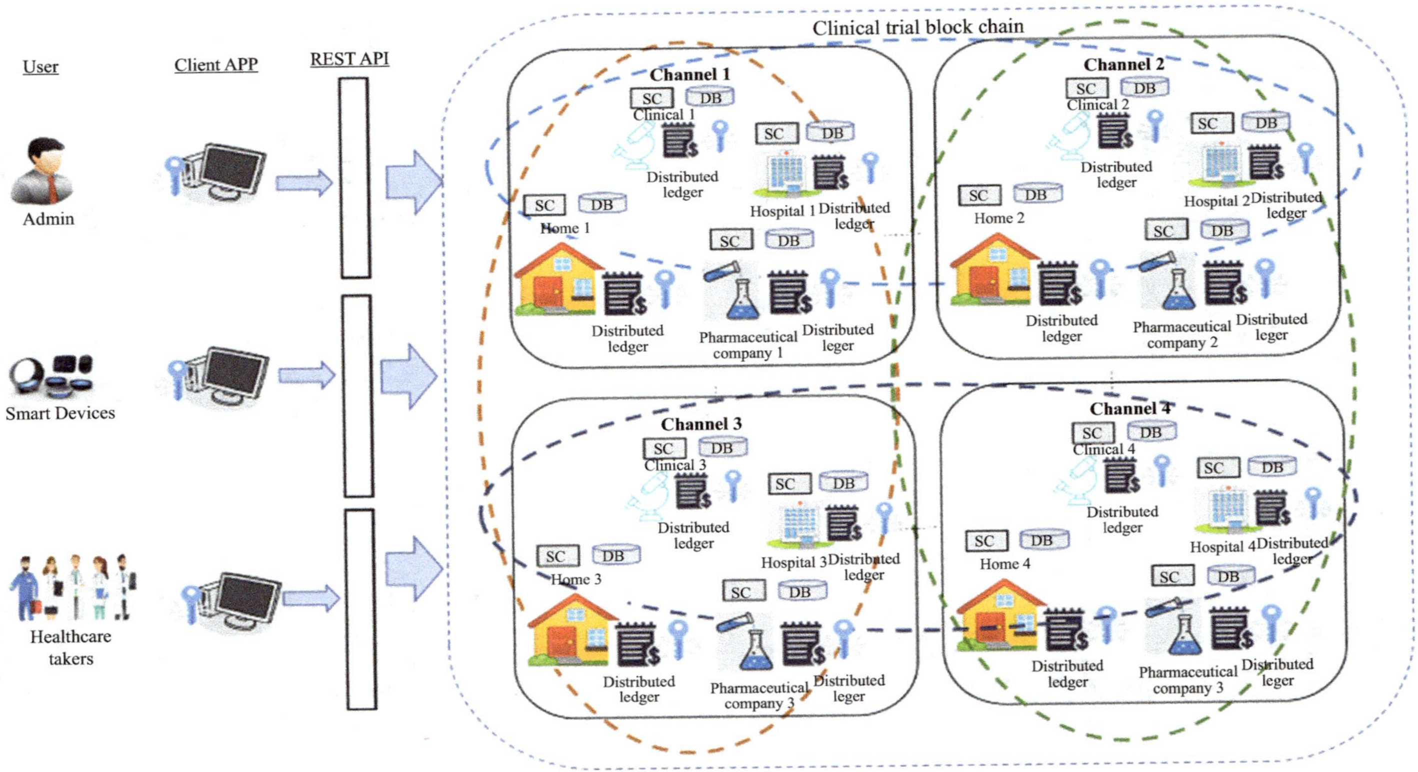

Figure 10.4 System architecture of the proposed BC-based clinical trial platform

The proposed clinical trial BC platform is encapsulated into two channels. Channels are private BC networks in which data is shared only between the network entities within a channel. Therefore, we use channels to represent various clinical trials. Each channel consists of multiple organizations and identities. As illustrated in Figure 10.5, they referred organization to a controlled collection of people, such as a pharmaceutical firm, a house, a hospital, and a contract research organization (CRO), all of which are participants in the clinical trials business network. A single MSP defines an organization's membership list. In this case study, the organizations oversee their members through many MSPs that represent various organizational groupings in different clinical trials. For example, the hospital organization is made up of two MSPs, ORG2.MSP and ORG6.MSP, which represents separate hospitals.

Eight organizations (R1, R2, R3, R4, R5, R6, R7, and R8) are selected and signed an agreement to use the fabric hyper ledger network, and authority to build the network's initial edition has been delegated to each organization R1 to R5, which are corresponded to pharmaceutical companies. Organizations R1, R2, R3, and R4, as well as organizations R5, R6, R7, and R8, require private communication within the overall network. Organization R1, like the other organizations in channel C1, has a client application that allows them to conduct commercial transactions via channel C1. Similarly, organizations in channel C2 have client applications that allow them to conduct business with channel C2. It is also worth noting that one institution, such as organizations R2 and R6, can have multiple client applications. Peer node P1, the other peer nodes to communicate in channel C1, keeps a duplicate of the ledger L1 linked to channel C1. Like the other peer nodes in channel C2, peer node P7 holds a copy of the ledger L2 connected with channel C2. The network is governed by policy rules stated in network configuration (NC), which is administered by R1 and R4. Channel C1 is regulated by channel policy CP1, whereas Channel C2 is governed by channel policy CP2. A network management point is provided by an ordering service and connects with the system via the system channel. Application channels C1 and C2, which are used to organize transactions into delivery blocks, are likewise supported by the ordering service. There is a favorite CA in each of the eight categories.

The proposed BC-based clinical trial platform's smart contract structure leads to the actual assets or intangible assets, services, or property that are recorded in registers and participants are part of a business network. Transactions are how parties exchange assets. Transaction functions are like implementing the transactions defined in the model files of the business network definition. Controlling who has access to which assets in the business network and under what conditions is made possible by access control policies. *Queries*: Queries are used to return information about the current status of the BC; they can hold assets and submit transactions on behalf of members in a professional network. Subject profiles can be managed by the PI/CRC. The PI/CRC can manage and configure device profiles (pillbox/bgm). Subjects' eCRF data is collected by the device (pillbox/bgm). The device's eCRF data can be viewed by the PI/CRC. The PI/CRC can manage eCRF PI consult data. The PI/CRC can manage eCRF lab test data. The CRA is capable of

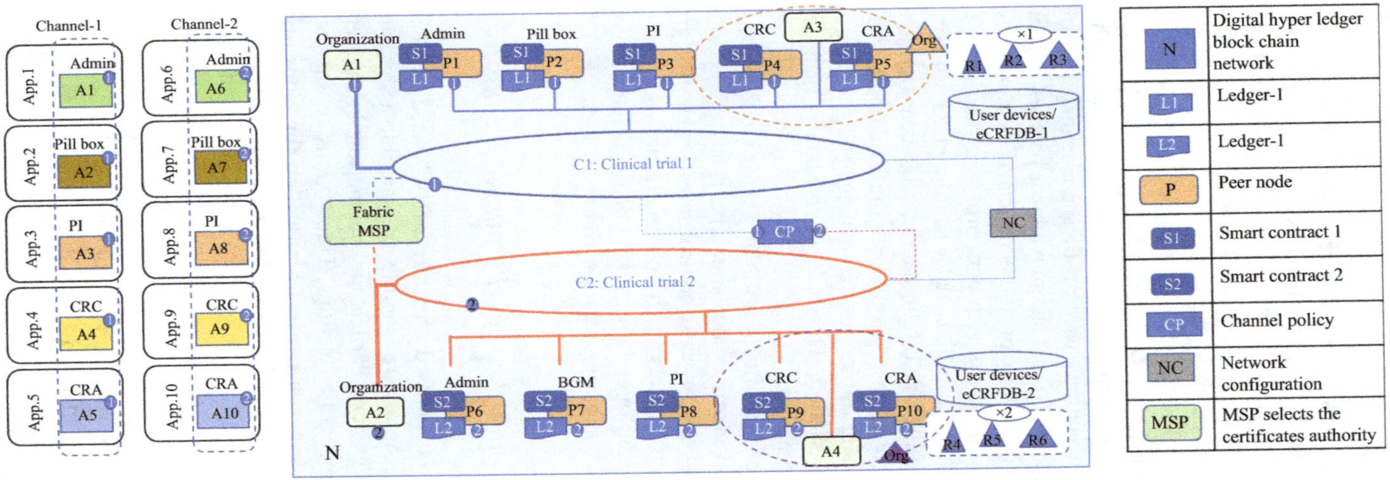

Figure 10.5 Healthcare clinical trials BC network topology

handling audit queries. The PI/CRC can read and update the PI consult data/lab test data by the CRA audit.

In a business network, assets can represent entire data including each asset must have a unique identifier and other assets or participants related to the assets. The clinical trial BC network's asset the ECRF pillbox asset, which is used to collect bio/non-bio data from subjects using the pillbox. Similarly, the eCRF BGM asset stores data gathered from the bgm. The ECRF PI consult asset stores the subject's consult data. The ECRF lab asset records the subject's lab test data and the audit data generated by CRA is recorded in the ECRF CRA audit asset.

Transactions are the means for participants to interact with assets. The clinical trial BC network transactions user profiles are managed using the user profile management transaction. To manage the subject profiles, the subject management transaction is used. The pillbox profile is managed using the device pillbox profile management transaction. The device bgm profile management transaction is similar. To create or read eCRF pillbox data, use the eCRF pillbox data management transaction. Similarly, the eCRF bgm data management transaction is also used to create or read eCRF bgm data. To manage the PI consult data, use the eCRF PI consult data management transaction. To manage eCRF lab data, use the eCRF lab data management transaction. The audit query data is managed using the eCRF CRA audit query transaction.

We construct cyber security rules in an access control list, which allows us to control domain model elements declaratively. A set of access control rules may exist in corporate networks. Access control rules enable fine-grained control over who has access to what assets in the corporate network and when. The access control rule supports four operations: read, update, delete, and read. The CRC, for example, has complete subject operation rights, whereas the CRA can only view the subject profile.

10.3.3 Smart devices for security purpose using BC technology

The proposed smart customised structure of BC-based clinical trial platform is shown in Figure 10.6. Assets are tangible or intangible assets, services, or property that are recorded in registers. Participants belong to a business network. They can possess assets and conduct transactions. Transactions are the mechanism by which assets are exchanged between parties. Transaction functions implement the transactions defined in the model files of the business network definition.

Access control rules: Access control rules provide for fine-grained control over who has access to which corporate network assets. *Queries*: Queries are used to obtain information about the BC's current status.

Participants form a professional network. They can own assets and make transactions. The PI/FDA can manage subject and device profiles (pillbox/bgm). The device configures itself accordingly. Subjects' eCRF data is collected by the device (pillbox/bgm). The device's eCRF data can be viewed by the PI/FDA. eCRF PI consult data and lab test data can be managed by the PI/FDA. The CRA is

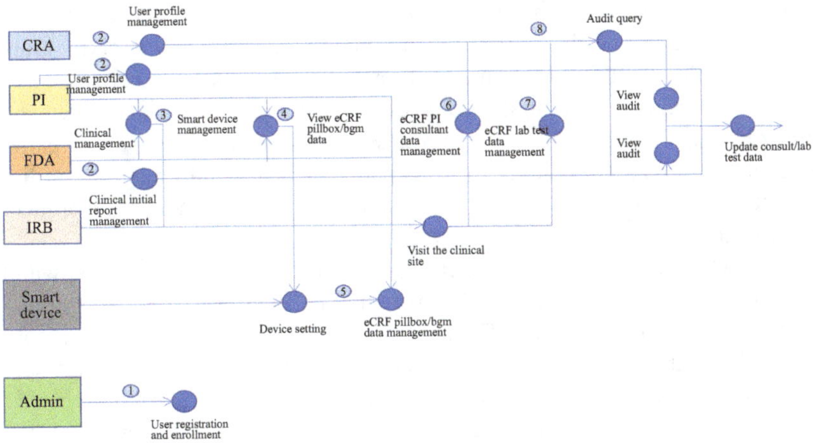

Figure 10.6 Implementation of a BC cyber security risk scenario on a healthcare clinical trial platform

capable of handling audit queries. According to the CRA audit, the PI/FDA can read and update the PI consult data/lab test data. Assets, which can be products, services, or property, are kept in registries. In a business network, assets can represent almost anything, and each asset must have a unique identifier. Other assets or participants may be related to the assets i.e., the clinical trial BC network's assets. The ECRF Pillbox asset is used to collect bio/non-bio data from subjects using the pillbox. Similarly, the eCRF BGM asset stores data gathered from the bgm. The ECRF PI consult asset saves the subject's consult data. The ECRF LAB asset records the subject's lab test data. The audit data generated by the ECRF CRA audit asset is recorded.

Transactions allow participants to interact with assets. The clinical trial BC network transactions on the user profile management transaction are used to manage user profiles. The subject management transaction is used to manage the subject profiles. The device pillbox profile management transaction is used to manage the pillbox profiles. The transaction for managing the device's bgm profile is similar. To create or read eCRF pillbox data, use the eCRF pillbox data management transaction.

To create or read eCRF bgm data, use the eCRF bgm data management transaction. Use the eCRF PI data operation transaction to manage the PI consult data. Use the eCRF lab data management transaction to manage eCRF lab data. The CRA eCRF creates cyber security rules in the form of an access control list, which provides critical accessibility control over domain model elements. A set of access control rules may exist in business networks. Access control rules enable fine-grained control over who has access to which corporate network assets and under what circumstances. Figure 10.6 describes the specific role of each access control rule. This figure displays some of the sample rules defined in the clinical trial BC network. The access control rule supports four operation types: read, update, delete, and read. The FDA has complete operational authorization on the topic, but the CRA can only examine the subject profile. Figure 10.7 depicts the cryptographic processes in the health chain, which are used to ensure data integrity and confidentiality and encrypt data in the IPFS off-chain storage system using specialized public-key cryptography. The IPFS provides a comprehensive view of the patient–physician relationship for accessing the patient's health information. A digital certificate verifies who owns a public key and issues certified network members to communications.

This method allows all parties to rely on the digital signature generated with the granted certificates' status and validity period. The physician seeks access to the patient's health record in the IPFS. Depending on the role and rule-based access control that has been authorized, the patient allows or grants requests from accepted users.

A client-side application is referred to as a system in this framework. Instead of transmitting the complete patient data, the system constructs a composite representation depending on the request. To gain data access, the system produces a session key, encrypts the combined viewpoint, and saves it in IPFS. Additionally, the system provides the physician with the encrypted session key composite view and supplies the patient with the encrypted session key. The session key is decrypted by the physician and inserts the amended record's composite view. Following that, the physician then uses the session key to encrypt the revised record in order to end the instance and delivers it to IPFS. The system notifies the patient of any modifications to the patient's record. The session key is decoded and used to update the composite view, and the encrypted medical record is decrypted using the patient's IPFS private key. The system then stores the modifications to the original record, encrypts it using the patient's public key, and uploads it to IPFS. The composite view and the session key both expire with a session.

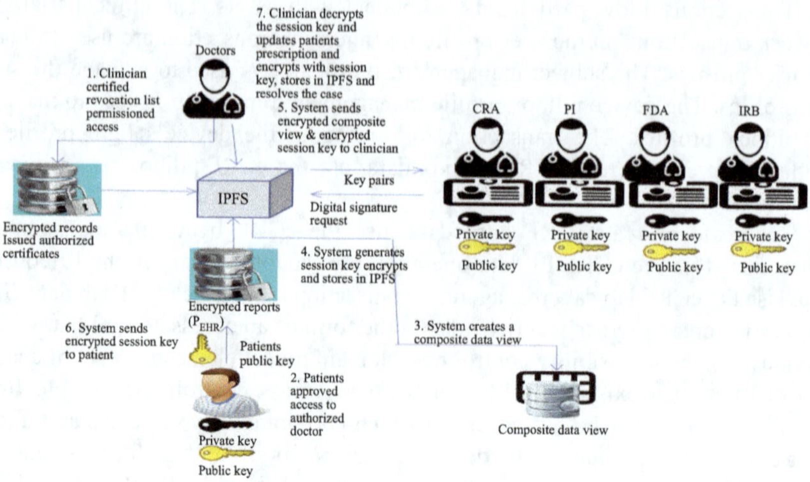

Figure 10.7 Cryptographic processes in the health chain

Figure 10.8 depicts the interaction of multiple stakeholders as well as various events that occur as a result of the smart contract's sequence of functions. Each network participant is assigned a hyper ledger Fabric Address, which allows them to communicate with one another by invoking smart contract operations. By completing the function, the Clinical Trail Sponsor is granted authorization to submit for an IND approval, which includes elements such as IPFS, trail phase scheduled, and estimated trial duration. The IPFS application is simply a hashed application document that the clinical trial sponsor enters into the system, from which the original file can be retrieved via IPFS.

A medical record is gathered using a variety of methods, including the patient using public key encryption to upload his data to IPFS. For the healthcare provider or interested party with record access from IPFS for a specific session, the system provides an automated composite data view that needs the session key for encryption. The encrypted public key of the physician will serve as the session key for secure transmission [32]. The physician comes and updates the medical records with the session key before saving them in IPFS. The patient is informed through the system when changes are made to the medical record is decrypted. After that, the patient encodes the modified record with the patient's public key and commits to modifications.

Figure 10.9 depicts physician and patient access permission retrieval of medical records. The clinician's public key will be used to encrypt IPFS' private key for secure transfer. When the doctor arrives, they make changes to the encrypts medical records which are using the session key before saving them to IPFS. When changes are made to a medical record, the system notifies a patient, and the new medical record is decrypted.

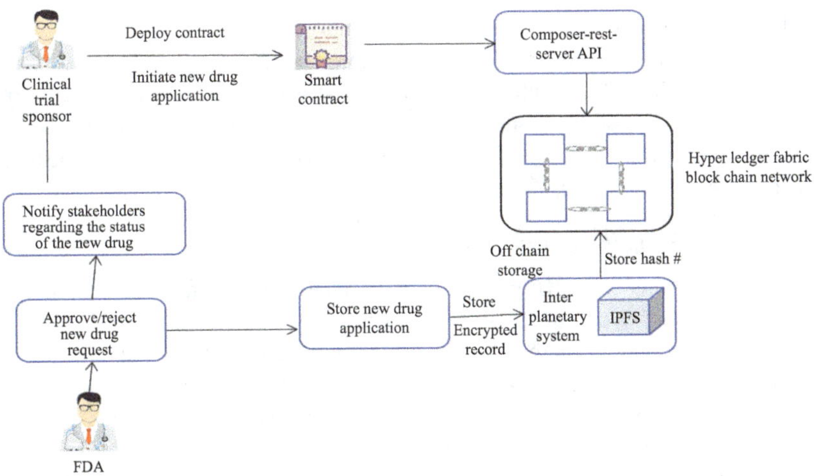

Figure 10.8 Clinical approval of physicians or drugs for healthcare

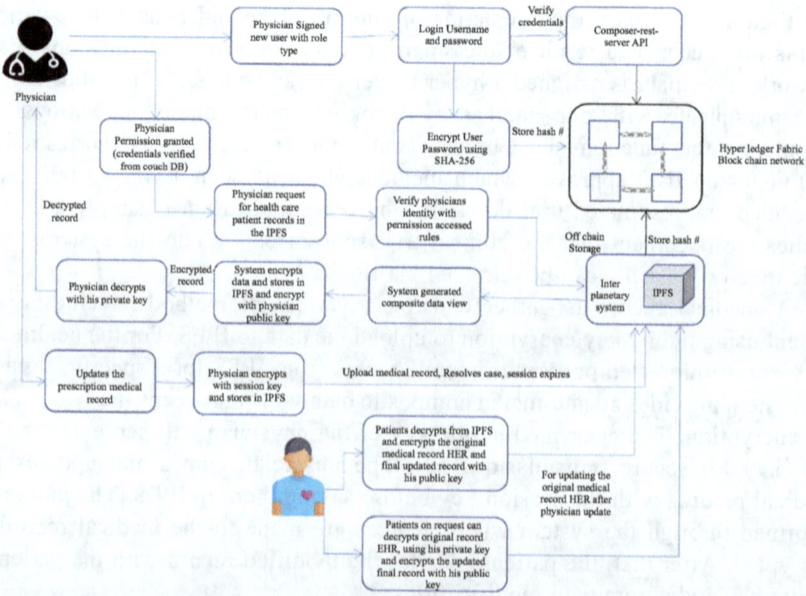

Figure 10.9 Physician and patient access permission retrieval of medical records to health chain

10.4 Implementation BC technology using hyper ledger

10.4.1 User profile management interface

The administrator can manage the identities of participants. The administrator has the authority to grant or revoke an identity for a specific user. The issued identity is a credential file that is stored locally. The administrator has control over the wallet, which stores each user's identity card. The admin can import or delete the identity card from the wallet. The identity information that is enrolled in the BC network is represented by the wallet dashboard. Figure 10.10 depicts the web-based user profile administration interface. The current identity that can interact with the BC network is represented by the default value.

10.4.2 eCRF pillbox data management

Figure 10.11 depicts an eCRF pillbox data management interface on the web. The eCRF PI consult data can be created and updated by both the PI and the CRC. A data lock field indicates whether the consulting data is locked. The CRA has access to the eCRF PI consult data. The CRC's consult data need to be confirmed by the PI. Figure 10.12 depicts the eCRF PI consult data management interface on the web. The consult data created by the PI, on the other hand, is confirmed at the same time it is created. Following confirmation, the data lock field is set to true, indicating that the consult data has been locked and cannot be edited.

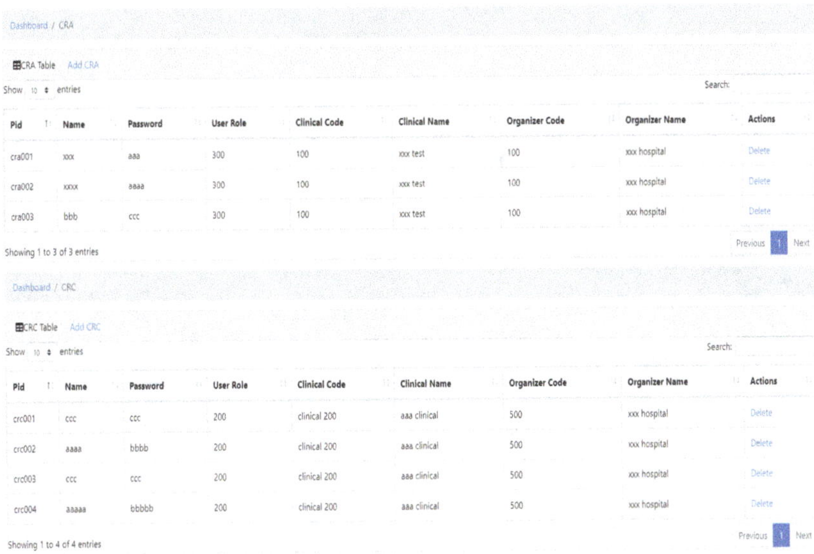

Figure 10.10 User profile management interface on the web

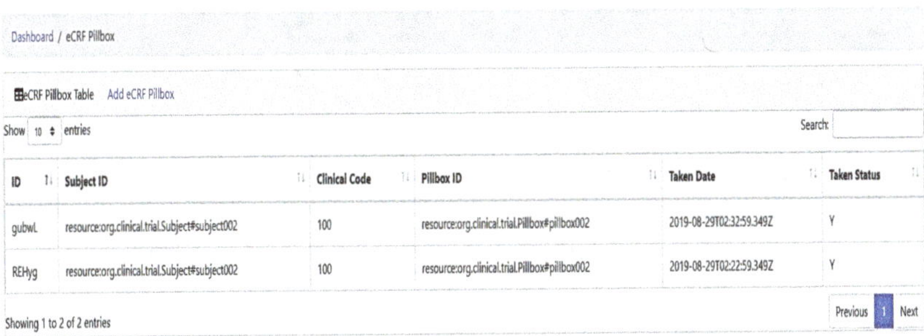

Figure 10.11 eCRF pillbox data management interface on a web

10.4.3 eCRF PI consult data management

Figure 10.13 depicts the web-based data management interface for eCRF Lab. The eCRF LAB test data can be created and updated by both the PI and the CRC. A data lock field indicates whether or not the LAB data is locked. The CRA has access to the eCRF LAB data. The PI needed to confirm the LAB data generated by the CRC. However, the PI's LAB data is confirmed at the same time it is created. Following confirmation, the data lock field is set to true, indicating that the consult data has been locked and cannot be edited.

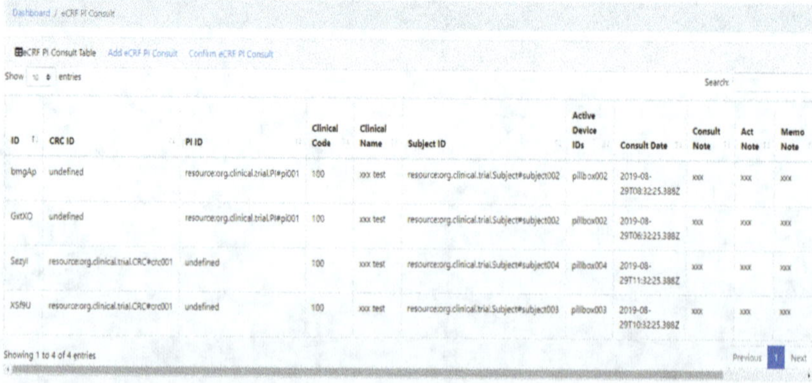

Figure 10.12 eCRF PI consult data management interface in the web

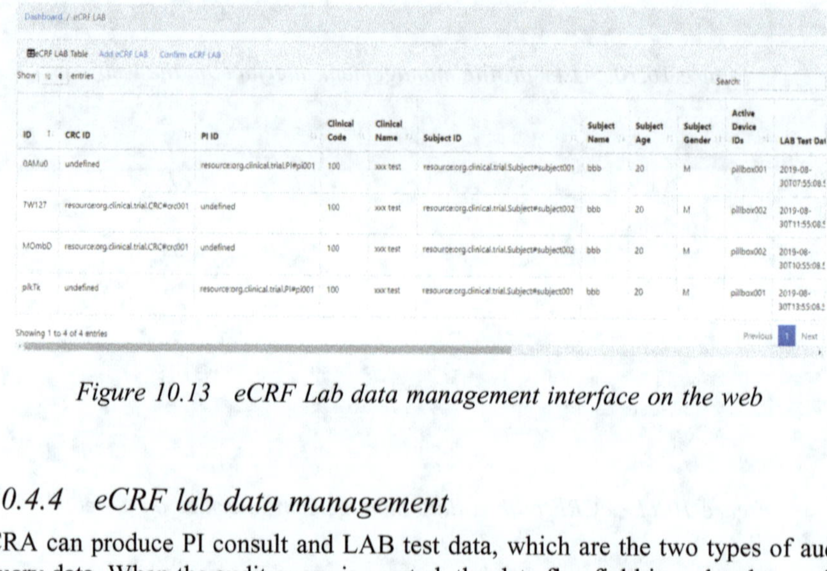

Figure 10.13 eCRF Lab data management interface on the web

10.4.4 eCRF lab data management

CRA can produce PI consult and LAB test data, which are the two types of audit query data. When the audit query is created, the data flag field is updated to notify the specified PI/CRC (PI Consult Flag or LAB Data Flag). The PI/CRC then updates the PI consults data or LAB test data, according to the active note. After completing the CRA-requested revision, the PI/CRC can update the audit query.

10.5 BCT applications in healthcare

The benefits of BCT are critical for medical research and healthcare applications; healthcare has emerged as one of the most important BC distributed ledger technology application areas in development. BC is a distributed ledger used to store healthcare-related data for stakeholders to share, trade, analyze, record, and validate.

When BC is used in healthcare, it creates a health information exchange (HIE) infrastructure, which allows for transactions between patients, providers, payers, and other organizations. These applications are further classified based on their primary goals for utilizing the data stored in the BC, which are improved medical record management, accelerated clinical/medical research sections, and enhanced insurance claim process. The advanced medical research/healthcare data ledger section displays applications that extend beyond HIE. For each type of application, analyze the large cases in greater detail and keep a record of the benefits of implementing BC technology.

10.5.1 Improved management of medical records

Many ongoing studies and projects are focusing on using BC to exchange patient care data being collected to improve medical record administration. Several well-known companies are also using BCT to store and manage healthcare data and records. The benefits of using blockchain to improve the insurance claim process are summarized in Table 10.1.

10.5.2 Insurance claim process improvements

The purpose of the insurance claim process is to validate improvements in claim transactions to alternative payment methods, smart contract-based automated claims for healthcare interoperability resources, and a smart health profile to manage Medicare beneficiaries' continual exit and re-entry due to eligibility

Table 10.1 Benefits and applications of BCT for the improvement of medical record management

Block chain: key benefit	**Applications to improvement of medical record management**
Distribution administration	*Records management*: The patient assumes the platform because they own and control to enter medical information. This removes any drawbacks in obtained medical records or carry them to a different healthcare provider [33]
Permanent audit trail	*Permanent files*: The data is stored in a private block chain cloud. Internally and natively, block chain may assure that no one, even physicians and patients, may alter medical data [34]
Data source	Records are verified and signed by the provider, allowing the validity of the records to be confirmed [35]
Robustness/accessibility	Patient records are less likely to be lost: Data is stored on a distributed network; no single organization can be vulnerabilities or hacked to access a substantial patient records [33]
Safety/privacy	*Improved medical record security*: Data is encrypted by the BC, and only the patient's private key can be used to decrypt it. There is no workable way to access patient data, even if a malicious actor infiltrates the network [33]

changes to support health plan claims. The benefits of using BC to improve the insurance claim process are summarized in Table 10.2.

10.5.3 Accelerated clinical/medical research

Several studies, including MedRec, data lake, health bank, and BC-based data sharing networks, suggest that clinical data, i.e. clinical and medical research investigations, can be used more quickly in the secondary stage. The model chain also used BC to improve the security and spanning several institutions, the scalability of distributed privacy-protecting healthcare predictive modeling. Table 10.3 summarizes the benefits and applications of using BCT to enhance clinical research.

10.5.4 Advanced medical research/healthcare data ledger

Many researchers have proposed that BC and transaction ledgers be used to archive data on biomarkers, clinical trial data, patient consent, provider/patient directories, treatment plans, genomics and precision medicine data, patient-related outcomes data, and other sorts of health care data (take, for instance, HIE). The advantages and applications of using BC to improve medical research/healthcare data ledgers are summarized in Table 10.4.

Table 10.2 Benefits and applications of BCT to improve the insurance claim process

Block chain: key benefit	**Applications to improve the insurance claim process**
Distribution administration	*Claim processing in real-time*: BCT able to eliminate intermediaries from a process that differentiate from other technologies. The solution will be able to promote online claims referring by substituting health plan with clear BCT [37]
Permanent audit trail	*Enhance fraud identification and assert auditing*: Audit facilitation and enhanced fraud identification are advantages for insurers as well as independent payers [36].
Data source	*Verifiable credentials for assert modification*: The many documents that patients, clinicians, and Medicaid managers use to make decisions are dispersed. The verification requirements are exactly the drawbacks that a distributed BC solution must overcome [38]
Robustness/accessibility	*Improved claim qualification records*: Patient data is accessible via many archives based on the BC virtual ledger, which benefits providers, services, and medical goods [37]
Safety/privacy	*Enhanced the safety of medical insurance information*: Benefit from a lower risk of financial information due to the BC mechanism [37]

Table 10.3 Benefits and applications of BCT in clinical and medical research

Block chain: key benefit	Applications of clinical and medical research
Distribution administration	*Enhance healthcare data exchange and analysis without attribution*: BC is a distributed, person-to-person, non-moderate system by design. While working with others for data exchange and analysis, each institution may keep total control over its computational resources [39]
Permanent audit trail	*Patient-generated data that is trackable and time stamped*: This notion may be further customized with BC, allowing researchers to track personal patient-generated health data in the study process with a time stamp [40,41]
Data source	*Source and evidence for research data*: The MedRec BC-based model type for EHR and research data includes decisive provenance properties [42,43]
Robustness/accessibility	*Higher healthcare data accessibility*: BC would provide continuous accessibility of real-time data to enhance healthcare coordination in medical emergency circumstances. Online data can also permit academics and public health workers to swiftly identify, isolate, and treat environmental factors that affect public health. Early detection and containment of epidemics is possible [44]
Safety/privacy	*Healthcare data exchange that is secure and protects patients' privacy*: Using the health block chain, millions of people, medical professionals and corporations, and healthcare researchers may securely and privately exchange massive sum of genetic, dietary, lifestyle, environmental, and health data [44]

Table 10.4 Block chain advantages and use cases to develop health data ledger and medical research

Block chain: key benefit	Develop data ledger in medical research
Distribution administration	*Distributed data backbone*: BC acts as digital health's backbone, integrating from healthcare technologies and the EMR to create a pool of information to authorized users, such as physicians and patients. There is no one entity that holds or has sole jurisdiction over any of the data that is stored in a distributed manner [40,45–47]
Permanent audit trail	*Immutable clinical research protocol log*: Recently, BC technology was shown to generate an unchangeable record of each step in process and that can be extended to fundamental and test model. The data, which is time-stamped and updated regularly, is accessible to all members of the network. It is virtually tamper-proof any changes, that predetermined data analysis [35,36]

(Continues)

Table 10.4 (Continued)

Block chain: key benefit	**Develop data ledger in medical research**
Data source	*Allow the transfer of ownership and the original producer in the particular supply chain*: The origin of the product and its components is discovered using BC, and any transfer of ownership is made explicit and accessible to all in each case. Forged, low-quality products are searched and identified
Robustness/ accessibility	*Increase robustness of pharmaceutical supply chain counterfeit drug prevention/detection systems*: Existing systems still have a central authority that can be hacked and documents can be faked. If the supply and dispensation system can be upgraded with BC-enabled anti-fiddling ability during manufacture, it has the potential to eradicate medication counterfeiting [48,49]
Security/privacy	*Increased patient trust in concurrent recording systems*: Patients can enter agreement declaration at any time during their treatment, knowing that the BC will keep them secure [50,51]

10.6 Advantages of block chain technology

10.6.1 Decentralized management

Electronic health records are more accessible, accurate, secure, and less expensive to create and manage (EHRs). BCT has several advantages in terms of safety, data protection, and ability to connect various systems used by user and research facilities. Medical researchers must share their findings, collaborate, and obtain permission to acquire and access data. Because BC technology does not store all data in one location, it is less vulnerable to external/internal threats.

10.6.2 Immutable audit trail

BCT provides the advantage of centralization without storing all data in one location, making it less vulnerable to external/internal threats [52,53].

10.6.3 Data provenance

The system administrator can modify digital assets while on the BC using cryptographic protocols, and the assets' origins are traceable, making them suitable for managing critical digital assets.

10.6.4 Robustness and availability

BC is based on distributed technology and is suitable for achieving high levels of data redundancy while preserving and continuously making records available.

10.6.5 Security and privacy

Protecting the data of a healthcare system from other threats and ensuring that supply chains are more dependable, easier to manage, and less expensive to run. Security and privacy are improved by using cryptographic algorithms and BC. To increase privacy, user addresses are additionally generated using a cryptographic hash function specified in the processing standards [54,55].

10.7 Challenges, limitations, and future perspectives

The design of a healthcare BC application must overcome various technological issues before it can be implemented. The main difficulties in developing a healthcare BC application are data protection and access to sensitive patient information and the implementation of cyber security to protect medical data from vulnerabilities. Electronic patient records systems have restrictions and obstacles in terms of BC technology and networking, such as accuracy. The cybersecurity BCT brings many challenges for intercommunication with the contribution of the rapidly growing medical field. The significant challenge in the IoMT domain web-based application, which contains secure user enrollment and authentication, medication data collection, case report form (CRF) management, and audit query, is also used to facilitate communication with the BC network. We demonstrated how smart contracts running on the hyper ledger Fabric BC may be utilized to increase data management transparency in clinical trials, security risk, threats, and attack in IoMT for healthcare clinical and diagnosis traits based on BC.

10.8 Conclusion

BC technology provides several advantages in terms of safety, data security, and the capacity to connect different systems used by the manufacturers, and research locations. This system benefits from centralization without keeping all data in a single area, making it less vulnerable to external/internal threats. In this chapter, a proposed BC-based framework for increasing data openness in clinical trials system was built on a permissioned network that ensures only authorized users have access. The BC records the entire clinical study life cycle implemented based on a clinical trial case study on hyper ledger fabric in the proof-of-concept with an experimental study. To facilitate interaction with the BC network, a web-based application with secure user enrollment and authentication, medication data collection, case report form management, and audit query are also implemented. The proposed system demonstrates the hyper ledger fabric of BC smart contracts that can be used to increase clinical trial data management and transparency.

References

[1] Nakamoto S. Bitcoin: a peer-to-peer electronic cash system. *Decentralized Business Review*, 2008, 21260.

[2] Gupta M. *Blockchain for Dummies* (2nd IBM Limited Edition). Hoboken (NJ): John Wiley & Sons, 2018, pp. 33–37.

[3] Kuo T.T., Kim H.E., and Machado O.L. Blockchain distributed ledger technologies for biomedical and health care applications. *J. Am. Med. Inform. Assoc.,* 2017;24(6):1211–20.

[4] Hussien H.M., Yasin S.M, Udzir S.N.I., Zaidan A.A., and Zaidan B.B. A systematic review for enabling of developing a blockchain technology in healthcare application: taxonomy, substantially analysis, motivations, challenges recommendation and future direction. *J. Med Syst.*, 2019;43(10):320.

[5] AI-Megrim S., Alabaman S., Altoaimy L., Alsalamah H., Soltanisehat L, and Almutairi E. Isopentane, Blockchain use cases in digital sectors: a review of the literature. In *2022 IEEE International Conferences on Internet of Things (iThings) and IEEE Green Computing & Communications (GreenCom) and IEEE Cyber, Physical & Social Computing (CPSCom) and IEEE Smart Data (SmartData)*, 2018, pp. 1417–1424.

[6] Tanwar S., Parekh K., and Evans R. Blockchain-based electronic healthcare record system for healthcare 4.0 applications. *J. Inf. Secur. Appl.*, 2020; 50:102407.

[7] Khan P., Byun Y.C., and Park N. A data verification system for CCTV surveillance cameras using blockchain technology in smart cities. *Electronics,* 2020;9(3):484.

[8] Jamil F., Ahmad S., Iqbal N., and Kim D.H. Towards a remote monitoring of patient vital signs based on IoT-based blockchain integrity management platforms in smart hospitals. *Sensors*, 2020;20(8):2195.

[9] Getz K. Insights into outsourcing practices and oversight effectiveness. *Appl. Clin. Trials,* 2018;1:27–12.

[10] Veeva. *Unified Clinical Operations Survey*. Pleasanton: Veeva Systems; 2018.

[11] Trial rove. *Clinical Trial Landscape: 2015 Trial Completions*. London: Informa UK Ltd.; 2016.

[12] Valkenhoef VG., Tervonen T., de Brock B., and Hillege H. Deficiencies in the transfer and availability of clinical trials evidence: a review of existing systems and standards. *BMC Med. Inf. Dec. Mak.* 2012; 12(1):95.

[13] Hume S., Sarnikar S., Becnel L., and Bennett D. Visualizing and validating metadata traceability within the CDISC standards. In *AMIA Summits on Translational Science Proceedings*, vol. 158, 2017.

[14] Zhuang Y., Lincoln S., Atkins A., Shyu CR., Shae Z., and Hsu C. A hybrid blockchain design for patient recruitments and persistent monitoring for clinical trials. In *IEEE Standards Association: Blockchain for Clinical Trials*, 2018, pp. 1–7.

[15] Nugent T., Upton D., and Cimpoesu M. Improving data transparency in clinical trials using block chain smart contracts. *F1000 Res.,* 2016;5:2541.

[16] Benchoufi M., Raphael P., and Philippe R. Block chain protocols in clinical trials: transparency and traceability of consent. *F1000 Res.,* 2017; 23(6):66.

[17] Shae Z. and Tsai J.P. On the design of a block chain platform for clinical trial and precision medicine. In *IEEE 37th International Conference on Distributed Computing Systems (ICDCS)*, Atlanta, GA, 2017, pp. 1972–1980.

[18] Rind D.M., Kohane I.S., Szolovits P., Safran C., Chueh H.C., and Barnett G. O. Maintaining the confidentiality of medical records shared over the Internet and the world wide web. *Ann Internal Med.*, 1997;127(2):138–141.

[19] Schoenberg R. and Safran C. Internet based repository of medical records that retains patient confidentiality. *Bmj* 2000;321(7270):1199–1203.

[20] Uckert F., Gorz M., Ataian M., and Prokosch H.U. Akteonline—an electronic healthcare record as a medium for information and communication. *Stud. Health Technol. Inform.*, 2002;90:293–297.

[21] Grant R.W., Wald J.S., Poon E.G., *et al.* Design and implementation of a web-based patient portal linked to an ambulatory care electronic health record: patient gateway for diabetes collaborative care. *Diabetes Technol. Therapeutics*, 2006;8(5):576–586.

[22] Gritzalis D. and Lambrinoudakis C. A security architecture for interconnecting health information systems. *Int. J. Med. Inform.*, 2004;73 (3):305–309.

[23] Bonacina S., Marceglia S., Bertoldi M., and Pinciroli F. A web-based system for family health record. In *Conference of the IEEE Engineering in Medicine and Biology Society*, 2007, pp. 3652–3656.

[24] Ibrahim L., Asim M., and Petkovic M. Secure management of personal health records by applying attribute-based encryption. In *6th International Workshop on Wearable, Micro, and Nano Technologies for Personalized Health*, 2009, pp. 71–74.

[25] Guntur SR., Gorrepati RR., and Dirisala VR. Internet of medical things remote healthcare and health monitoring perspective. In *Medical Big Data and Internet of Medical Things: Advances, Challenges, and Applications.* Boca Raton (FL): CRC Press Taylor & Francis Group, 2018.

[26] Guntur S.R., Gorrepati R.R., and Dirisala V.R. Robotics in healthcare: an Internet of Medical Robotic Things (IoMRT) perspective. In *Machine Learning in Biosignal Analysis and Diagnosis Imaging*. Amsterdam: Elsevier, 2019.

[27] Gorrepati R.R. and Guntur S.R. DroneMap: an IoT network security in Internet of Drones. In *Development and Future of Internet of Drones (IoD): Insights, Trends and Road Ahead.* New York, NY: Springer, 2021, pp. 251–268.

[28] Azaria A., Ekblaw A., Vieria T., and Lippman A. MedRec: using block chain for medical data access and permission management. In *International Conference on Open and Big Data (OBD)*. New York, NY: Springer, 2016, pp. 25–30.

[29] Dagher G.G., Mohler J., Milojkovic M., and Marella P.B. Ancile: privacy-preserving framework for access control and interoperability of electronic health records using block chain technology. *Sustain. Cities Soc.*, 2018;39: 283–297.

[30] Li H., Zhu L., Shen M., Gao F., Tao X., and Liu S. Blockchain-based data preservation system for medical data. *J. Med. Syst.*, 2018;42(8):141.

[31] Bocek T., Rodrigues BB., Strasser T., and Stiller B. Block chains everywhere-a use-case of block chains in the pharma supply-chain. In *IFIP/IEEE Symposium on Integrated Network and Service Management (IM)*, Lisbon, Portugal, 8–12 May 2017, pp. 772–777.

[32] Ivan D. Moving toward a block chain-based method for the secure storage of patient records. In *ONC/NIST Use of Block Chain for Healthcare and Research Work Shop*. Gaithersburg, MD: ONC/NIST, 2016.

[33] Yue X., Wang H., Jin D., Li M., and Jiang W. Healthcare data gateways: found healthcare intelligence on block chain with novel privacy risk control. *J. Med. Syst.* 2016;40(10):218.

[34] Blough D., Ahamad M., Liu L., and Chopra P. Med Vault: ensuring security and privacy for electronic medical records. In *NSF Cyber Trust Principal Investigators Meeting, 2008*, 2016.

[35] Attili S., Ladwa S.K., Sharma U., and Trenkle A.F. Blockchain: the chain of trust and its potential to transform healthcare – our point of view. In *ONC/NIST Use of Blockchain for Healthcare and Research Workshop*. Gaithersburg, MD: ONC/NIST, 2016.

[36] Use of Blockchain for Healthcare and Research Workshop. Gaithersburg, MD: ONC/NIST, 2016.

[37] Culver K. Blockchain technologies: a whitepaper discussing how the claims process can be improved. In *ONC/NIST Use of Blockchain for Healthcare and Research Workshop*. Gaithersburg, MD: ONC/NIST, 2016.

[38] Vian K., Voto A., and Haynes-Sanstead K. A blockchain profile for medicaid applicants and recipients. In *ONC/NIST Use of Blockchain for Healthcare and Research Workshop*. Gaithersburg, MD: ONC/NIST, 2016.

[39] Kuo TT., Hsu CN., and Ohno-Machado L. Model chain: decentralized privacy preserving healthcare predictive modeling framework on private blockchain networks. In *ONC/NIST, Use of Block Chain for Healthcare and Research Workshop*. Gaithersburg, MD: ONC/NIST, 2016.

[40] Mettler M. Blockchain technology in healthcare: the revolution starts here. In *IEEE Conference on e-Health Networking, Applications and Services (Healthcom)*. Munich, Germany: IEEE, 2016, pp. 1–3.

[41] Healthbank.coop. *Health Bank*, https://www.healthbanl.coop/. Accessed December 20, 2016.

[42] Ekblaw A., Azaria A., Halamka J.D., and Lippman A. A case study for blockchain in healthcare: "MedRec" prototype for electronic health records and medical research data. In *Use of Block chain for Healthcare and Research Workshop*, 2016.

[43] Azaria A., Ekblaw A., Vieira T., and Lippman A. MedRec: using blockchain for medical data access and permission management. In *International Conference on Open and Big Data (OBD)*. Vienna, Austria: IEEE; 2016, pp. 25–30.

[44] Linn L.A. and Koo M.B. Blockchain for health data and its potential use in health IT and health care related research. In *ONC/NIST Use of Blockchain for Healthcare and Research Workshop*. Gaithersburg, MD: ONC/NIST, 2016.

[45] Goldwater J. The use of a blockchain to foster the development of patient-reported outcome measures. In *ONC/NIST Use of Blockchain for Healthcare and Research Workshop*. Gaithersburg, MD: ONC/NIST, 2016.

[46] Topol EJ. Money back guarantees for non-reproducible results. *BMJ.* 2016; 353:i2770.

[47] Taylor P. Applying blockchain technology to medicine traceability. *Securing Industry.com*. https://www.securingindustry.com/pharmaceuticals/applying-blockchain technology-to-medicine-traceability/s40/a2766/-.WFnPQ7GZNzg, 2016.

[48] Bhavin M., Tanwar S., Sharma N., Tyagi S., and Kumar N. Blockchain and quantum blind signature-based hybrid scheme for healthcare 5.0 applications. *J. Inf. Secur. Appl.*, 2021;56:102673.

[49] Gupta R., Shukla,A., and Tanwar S. BATS: a blockchain and AI-empowered drone-assisted telesurgery system towards 6g. *IEEE Trans. Netw. Sci. Eng.*, 2021;8(4):2958–2967.

[50] Tayal A., Solanki A., Kondal R., Nayyar A., Tanwar S., and Kumar N. Blockchain-based efficient communication for food supply chain industry: transparency and traceability analysis for sustainable business. *Int. J. Commun. Syst.*, 2021;34(4):e4696.

[51] Vora J., Nayyar A., Tanwar S., *et al.* BHEEM: a blockchain-based framework for securing electronic health records. In *2018 IEEE Globecom Workshops (GC Wkshps)*, December, 2018, pp. 1–6.

[52] Gupta R., Bhattacharya P., Tanwar S., Kumar N., and Zeadally, S. GaRuDa: a blockchain-based delivery scheme using drones for healthcare 5.0 applications. *IEEE Internet of Things Mag.*, 2021;4(4):60–66.

[53] Kapadiya K., Patel U, Gupta R., *et al.* Blockchain and AI-empowered healthcare insurance fraud detection: an analysis, architecture, and future prospects. *IEEE Access*, 2022; 10:79606–79627.

[54] Gupt R., Shukla A., and Tanwar S. AaYusH: a smart contract-based telesurgery system for healthcare 4.0. In *2020 IEEE International Conference on Communications Workshops (ICC Workshops)*, Dublin, Ireland, 2020, pp. 1–6.

[55] Bodkhe U., Tanwar S., Bhattacharya P., and Verma A. Blockchain adoption for trusted medical records in healthcare 4.0 applications: a survey. In *Proceedings of Second International Conference on Computing, Communications, and Cyber-Security*, Springer, Singapore, 2021, pp. 759–774.

Chapter 11

A recent study based on prediction approach over Coronavirus disease-2019 cases in India using regression modeling of artificial intelligence

Shishupal Kumar[1], *Kumari Nidhi Lal*[2] *and Aman Singh*[3]

Abstract

The advent of severe acute respiratory syndrome coronavirus 2 (SARS-CoV-2; formerly named coronavirus virus or 2019-nCoV) disease (COVID-19) in China at the end of 2019 has caused a large global outbreak and is a major public health issue. As of February 20, 2021, data from the World Health Organization (WHO) have shown that more than 10,977,387 confirmed cases and 156,240 deaths have been identified in India. On January 30, 2020, the WHO declared COVID-19 as the sixth public health emergency of international concern. In India, there are 81,990 numbers of confirmed cases that have been identified as on May 15, 2020. In this regard, the Indian Council of Medical Research (ICMR) takes an initiative for providing intensive care to COVID-19 patients and to fulfill their needs. The percentage of patients infected with COVID-19 reported daily in India between April 1 and May 14, 2020 has been consistently between 20.64% and 30.81% of patients who are actively infected. Currently, India is sustaining in the pandemic situation. The number of patients infected since May 14 in India closely follows an exponential trend. If this trend continues for 1 more week, there will be 1–13 lakhs infected patients till mid-March and India will enter into a serious situation. Intensive care units are then at maximum capacity; up to 710,761 hospital isolation beds will be needed by mid-March, 2020. In this chapter, we discuss the preparations related to COVID-19 and its impact on the second highest populated country in the world. A prediction model is proposed to forecast the COVID-19 confirmed cases with high accuracy. The analysis might help political leaders and health authorities to

[1]Department of Computer Science & Engineering, Indian Institute of Information Technology, Nagpur (IIIT Nagpur) India
[2]Department of Computer Science & Engineering, VNIT, Nagpur, India
[3]Department of Computer Engineering, Universidad Internacional Iberoamericana, USA

allocate enough resources, including personnel, beds, and intensive care facilities, to manage the situation in the next few days and weeks.

Keywords: SARS-CoV-2; COVID-19; Machine learning; Epidemic; Predictive modeling

11.1 Introduction

At the end of December 2019, when all of the countries of the world were engaged in preparing to celebrate the new year 2020, the residents of a city named Wuhan in China started to suffer from a novel COVID-19 disease. The common symptoms of COVID-19 are dry cough, throat pain, high-grade fever, sneezing, and breathing problems. Due to its high transmission efficiency, the virus had been spread into the community of China very fast. It leads to a death-to-case ratio between 17.3% and 0.6%. This virus can be spread in the air in the form of droplets that are produced by an infected person [1]. Due to its high survival rate on the earth's surface, most of the population got affected that are close contact under the area of 1 or 2 m of an infected person [2]. The effect of the virus çan be seen in Italy and France in the worst condition rather than in China. As we know that India is the second most populated country in the world, the spreading of the virus will take a lot of lives and make worsen its economic situation. In the case of spreading the virus in the local community of the people, India will be going back more than 21 years as stated by the government [3]. Currently, the population of India is 133.92 crores. The first COVID-19 case was detected on 30 January 2020. Afterward, no infections were found in any city in India till February 2020. The suspected increasing number of cases of this virus began on February 15, 2020. On March 15, the cases jumped to 82 counts and all the schools/colleges got closed for preventive measures. When the number of cases crossed up to 194, then the government of India introduced a term called "Janta curfew" in which people have to remain inside the house. It was also called the lockdown of India to prevent the spreading of the virus across the community [4].

It was a very admirable decision taken by the Prime Minister of India. During this lockdown, all the public gatherings got canceled or postponed personally. All the domestic and international flights along with the Indian railway system were completely shutting down. The government takes favorable steps regarding providing food for poor people of India free of cost. In this situation of lockdown, only facilities such as groceries, milk bars, banks, and hospitals were available for people. However, the COVID-19 cases in India are increasing day by day. The total confirmed cases are 81,990 on May 14 with 2,649 deaths. However, the question is: Will lockdown help to win the war against COVID-19? The rising of cases provides the answer.

In spite of this panic situation, the people of India have concepts regarding spiritual thinking. Lord Shiva and Lord Ram made the land of India and according to this concept, COVID-19 can be easily defeated by the Indian people. However, due to increasing number of deaths and cases of infection leads to re-think about the future condition of India being worst. In this infection, health plays an important role in being a victim of the disease. According to the victim's health condition, it can be said that COVID-19 is directly proportional to the immune system of a person. However, it is observed from Figure 11.1 that in the age group of 0–9, it contains 3.18% victim, 10–19 contains 3.90% victims, 20–29 contains 24.8% victims, 30–39 contains 21.7% victims, 40–49 contains 16.18% victims, 50–59 contains 11.1% victims, 60–69 contains 12.86% victims, 70–79 contains 4.05% victims, 80–89 contains 1.45% victims, and 90–99 contains 1.30% of victims [5]. Also, 73% males and 27% of females are victims of COVID-19 [6]. According to these values, youngsters of India are more prone to being affected by this disease. However, the death toll rate is arising over the

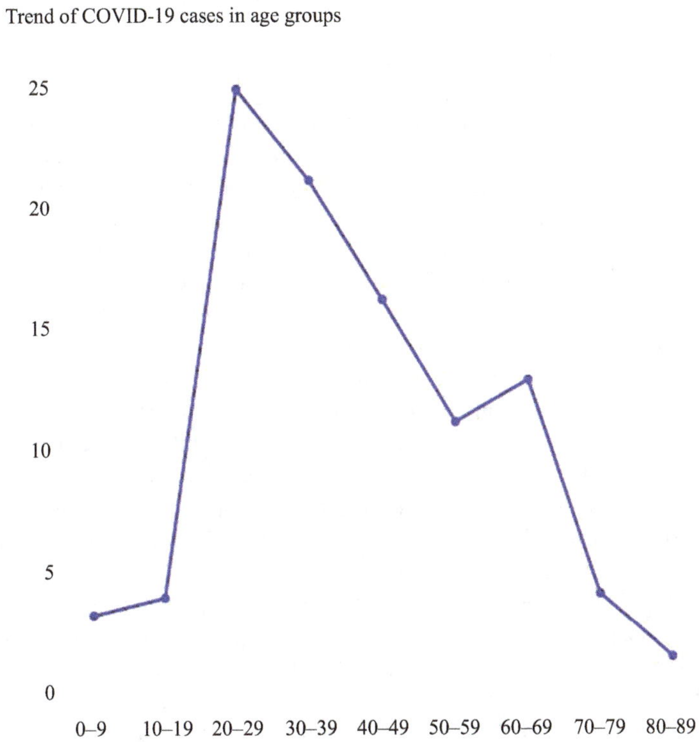

Figure 11.1 COVID-19 patients with respect to age group to this question in a negative manner. Every person in India thinks about the situation of spreading the virus among 133.92 crores people as a local community transmission.

age group greater than 50 and youngsters are recovering in a speedy way [7]. The Objectives of the chapter are:

1. To propose a machine learning (ML) model to predict the COVID-19 cases accurately with a high precision score;
2. And, to utilize the proposed model, to forecast COVID-19 situation in future and be able to prepare plans to fight against the medical emergency.

11.1.1 Organization of chapter

The rest of the chapter is organized as: Section 11.2 enlightens the literature review of ML-based works for COVID-19 prediction to provide safety. Section 11.3 illustrates the current situation of COVID-19 in India that collaborates with the real-time capturing of healthcare conditions. Sections 11.4 and 11.5 highlights the contribution that illustrates the novelty of the proposed work with reference to research gap and proposed model of ML respectively. Simulation results are covered in Section 11.6 to demonstrate the graphical analysis regarding achieved improvement used by the proposed work and Section 11.7 concludes the chapter with future scope.

11.2 Related work

In [8], a ML-based approach was proposed to identify the high-risk patients with abnormalities of COVID-19. The authors stated that it can be used to understand the nature of this virus and further predict the upcoming issues. This work provides accurate and useful features rather than a traditional explicitly calculation-based method. However, effective utilization of the ML concept is missing in the paper in terms of analysis and prediction. In [9], for the prediction of COVID-19 cases, a hybrid ML approach was proposed by the authors. This hybrid model used an Adaptive Network-based Fuzzy Inference System (ANFIS) and Multi-Layered Perceptron-Imperialist Competitive Algorithm (MLP-ICA) that are proposed to predict the time series of infected individuals and mortality rate. However, the drawback of this scheme was the model that maintains its accuracy as long as no significant interruption occurs. In an examination work proposed in [10], creators proposed an autoregressive coordinated moving normal (ARIMA) model to foresee the spread of COVID-2019. In this paper, authors estimated the different boundaries for the following two days based on the investigation of the pervasiveness and occurrence of the COVID-19. This research work likewise exhibits the correlogram and ARIMA conjecture chart for the plague rate and commonness. In [11], the authors proposed a period arrangement technique to break down rate design and the assessed proliferation number of COVID-19 episodes. They performed factual examination to investigate the patterns of the flare-up to feature the present epidemiological phase of a locale so that different approaches can be recognized to address the COVID-19 pandemic in various nations [7]. As per the current circumstance, it is fundamental to comprehend the early spread examples of the disease to plan and control the viable security measures [12,13].

In [14], the authors proposed a novel support vector regression (SVR) technique to investigate five distinct errands identified with the novel COVID-19. In this work, rather than straightforward relapse line, authors utilized the upheld vectors likewise to improve characterization precision. This methodology is assessed and contrasted and other notable relapse models on standard accessible datasets [11,15–46]. In [47], an improved mathematical model was proposed by the authors to analyze and predict the growth of the epidemic. An ML-based improved model was applied to predict the potential threat of COVID-19 in countries worldwide. The authors showed that by using iterative weighting for fitting Generalized Inverse Weibull distribution, a better fit can be obtained to develop a prediction framework.

11.3 Current situation of India for COVID-19

On June 22, 2022, there are 4.3 crore total confirmed cases of COVID-19 in India. The value is low in comparison with the active infections inside Italy and USA [48]. India is the second most populated country in the world after China having a 133.92 crores population. In India, the highest populated states are Uttar Pradesh, Bihar, Maharashtra, Andhra Pradesh, and Delhi. In this regard, it can be said that the spreading will affect to highest count value because it is starting to threaten the most populated cities of India including urban and rural communities as shown in Figure 11.2. The confirmed COVID-19 cases are represented in Figure 11.3 in all the states. These values among the highly populated cities give an idea of forecasting COVID-19 infection transmission. If the condition remains the same, then the counting of confirmed cases will get higher in comparison to Italy, Spain, and the USA.

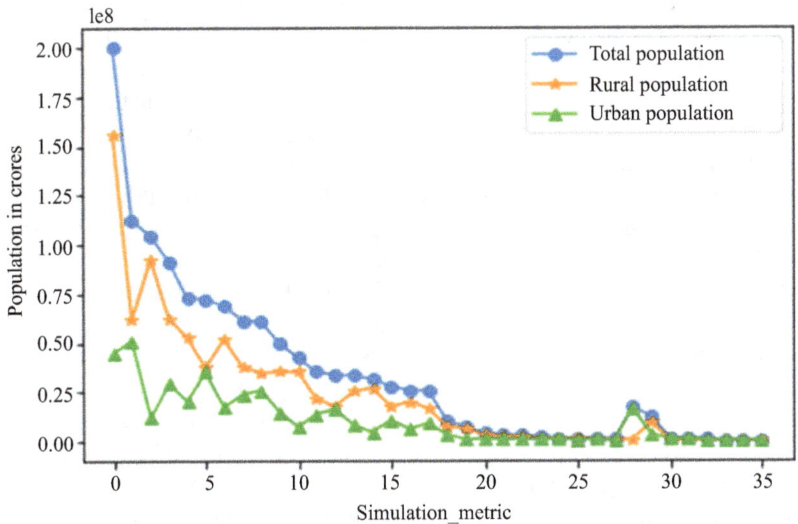

Figure 11.2 Population in India

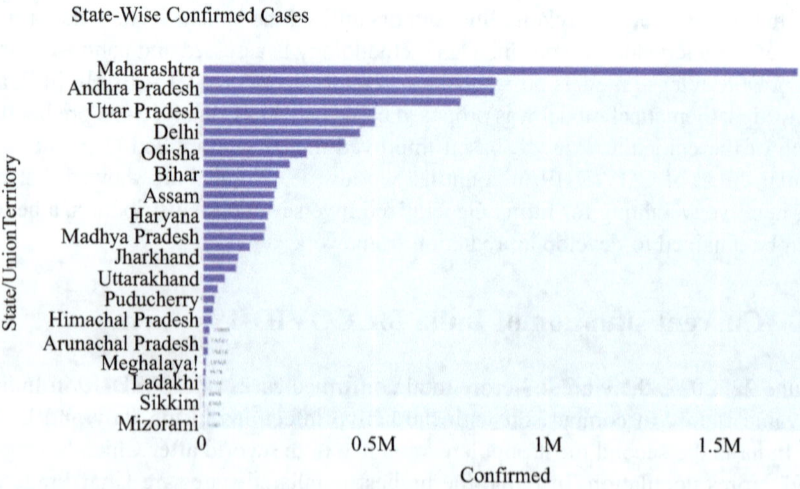

Figure 11.3 Confirmed COVID-19 cases in all the states of India

In [49], it was mentioned that humans having some autoimmune disease like cancer, HIV-AIDS, diabetes, etc. are more prone to be a victim of COVID-19. In India, out of a total of 118 deaths, above 80% are the person having some disorders of autoimmune disease. In this regard, one's assumption is to eat particular foods that are used to boost the immune system. In this way, India can win the battlefield against COVID-19 [50]. However, if this is a true case, then what about other developed countries like Italy and Spain? What is the reason that the death toll rate is high in Italy and Spain compared to India? Are the residents of Italy and the USA do not include immune booster-related foods? Along with food, another important term that plays a vital role in the spreading of COVID-19 is temperature. In India, there are mainly three weather seasons: summer, rainy, and winter. The summer season lies from April to June with a maximum temperature of 45 degrees Celsius. Therefore, another discussion point is the impact of temperature over COVID-19. Viruses can indeed be active more hours in cold places rather than hot environments. However, there is no research-related proof over it that can assure that COVID-19 does not show its full progression inside hot countries [51].

Along with this, there are various factors that can play a huge role in increasing the count of COVID-19 patients in India. Lack of support from the community upon the lockdown decision taken by the government provides a great impact on this situation. If the population is high, then there will be numerous problems that occur in the management. In India, people are not following the restrictions of public gatherings due to their spiritual thinking. People are not able to understand the seriousness and caution regarding this disease. The outcome will be just because of a few hundred or thousand humans may suffer and be infected. However, the government has done many advertisements and online programs to distribute awareness of COVID-19. Another factor that plays a role in the spreading of COVID-19 is that people hide after being suspected or confirmed of COVID-19 after testing. They are not following the quarantine or isolation procedures. Also, some of them tried to hide their travel history or listing of visited places and persons. In this regard, the government of India is taking strict actions in favor of fines or imprisonment against these people. In these stated measures, the only panic concern is: how to prevent or stop the spreading of infection of COVID-19?

Along with these conditions, the growth of confirmed cases in India is lower than in other countries like the USA, Italy, France, and Spain. In this regard, WHO admires the Prime Minister of India regarding taking preventive steps like lockdown, use of sanitizers, wearing of a mask, and suspending local gatherings at the start of the COVID-19 cases. However, some aspects arise over it like random testing of samples, and few testing processes are done by medical authorities [52,53]. For more accurate identification, private labs are involved in the testing process of COVID-19 as instructed by the government [54].

Currently, ICMR describes four stages of COVID-19 spreading in India. In stage 1, only the first starting cases are diagnosed that having a travel history in one of the infected countries. In stage 2, after the virus reaches the destination country through the body of the victim and spreads over his/her close relatives and friends. In stage 3, the local transmission of the virus begins, and sometimes it is called community transmission. In this stage, the history of the origination of the virus could not be traced. In stage 4, the condition of the spreading of the virus becomes uncontrollable. In the current situation, India lies in stage 3 and the origination of virus spreading is not easily accessible. Regarding this, the government of India needs the preparation of medical facilities to be sustained into stage 3. The interventions implemented throughout India include complete lockdown of cities, active case surveillance, rapid investments in increased testing capacity, isolation of cases, ventilators, treatment of severe cases, quarantine of cases and high-risk groups, and behavioral risk-reduction strategies, such as the compulsory use of masks in the general population [55,56].

India is a developing country with the highest population in which the facilities are not up to the remarkable way. The government hospital will need to increase the number of beds and isolation wards to treat COVID-19 patients. A dataset is collected that consists of the data of COVID-19 cases of India up to May 14, 2020 [12]. The collective measures of the dataset are presented in Figures 11.4–11.6. The government of India provides free testing services and treatment to the infected COVID-19 patients by the public sector hospitals [57]. However, the facilities regarding isolation wards and cleaning of washrooms are not up to the level. There are some cases in which the patients have attempted to run away from the hospitals without informing medical authorities. Moreover, the treatment facility is provided by private hospitals with high charges. This high cost cannot be borne by the common middle or low-class people of India. Currently, India has 29,899 total hospitals with a capacity of 7,39,024 beds. Out of the total beds, only 279,588 reside in the rural area and 431,173 beds reside in an urban area. Besides these, to cope with the lakhs of COVID-19 cases, the government of India has prepared

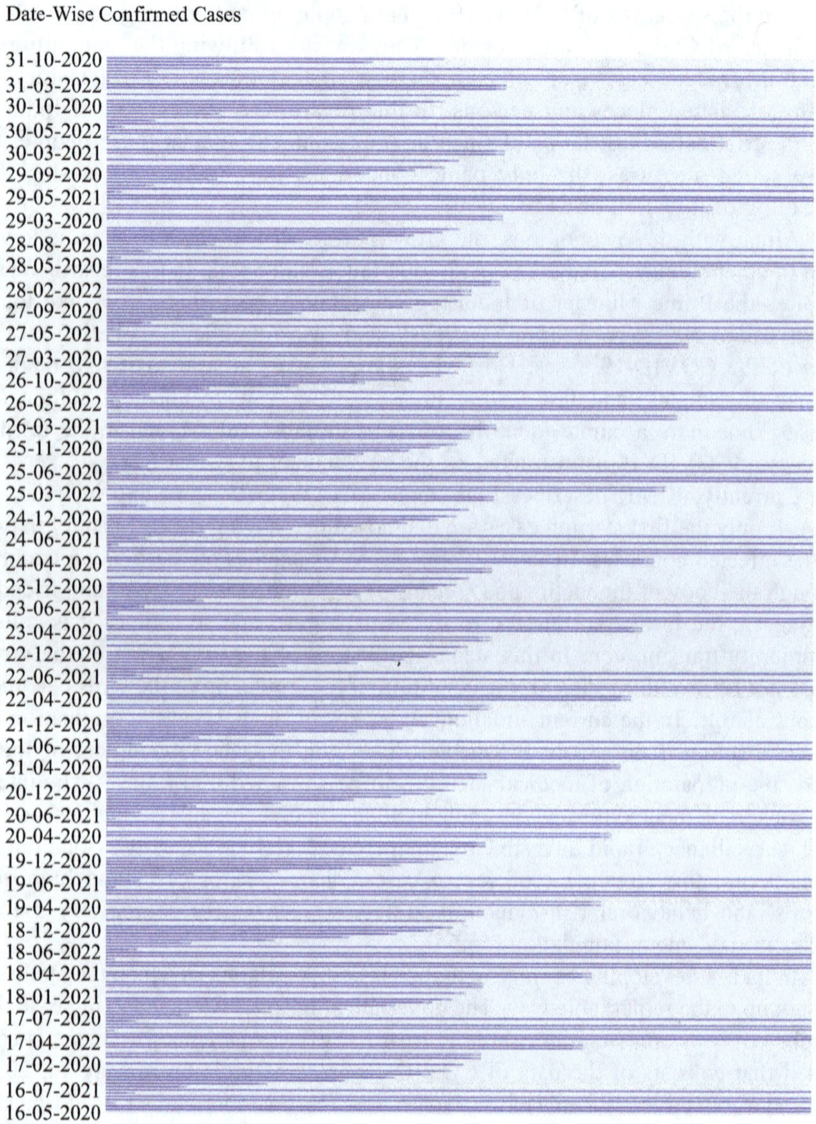

Figure 11.4 Confirmed cases growth of COVID-19 in India

railway coaches to act as isolation wards with having a capacity of thousands per train. Besides, the government also provides extensive care and protection to the working team and their family that including doctors and medical staff. The Prime Minister of India provided an emergency relief fund where people can contribute and help in the government's fight against coronavirus [58]. A Reuters report has

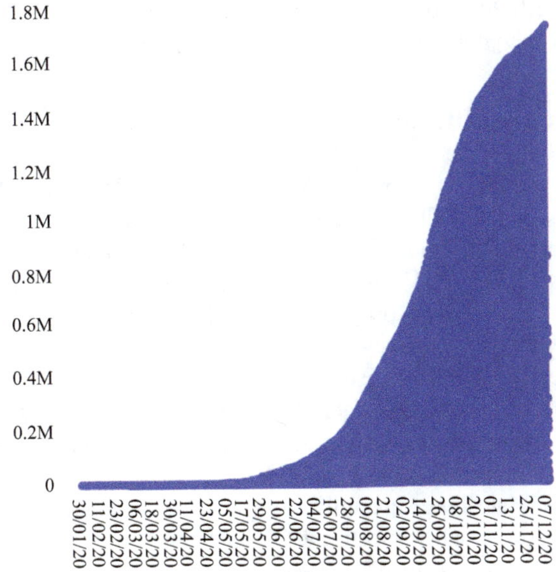

Figure 11.5 Cured cases of COVID-19 in India

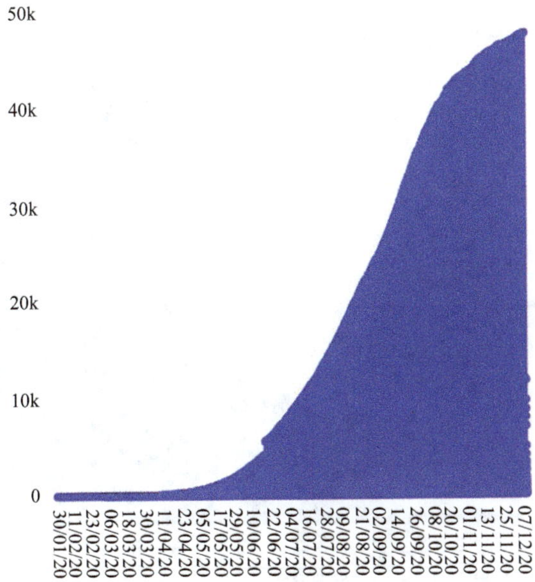

Figure 11.6 Death cases of COVID-19 in India

now said that the Indian government may bring an INR 2.3 Tn relief fund for the poor sections of the population and businesses [59].

11.4 Our contributions

In this chapter, the aim is to build a ML-based prediction model to forecast the confirmed cases of COVID-19 in India [60,61]. In this way, the model can be utilized by the politicians and health service providers to take some necessary and preventive steps against COVID-19-related current situation of India as indicated by Figure 11.7. Therefore, we proposed a model that predicts the confirmed cases of COVID-19 with 96% accuracy. To build the model, we utilized a dataset provided by Kaggle [12]. In this dataset, the correct and ongoing measure of COVID-19 cases can be acquired. The dataset contains various attributes like “Date,” “Time,” “State,” “Confirmed,” “Deaths,” “Cured,” “Confirmed Indian National,” etc. To build the proposed model, only “Date,” “Confirmed,” “Deaths,” “Avg-Age,” and “Cured” attributes are utilized. The attribute named “Date” is used to indicate the date on which respective COVID-19 measures cases were captured. The attribute named “Confirmed” is used to denote the total number of confirmed COVID-19 cases on a particular date. The attribute named “Deaths” is used to indicate the total number of death cases that occurred due to COVID-19 on that particular date. Like ways, the attribute named “Cured” is used to indicate the

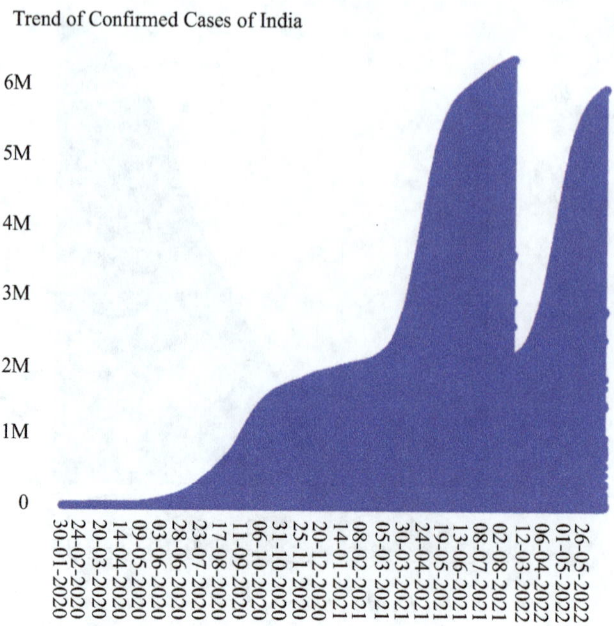

Figure 11.7 Current condition of India

survival cases against COVID-19 on that particular date. Like ways, the attribute named "Avg-Age" is used to indicate the average age of patients reported on that particular date. The dataset contains the measure cases of COVID-19 from January 30, 2020 to December 31, 2020. In this way, it can be stated that the proposed work is based on a recent trend and a novel approach to making predictions.

11.5 Proposed work

11.5.1 Predictive modeling of COVID-19 using ML

Fundamentally, ML is an approach that gives learning ability to the machines in the form of training and testing. Afterward, the machines can able to make a decision and give prediction values. To give the ability to learn to the machine, there is a need for the algorithm. According to the prediction requirements, the approaches of ML are divided into three parts: supervised learning, unsupervised learning, and reinforcement learning [62,63]. In this chapter, a supervised learning approach is utilized by the proposed prediction model. The work of these algorithms is to make the relationship between the predictor variable (independent variable) and the prediction variable (dependent variable).

Selection of the algorithm for ML is dependent on the nature of the prediction variable. According to the application, this variable is of two types of nature: continuous and discrete. In this chapter, the prediction variable is the "Confirmed" attribute of the dataset with the nature of continuous. However, if the need for predicting the values of the prediction variable in terms of 0 or 1 (binary), then it is said to be discrete. In the supervised learning approach, the linear regression (LR) model is utilized by the proposed work as shown in Figure 11.8(a). The LR model is one of the simplest and easier ML algorithms to make predictions. Despite its simplicity, another reason for selecting the LR algorithm in the proposed work is regression. The line plotted for making predictions is called the regression line. Mathematically, equation for the regression line is represented by (11.1):

$$y = \theta_1 + \theta_2 * x \tag{11.1}$$

Here, y denotes the dependent or prediction variable, x denotes the independent or predictor variable, θ_1 denotes intercept of the line, and θ_2 denotes the slope of the line.

These parameters are also known as the regression coefficient. The values on the regression line are understood as predicted values.

Figure 11.8(b) shows the model of LR. In this figure, X-axis is used to denote the independent variable, and Y-axis is used to denote the dependent variable. The scatter plot is the actual values from the data in the used dataset. The straight line is known as the regression line and it is used to predict the values. This line is constructed by considering the mathematical equation (11.1). The actual values are denoted by y_i and predicted values are denoted by y_i. Now, there can be as many lines passing through the given scatter points in the plot. To find the best fit line according to the dataset, the least square method is used. The least square method is

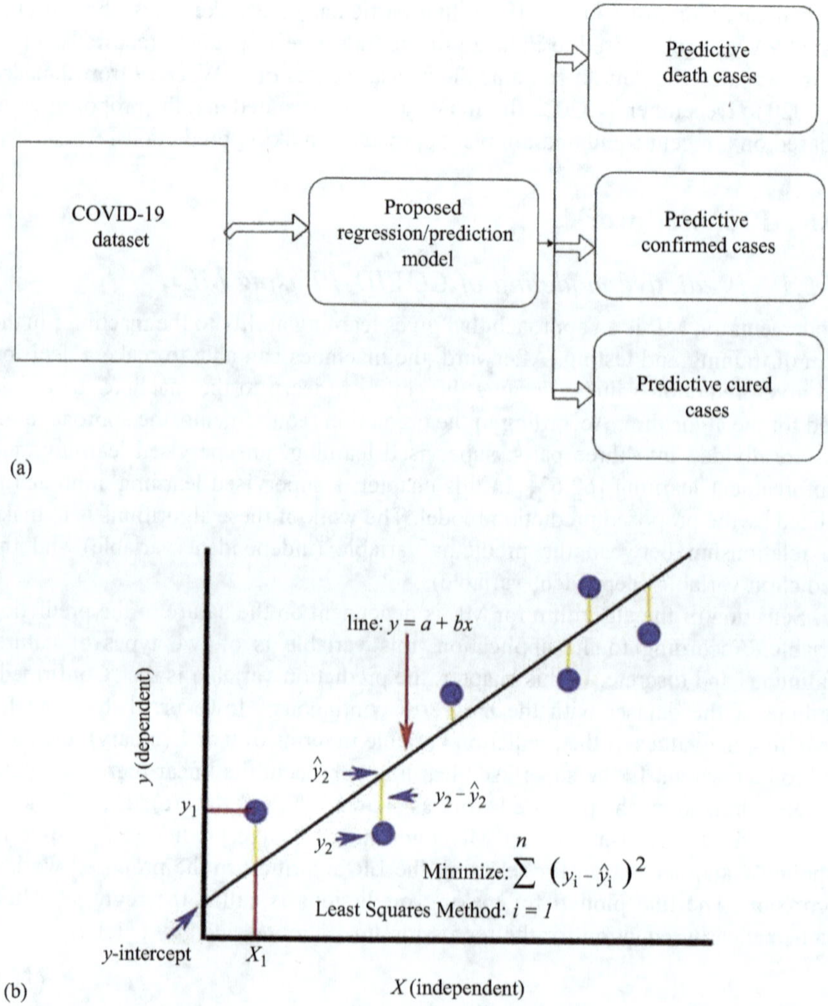

Figure 11.8 (a) Proposed action model for COVID-19. (b) LR model for COVID-19.

a mathematical regression analysis modeling. This method is used to get the best fit of the line for the given data or set of data values. Besides, it provides a visual demonstration of the relationship between given data set scatter points and represents the relationship between an independent variable and a dependent variable.

$$i = 1(y_i - y_i^{\cap})n \tag{11.2}$$

In (11.2), n is used to represent all data points, y_i is used to denote the value of actual data, and $y_i^{\cap}$ is used to represent the value of prediction. In the proposed prediction model for COVID-19 cases in India, an attribute named "Confirmed" in

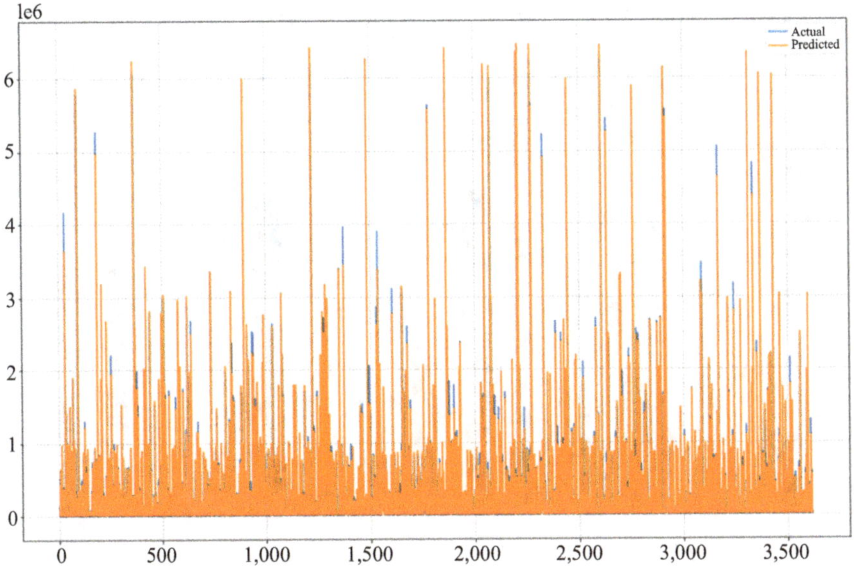

Figure 11.9 Actual and predicted values generated by proposed model

the dataset acted as *Y*, and "Date," "Cured," and "Deaths" acted as *X*. For modeling and simulation, Python3 language is used with the Jupyter Notebook. The utilized packages are Scikit learn, pandas, matplotlib, and NumPy. After initializing the *X* and *Y* values for the proposed model, the used data values are needed to split into two parts: training and testing. The data values associated with training are used to train the proposed model and cover 80% of the data. The rest of the data values (20%) are used to test the modeling after the training process.

11.6 Results discussion and analysis

After the phase of training and testing of the proposed model, the prediction of the confirmed cases of COVID-19 has to be measured by plotting the regression line [64]. For the proposed working model, the intercept is calculated as 4,553 and the regression coefficient comes out to be 2.7. After extracting these values, the prediction value can be calculated for COVID-19 cases. In Figure 11.9, the actual and predicted values are shown for COVID-19 confirmed cases. In the figure, the actual represents the values that already exist in the dataset. The predicted values are used to represent the values that are generated by the proposed predictive modeling approach [65].

In Figure 11.10, a bar chart is shown for the top 15 values of actual and predicted. From the figure, it can be seen that the proposed model predicts the values of COVID-19 confirmed cases nearby to actual cases. More descriptively, Figure 11.11 is plotted to show the outcome of the proposed model for the last 15

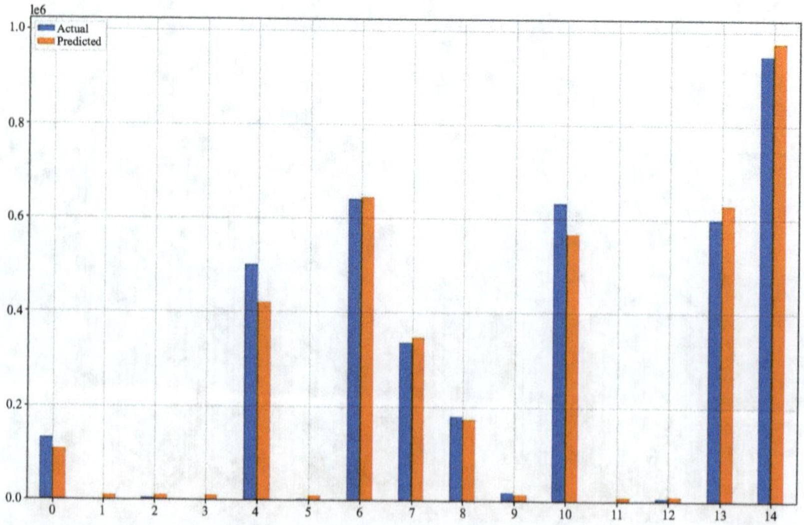

Figure 11.10 Actual and predicted values for the last 15 values

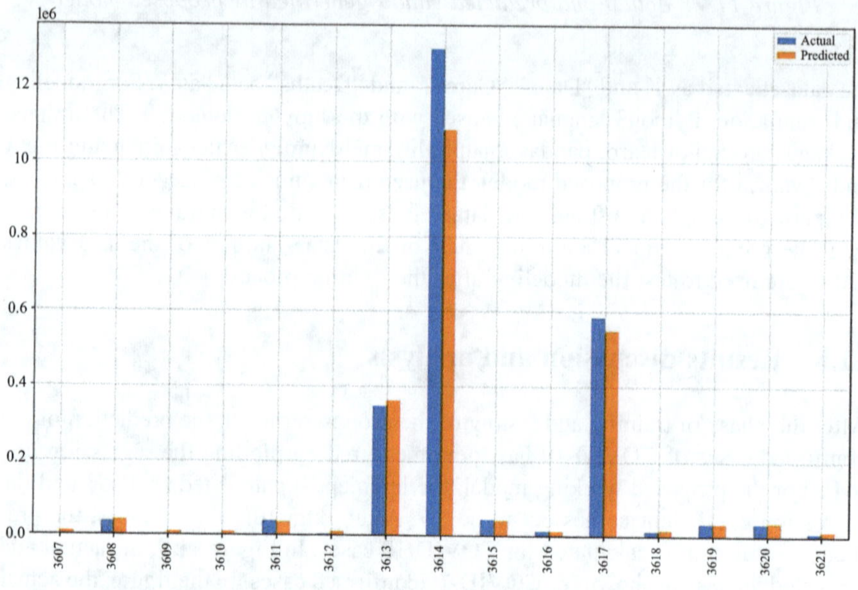

Figure 11.11 Outcome of proposed model for the last 15 values after training and tests phase

values after the training and testing phase. In this figure, it can be seen that the proposed model successfully predicts the values nearby to actual values of confirmed cases. In this way, it can say that the proposed model can be utilized to forecast the cases of COVID-19. The proposed model can be utilized to predict death as well as cured cases of COVID-19. However, in this chapter, our focus is restricted to only making the forecasting confirmed cases in India. The forecasting values are denoted by predicted values as shown in Table 11.1. We have used two datasets for simulation. The first dataset is taken up to the values collected in May 2020 and the second dataset belongs to the values collected latest by December 2020. In Table 11.1, the first two column pairs "actual" and "predicted" are used to indicate the tail 50 values and head 50 values, respectively. It can be seen from Table 11.1 that the proposed model achieves high accuracy by predicting the nearest value to the actual value in the forecasting of COVID-19 cases [66,67].

To evaluate the performance of the proposed LR model, certain performance metrics are considered. These metrics are used to find the suitability and efficiency in the usage of the proposed model to predict the confirmed cases. These performance metrics are *R*-squared score, MAE, and RMSE as shown in Table 11.2.

R-squared is a goodness-of-fit measure for LR models. This statistic indicates the percentage of the variance in the dependent variable to that of the independent variables. *R*-squared measures the strength of the relationship between the proposed model and the dependent variable on a convenient 0–100% scale. In this chapter, the confirmed cases of COVID-19 are acted as a dependent variable. The *R*-squared method is used to find the difference between actual values that exist in the dataset and predicted values forecasted by the proposed model. It is also stated by the smallest sum of squared residuals. The value of *R*-squared always lies between 0% and 100%. If the model generates a high

Table 11.1 Actual and predicted outcome values

Dataset up to May 5, 2020				Dataset up to June 20, 2022			
Actual	**Predicted**	**Actual**	**Predicted**	**Actual**	**Predicted**	**Actual**	**Predicted**
100	74	35	36	22,333	18,303	3,081	5,293
33	78	5	13	1,077,374	929,893	3,078	4,749
63	51	86	96	5,444	9,502	3,182	5,137
70	53	20	25	79,451	58,044	22,918	20,674
77	87	40	44	4,563	5,082	80,476	77,217
437	311	120	114	16,806	16,163	79,667	62,696
35	54	1952	1729	166,586	102,066	4,238	5,841
100	107	328	200	97,920	78,572	2,089	5,316
30	33	100	101	132,364	122,370	23,018	26,361
40	42	13	50	182,728	178,289	1,324	5,012
3061	3872	43	107	13,643	16,116	7,642	5,410
2	4	70	42	748,225	783,992	5,087	5,215
2,626	1,998	245	349	262,011	248,340	18,138	14,728
120	137	42	61	72,777	64,667	2,066	6,148
184	192	30	55	3,718	6,945	187,240	196,841
70	72	500	1,001	2,018	6,231	5,409	5,408

(Continues)

Table 11.1 (*Continued*)

Dataset up to May 5, 2020				Dataset up to June 20, 2022			
Actual	**Predicted**	**Actual**	**Predicted**	**Actual**	**Predicted**	**Actual**	**Predicted**
50	53	121	122	5,672	9,146	8,239	8,311
70	71	319	400	59,591	59,598	698	679
50	58	414	245	3,093	7,571	22,208	25,554
363	121	15	69	12,905	15,370	23,454	23,065
43	41	67	70	154,418	119,401	5,587	5,603
169	102	1,852	1,592	4,987	5,156	8,281	8,300
91	122	969	410	4,077	5,548	37,497	37,923
1,229	566	12	48	10,852	14,946	166,666	174,405
2,364	2,315	3	19	8,322	9,551	5,445	9,198
182	115	61	111	121,947	106,165	1,609	6,219
60	69	2	67	67,706	73,234	10,220	12,193
27	55	3	64	241,059	241,985	59,711	49,269
657	205	19	22	13,944	15,718	462	5,478
13	31	96	164	243,896	246,472	10,108	8,386
23,401	21,431	1,731	723	10,511	14,900	893,006	923,128
91	119	557	809	2,918	6,735	269,223	269,342
40	43	451	787	26,680	33,546	30,152	22,898
44	54	20	22	6,004	6,034	4,900	4,914
3	3	49	49	687	552	4,956	4,918
45	46	700	715	4,888	4,988	7,643	73,648
7	14	39	69	829,640	837,612	3,416	4,972
70	55	19	22	3,289	5,453	3,689	5,457
38	33	19	28	383	5,015	8,969	13,804
20	26	33	43	8,607	8,651	385,352	366,902
2,376	2,371	7	14	72,300	50,455	107,001	73,417
23	12	387	508	14,747	17,456	2,998	5,608
305	100	353	386	2,137	7,106	3,419	7,554
23	24	39	41	5,001	5,535	4,835	6,886
42	55	5	74	4,445	7,909	5,087	5,312
100	104	1,259	2,247	5,089	5,056	5,172	5,192
16	23	30	31	5,100	5,420	3,148	4,967
70	79	185	168	31,622	36,794	191,246	199,465
213	210	19	40	5,643	5,322	10,747	10,147
42	53	11	58	92,175	84,258	4,352	57,092

Table 11.2 *Result outcome values*

Parameter name	Dataset up to May 5, 2020	Dataset up to June 20, 2022
RMSE	309	17,387
Mean absolute error (MAE)	136	8,844
R^2-score	0.96	0.99
Intercept	−2.23	4,553
Regression-coefficient	0.755	2.707

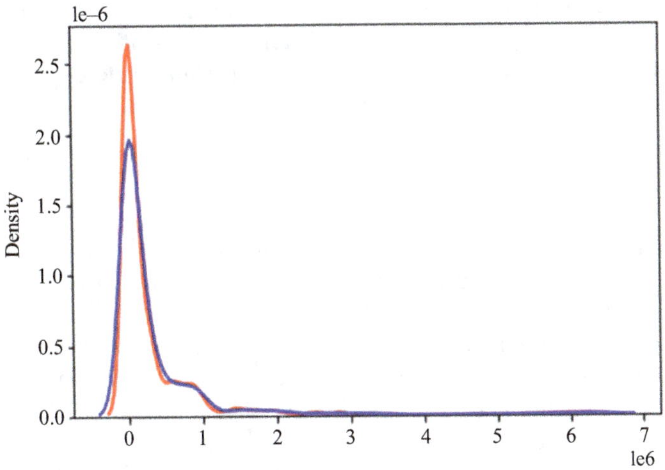

Figure 11.12 Actual vs. predicted values for RMSE

R-squared value, then it better fits on calculated predictions and is said to be accurate. For the proposed working model, the *R*-squared value comes out to be 99%. In this way, it can say that the proposed prediction model can forecast the confirmed cases of COVID-19 with high accuracy. The RMSE is the square root of the variance of the residuals. It indicates the absolute fit of the model to the data – how close the observed data points are to the model's predicted values. Whereas *R*-squared is a relative measure of fit, RMSE is an absolute measure of fit. For the proposed approach, the RMSE value comes out to be 17,387 as indicated in Figure 11.12.

11.7 Conclusion and future scope

India is the second most populated country in the world after China with technologies and facilities that are not up to a remarkable point. If India could not able to stop the spreading of COVID-19 infection into stage 3, then it will be very harmful, and a worse situation occurs in India that cannot be easily stopped. In this regard, India will automatically enter into stage 4 with a high count in loss of lives and economic conditions. It is a bitter truth that COVID-19 becoming a medical emergency in all the nations including India. The preparations and effective decisions will save numerous lives in India. This disease gives us a message to maintain social distancing and to be careful to save our lives. Therefore, our aim in this chapter is to propose a ML model to predict the COVID-19 cases accurately with a high precision score. By the utilization of this model, the healthcare workers and politicians can analyze the forecasting and be able to prepare the plans to fight against this medical emergency. However, we have used only a simple regression model to predict the confirmed cases of COVID-19 and achieve 99% accuracy.

This model can be improved by utilizing other models like SVR. Besides, the proposed model can be utilized over the COVID-19 vaccination data to predict its success rate. In future, the work will be analyzed using different ML and deep learning algorithms.

References

[1] Huang, C., Wang, Y., Li, X., *et al.* "Clinical features of patients infected with 2019 novel coronavirus in Wuhan, China." *The Lancet* 395, no. 10223 (2020): 497–506.

[2] Zhao, S., Lin, Q., Ran, J., *et al.* "Preliminary estimation of the basic reproduction number of novel coronavirus (2019-nCoV) in China, from 2019 to 2020: a data-driven analysis in the early phase of the outbreak." *International Journal of Infectious Diseases* 92 (2020): 214–217.

[3] Singhal, T. "A review of coronavirus disease-2019 (COVID-19)." *The Indian Journal of Pediatrics* 87 (2020): 1–6.

[4] Pulla, P. "Covid-19: India imposes lockdown for 21 days and cases rise." *BMJ* 368 (2020): m1251.

[5] Singh, R. and Adhikari, R., "Age-structured impact of social distancing on the COVID-19 epidemic in India." 2020. *arXiv preprint arXiv:2003.12055.*

[6] Ji, Y., Ma, Z., Peppelenbosch, M.P., and Pan, Q. "Potential association between COVID-19 mortality and health-care resource availability." *The Lancet Global Health 8*, no. 4 (2020): e480.

[7] Sohrabi, C., Alsafi, Z., O'Neill, N., *et al.* "World Health Organization declares global emergency: a review of the 2019 novel coronavirus (COVID-19)." *International Journal of Surgery* 78, (2020): 185–193.

[8] Yadav, M., Perumal, M., and Srinivas, M. "Analysis on novel coronavirus (COVID-19) using machine learning methods." *Chaos, Solitons & Fractals* 139 (2020): 110050.

[9] Tuli, S., Tuli, S., Tuli, R., and Gill, S.S. "Predicting the growth and trend of COVID-19 pandemic using machine learning and cloud computing." *Internet of Things* 11 (2020): 100222.

[10] Github Repository. 2019 Novel Coronavirus COVID-19 (2019-nCoV) Data Repository by Johns Hopkins CSSE. Retrieved March 31, 2020 from https://github.com/CSSEGISandData/COVID-19.

[11] WHO coronaviruses (COVID-19). Retrieved March 30, 2020 from https://www.who.int/emergencies/diseases/novel-coronavirus-2019.

[12] Stoecklin, S.B., Rolland, P., Silue, Y., *et al.* "First cases of coronavirus disease 2019 (COVID-19) in France: surveillance, investigations and control measures, January 2020." *Eurosurveillance* 25, no. 6 (2020): 2000094.

[13] Mistry, C., Thakker, U., Gupta, R., *et al.* "MedBlock: an AI-enabled and blockchain-driven medical healthcare system for COVID-19." In *ICC 2021 – IEEE International Conference on Communications,* Montreal, *QC*, Canada, 2021, pp. 1–6.

[14] Benvenuto, D., Giovanetti, M., Vassallo, L., Angeletti, S., and Ciccozzi, M. "Application of the ARIMA model on the COVID- 2019 epidemic dataset." *Data in Brief* 29, (2020): 105340.

[15] Leppard, G.G. "Evaluation of electron microscope techniques for the description of aquatic colloids." In *Environmental Particles*. CRC Press, London, 2019, pp. 231–289.

[16] Jia, L., Li, K., Jiang, Y., and Guo, X. Prediction and analysis of Coronavirus Disease 2019. 2020. *arXiv preprint arXiv:2003.05447*.

[17] Kalipe, G., Gautham, V., and Behera, R.K. "Predicting malarial outbreak using machine learning and deep learning approach: a review and analysis." In *2018 International Conference on Information Technology (ICIT)*, December 2018. IEEE, pp. 33–38.

[18] Singh, R., Singh, R., and Bhatia, A. "Sentiment analysis using machine learning technique to predict outbreaks and epidemics." *International Journal of Advanced Science and Research* 3, no. 2 (2018): 19–24.

[19] Ting, D.S.W., Carin, L., Dzau, V., and Wong, T.Y. "Digital technology and COVID-19." *Nature Medicine* 26, no. 4 (2020): 459–461.

[20] Benvenuto, D., Giovanetti, M., Vassallo, L., Angeletti, S., and Ciccozzi, M. "Application of the ARIMA model on the COVID-2019 epidemic dataset." *Data in Brief* 29 (2020): 105340.

[21] Menni, C., Valdes, A.M., Freidin, M.B. *et al.* "Real-time tracking of self-reported symptoms to predict potential COVID-19." *Nature Medicine* 26 (2020): 1037–1040. https://doi.org/10.1038/s41591-020-0916-2.

[22] "Question and answers hub." https://www.who.int/emergencies/diseases/novel-coronavirus-2019/question-and-answers-hub

[23] Centers for Disease Control and Prevention. Symptoms of Coronavirus, 2020, https://www.cdc.gov/ coronavirus/2019-ncov/ symptoms-testing/ symptoms.html (last accessed: 11.05.2020).

[24] Yang, P. and Wang, X. "COVID-19: a new challenge for human beings." *Cellular & Molecular Immunology* 17, (2020): 555–557. https://doi.org/10.1038/s41423-020-0407-x

[25] Coronavirus Disease 2019 (COVID-19), https://www.mayoclinic.org/diseasesconditions/coronavirus/diagnosis-treatment/drc20479976 (last accessed 12.05.2020).

[26] Advice on the Use of Point-of-Care Immunodiagnostic Tests for COVID-19, https://www.who.int/newsroom/commentaries/detail/advice-on-the-use-of-pointof-care-immunodiagnostic-tests-for-covid-19 (last accessed 12.05.2020).

[27] Centers for Disease Control. Viral Testing Data in the U.S., 2020, https://www.cdc.gov/coronavirus/2019-ncov/casesupdates/testing-in-us.html (last accessed 12.05.2020).

[28] ICMR, COVID-19 Testing, https://main.icmr.nic.in/sites/default/files/upload_documents/Revised_Advisory_Rapid_Anibody _blood_tests.pdf (last accessed 12.05.2020).

[29] COVID-19 in India, https://www.kaggle.com/ sudalairajkumar/covid19-inindia?select=AgeGroupDetails.csv (last accessed 12.05.2020). COVID-

19 in India, https://www.kaggle.com/sudalairajkumar/covid19-inindia?select=covid_19_india.csv (last accessed 12.05.2020).

[30] Novel Corona Virus 2019 Dataset, https://www.kaggle.com/sudalairajkumar/novelcorona-virus-2019-dataset?select=covid_19_data.csv (last accessed 12.05.2020).

[31] Chelliah, B.J., Kalaiarasi, S., Anand, A., Janakiram, G., Rathi, B., and Warrier, N.K. "Classification of mushrooms using supervised learning models," *IJETER* 6, no. 4 (2018): 229–232. ISSN 2454-6410.

[32] Shona, D. and Shobana, A. "Fast and effective network intrusion detection technique using hybrid revised algorithms." *IJETER* 4, no. 11, (2016): 42–46. ISSN 2454-6410.

[33] Dharmapuri, S.L., Dandamudi, P.K., Botcha, V.M., and Kolla, B.P. "Detecting central nervous system disorder using machine learning technique (XGB classifier)." *IJETER* 8, no. 4, (2020): 1142–1147. ISSN 2454-6410.

[34] Prakash, K.B. "Information extraction in current Indian web documents." *International Journal of Engineering and Technology (UAE)* 7, no. 2.8, (2018): 68–71. https://doi.org/10.14419/ijet.v7i2.8.10332

[35] Kolla, B.P., Dorairangaswamy, M.A., and Rajaraman, A. "A neuron model for documents containing multilingual Indian texts." In *2010 International Conference on Computer and Communication Technology, ICCCT2010*, 2010, pp. 451–454.

[36] Kolla, B.P. and Raman, A.R. "Data engineered content extraction studies for Indian web pages." *Advances in Intelligent Systems and Computing* 711, (2019): 505–512.

[37] Prakash, K.B. and Dorai Rangaswamy, M.A. "Content extraction studies using neural network and attribute generation." *Indian Journal of Science and Technology* 9, (2016): 1–10.

[38] Prakash, K.B., Dorai Rangaswamy, M.A., and Raman, A.R. "Text studies towards multi-lingual content mining for web communication." In *Proceedings of the 2nd International Conference on Trends in Information Sciences and Computing, TISC-2010* 2010, pp. 28–31. https://doi.org/10.1109/TISC.2010.5714601.

[39] Prakash, K.B. and Rangaswamy, M.A.D. "Content extraction of biological datasets using soft computing techniques." *Journal of Medical Imaging and Health Informatics* 6 (2016): 932–936.

[40] Prakash, K.B., Rajaraman, A., Perumal, T., and Kolla, P. "Foundations to frontiers of big data analytics." In *Proceedings of the 2016 2nd International Conference on Contemporary Computing and Informatics, IC3I 2016*, 2016, p. 242. https://doi.org/10.1109/IC3I.2016.791796.

[41] Kolla, B.P. and Raman, A.R. "Data engineered content extraction studies for Indian web pages." *Advances in Intelligent Systems and Computing* 711 (2019): 505–512.

[42] Prakash, K.B. "Content extraction studies using total distance algorithm." In *Proceedings of the 2016 2nd International Conference on Applied and*

Theoretical Computing and Communication Technology, iCATccT 2016, 2017, p. 673.

[43] Prakash, K.B., Kumar, K.S., and Rao, S.U.M. "Content extraction issues in online web education." In *Proceedings of the 2016 2nd International Conference on Applied and Theoretical Computing and Communication Technology, iCATccT 2016*, 2017, p. 680. https://doi.org/10.1109/ICATCCT.2016.7912086.

[44] Prakash, K.B., Rajaraman, A., Perumal, T., and Kolla, P. "Foundations to frontiers of big data analytics." In *Proceedings of the 2016 2nd International Conference on Contemporary Computing and Informatics, IC3I 2016*, 2016, p. 242.

[45] Ismail, M., Prakash, K.B., and Rao, M.N. "Collaborative filtering-based recommendation of online social voting." *International Journal of Engineering and Technology (UAE)* 7, no. 3 (2018): 1504–1507.

[46] Prakash, K.B., Rajaraman, A., and Lakshmi, M. "Complexities in developing multilingual on-line courses in the Indian context." In *Proceedings of the 2017 International Conference on Big Data Analytics and Computational Intelligence, ICBDACI 2017*, 2017, p. 339.

[47] Deb, S. and Majumdar, M. "A time series method to analyze incidence pattern and estimate reproduction number of COVID-19." *arXiv preprint arXiv:2003.10655* (2020).

[48] Remuzzi, A. and Remuzzi, G. "COVID-19 and Italy: what next?" *The Lancet* 395, no. 10231 (2020): 1225–1228.

[49] Fang, L., Karakiulakis, G., and Roth, M. "Are patients with hypertension and diabetes mellitus at increased risk for COVID-19 infection?" *The Lancet Respiratory Medicine* 8, no. 4(2020): e21.

[50] Legido-Quigley, H., Asgari, N., Teo, Y.Y., *et al.* "Are high-performing health systems resilient against the COVID-19 epidemic?" *The Lancet* 395, no. 10227 (2020): 848–850.

[51] Dhama, K., Khan, S., Tiwari, R., *et al.* "Coronavirus disease 2019–COVID-19." *Clinical Microbiology Reviews* 33, no. 4 (2020): e00028-20.

[52] Oxford Analytica. "India faces huge COVID-19 burden absent more testing." *Emerald Expert Briefings* (oxan-db).

[53] Oxford Analytica. "India will come under heavy strain over COVID-19." *Emerald Expert Briefings* (oxan-es).

[54] Tanne, J.H., Hayasaki, E., Zastrow, M., Pulla, P., Smith, P., and Rada, A.G. "Covid-19: how doctors and healthcare systems are tackling coronavirus worldwide." *BMJ* 368 (2020): m1090.

[55] John, T.J. "How prepared is India to control the COVID-19 pandemic?" *Economic & Political Weekly* 55, no. 11 (2020): 13.

[56] Pinter, G., Felde, I., Mosavi, A., Ghamisi, P., and Gloaguen, R. "COVID-19 pandemic prediction for Hungary: a hybrid machine learning approach." *Mathematics* 8, no. 6 (2020): 890.

[57] Rodriguez-Morales, A., Tiwari, R., Sah, R., and Dhama, K. "COVID-19, an emerging coronavirus infection: current scenario and recent developments – an overview." *Journal of Pure and Applied Microbiology* 14 (2020): 6150.

[58] Kushwaha, S., Bahl, S., Bagha, A.K., *et al.* "Significant applications of machine learning for COVID-19 pandemic." *Journal of Industrial Integration and Management* 5, no. 4 (2020): 453–479.

[59] Jha, V., Dinesh, T.A., and Nair, P. "Are we ready for controlling community transmission of COVID 19 in India?." *Epidemiology International* 5, no. 1 (2020): 10–13.

[60] Tanwar, S., Ramani, T., and Tyagi, S. "Dimensionality reduction using PCA and SVD in big data: a comparative case study." In *International Conference on Future Internet Technologies and Trends,* 2017, August. Springer, Cham, pp. 116–125.

[61] Chauhan, K., Jani, S., Thakkar, D., *et al.*, "Automated machine learning: the new wave of machine learning." In *2020 2nd International Conference on Innovative Mechanisms for Industry Applications (ICIMIA)*, Bangalore, India, 2020, pp. 205–212.

[62] Shukla, A., Ramdasani, U., Vinzuda, G., Obaidat, M. S., Tanwar, S., and Kumar, N. "BCovX: blockchain-based COVID diagnosis scheme using chest X-ray for isolated location." In *ICC 2021 – IEEE International Conference on Communications*, Montreal, QC, Canada, 2021, pp. 1–6.

[63] Tanwar, S., Kumari, A., Vekaria, D., Kumar, N., and Sharma, R. "An AI-based disease detection and prevention scheme for COVID-19." *Computers and Electrical Engineering* 103 (2022): 108352.

[64] Jadav, D., Patel, D., Gupta, R., Jadav, N. K., and Tanwar, S. "BaRCODe: a blockchain-based framework for remote COVID detection for healthcare 5.0." In *2022 IEEE International Conference on Communications Workshops (ICC Workshops)*, Republic of Seoul, Korea, 2022, pp. 782–787.

[65] Mankodiya, H., Palkhiwala, P., Gupta, R., *et al.* "A real-time crowdsensing framework for potential COVID-19 carrier detection using wearable sensors." *Mathematics* 10, no.16 (2022): 2927.

[66] World Health Organization. "Coronavirus Disease 2019 (COVID-19): Situation Report: 67." (2020).

[67] Ardabili, S.F., Mosavi, A., Ghamisi, P., *et al.* "Covid-19 outbreak prediction with machine learning." *Algorithms* 13, no. 10 (2020): 249.

Index